Advances in Molecular Genetics of Brain Tumors

Advances in Molecular Genetics of Brain Tumors

Editors

Nives Pećina-Šlaus
Ivana Jovčevska

MDPI • Basel • Beijing • Wuhan • Barcelona • Belgrade • Manchester • Tokyo • Cluj • Tianjin

Editors
Nives Pećina-Šlaus
Department of Biology and Croatian Institute for Brain Research, School of Medicine
University of Zagreb
Croatia

Ivana Jovčevska
Institute of Biochemistry and Molecular Genetics, Faculty of Medicine, University of Ljubljana
Slovenia

Editorial Office
MDPI
St. Alban-Anlage 66
4052 Basel, Switzerland

This is a reprint of articles from the Special Issue published online in the open access journal *International Journal of Molecular Sciences* (ISSN 1422-0067) (available at: https://www.mdpi.com/journal/ijms/special_issues/Molecular_Genetics_Brain_Tumors).

For citation purposes, cite each article independently as indicated on the article page online and as indicated below:

LastName, A.A.; LastName, B.B.; LastName, C.C. Article Title. *Journal Name* **Year**, *Volume Number*, Page Range.

ISBN 978-3-0365-6644-3 (Hbk)
ISBN 978-3-0365-6645-0 (PDF)

Cover image courtesy of Ivana Jovčevska
glioblastoma stem cells

Contents

About the Editors

Nives Pećina-Šlaus

Nives Pećina-Šlaus is a tenured professor at the Department of biology and Head of the Laboratory of Neuro-oncology Croatian Institute for Brain Research at the School of Medicine of the University of Zagreb. She graduated in 1990. and obtained M.S. in 1992. from the Faculty of Sciences, and her Ph.D. in the field of molecular oncology in 1998 from the School of Medicine at the University of Zagreb. She was trained at Cold Spring Harbor Laboratory, New York, and at Georgetown University, Washington DC, USA. She was granted as principal investigator for six scientific projects and is/was a collaborator on three international projects funded by EU. Her research has led to the publication of more than 100 publications—75 scientific papers, a book, abstracts, 7 book chapters and many professional papers. Her main fields of research are cancer genetics, Wnt signaling pathway, brain tumorigenesis, genomic instability, and molecular profiles of brain tumors. She was an invited lecturer in 29 scientific meetings. She organized several scientific workshops and often participated in the population of science. She was cited 2251 in Google Scholar (H index 24), 1130 times in Web of Science (H index 17), and 1358 in Scopus (H index 19). She acts as a reviewer for many scientific projects and indexed journals and is an editorial board member of Cancer Cell International, Frontiers in Bioscience, Acta Clinica Croatica, the Croatian Medical Journal in 2011. In 2021, she edited the Special Issue "Advances in Molecular Genetics of Brain Tumors" in *IJMS* (ISSN 1422-0067). She teaches medical biology and many doctoral courses for medical and dentistry students and was a mentor on numerous theses (29 defended theses) including seven papers rewarded by Rector's award. She is a member of professional associations and councils (European Society for Human Genetics, European Association for Cancer Research, Croatian Society for Neuroscience, Council for neuroscience and brain disease CASA (HAZU), Council of School of Medicine at the University of Zagreb). She was awarded four scientific awards by the Croatian Medical Association and the Academy of Medical Sciences in 2011. she received National Science Award for outstanding scientific research from the Ministry of Science, Education and Sports and the award for publication activity from the Medical School University of Zagreb.

Ivana Jovčevska

Ivana Jovčevska is a researcher at the Faculty of Medicine, University of Ljubljana in Ljubljana, Slovenia. She was born and raised in Skopje, R. N. Macedonia, where at the Ss Ss Cyril and Methodius University she finished her Bachelor's studies in analytical biochemistry in 2011. Since October 2011 she lives in Ljubljana, Slovenia, where she completed her PhD studies in biochemistry and molecular biology, and was awarded a doctoral degree in 2015. She was trained at the Vrije Universiteit Brussel in Brussels, Belgium and Max Delbrück Center for Molecular Medicine (MDC) in Berlin, Germany. Her main fields of research are molecular changes in glioblastoma, proteomics, nanobodies, and gliomagenesis. So far, during her research career, she has been granted one international CELSA grant, as well as one post-doctoral grant and two smaller bilateral projects from SRA. She was also awarded several scholarships in the course of her Bachelor's and PhD studies, and her post-doctoral training. She has also cooperated on two Interreg projects. Her research led to the publication of 24 papers, a book chapter and a review in Slovenian. She was an invited lecturer at two scientific meetings, organized the one-day conference with international participation "Brain tumors: From bench to clinic" in 2019, and is constantly working on the popularization of science. Her research has been cited over 300 times, her H index is 11. She acts as a reviewer for many peer-reviewed journals. In 2021, she edited the Special Issue "Advances in Molecular Genetics of Brain Tumors" for the *International Journal of Molecular Sciences* (ISSN 1422-0067). She has one elective course in the PhD programme Biomedicine at the UL Faculty of Medicine and is currently mentoring two PhD students and one MSc student. She is a member of FEBS and the Slovenian Biochemical Society, the latter awarded her the Lapanje recognition for extraordinary achievements in the field of biochemical sciences in 2022. The award is given to young members for significant contributions to the development of biochemical sciences in the Slovenian and international research society.

Preface to "Advances in Molecular Genetics of Brain Tumors"

It is our great pleasure and joy to print the Special Issue "Advances in Molecular Genetics of Brain Tumors" in the form of a book. When planning this Special Issue, our aim was to provide readers with studies dealing with different aspects of brain tumors. Since the Special Issue contains manuscripts covering various features of brain tumors, we believe we have shown how they fit together and complement each other to accelerate the design and development of more promising diagnostic and treatment methods. The result, we hope, is a solid set of manuscripts that contribute to filling in the gap of knowledge in the biology of brain tumors.

The work that our authors, the *IJMS* editorial office and ourselves conducted during the past two years will hopefully provide valuable information for the scientific community on the molecular characteristics of brain tumors and their prospective treatment options. This unique editing experience during the work on the interesting papers that we received, provided us with valuable knowledge and we learned a great deal about novel approaches and ideas behind the formidable studies. We would like to thank and acknowledge the contributing authors for their great research studies and literature reviews. Our heartfelt thanks to Mrs Clytie Yin and the *IJMS* editorial office for encouraging us to edit the Special Issue and providing great assistance in the making of this project.

As guest editors, we are very excited about the book coming to fruition and we sincerely hope that the readers are going to find the book useful and interesting.

Nives Pećina-Šlaus and Ivana Jovčevska
Editors

International Journal of *Molecular Sciences*

MDPI

Editorial

Different Approaches to Study Molecular Blueprint and Biological Behavior of Brain Tumors: Editorial to the Special Issue "Advances in Molecular Genetics of Brain Tumors"

Nives Pećina-Šlaus [1,2,*] and Ivana Jovčevska [3]

1 Laboratory of Neuro-Oncology, Croatian Institute for Brain Research, School of Medicine University of Zagreb, Šalata 12, 10000 Zagreb, Croatia
2 Department of Biology, School of Medicine, University of Zagreb, Šalata 3, 10000 Zagreb, Croatia
3 Center for Functional Genomics and Biochips, Institute of Biochemistry and Molecular Genetics, Faculty of medicine, University of Ljubljana, Zaloška cesta 4, 1000 Ljubljana, Slovenia
* Correspondence: nina@mef.hr; Tel.: +385-1-45-66-952 or +385-14566777

Cancer remains one of the leading causes of mortality worldwide. In this context, brain tumors are also characterized by poor prognosis and short survival [1,2]. Brain tumors demonstrate the great heterogeneity and inherent molecular variability of tumor cells even within a specific subtype. Therefore, for precise diagnosis and successful treatment it is important to understand their biology, which, at this point, remains largely unexplained. Understanding, preventing, and treating brain tumors are among the most important ongoing challenges of neuro-oncology and a crucial public health issue.

The aim of the Special Issue "Advances in Molecular Genetics of Brain Tumors" was to present a collection of selected relevant papers in the rapidly expanding field of brain tumor genetics. The topics covered molecular genetics, signaling pathways governing the mechanisms of tumor formation and invasion, DNA methylation, predictive biomarkers, diagnostic improvement, potential therapeutic targets, metastasis and recurrence, as well as the difference in the therapeutic response to specific clinical subtypes. In this issue, we published a total of 14 papers, particularly 10 novel innovative studies on the genetics of brain tumors, as well as 1 communication and 3 reviews providing perspectives that can shed light on novel research directions. The contributing authors report a range of different and versatile molecular studies, but also offer insight into the state-of-the-art progress in several critical reviews.

The original work on the genetics of brain tumors consists of diverse studies ranging from profiling of glial cancers pathway-related genes [3] on one end, to investigating the migratory properties of glioblastoma cells [4] and proteins involved in the process [5], and identification of proteins crucial for stemness and self-renewal of glioblastoma cells [6] on another end. In more detail, in the comparative study by Majercikova et al. [3] the authors performed transcriptomic profiling of cancer pathway-related genes by combining the PCA method and multi-criteria decision making in the analysis of gene expression. They observed changes in the expression of 26 genes compared to the average expression value of three different controls belonging to various pathways including cellular senescence, metabolism, angiogenesis, apoptosis, DNA damage and repair, epithelial to mesenchymal transition, and telomeres and telomerase. Migratory glioma cell properties were examined in two studies: one by Odrzywolski et al. [4] analyzed the correlation of doublecortin (DCX) expression and glioblastoma cell migration using immunohistochemistry and single-cell RNA-seq. Along with DCX, the analysis also included the expression patterns of Nestin (NES) and Oligodendrocyte lineage transcription factor 2 (OLIG2). The findings indicated that there was a set proportion of cells expressing DCX/NES/OLIG2, regardless of treatment, guided by tumor plasticity.

Citation: Pećina-Šlaus, N.; Jovčevska, I. Different Approaches to Study Molecular Blueprint and Biological Behavior of Brain Tumors: Editorial to the Special Issue "Advances in Molecular Genetics of Brain Tumors". *Int. J. Mol. Sci.* **2023**, *24*, 948. https://doi.org/10.3390/ijms24020948

Received: 18 November 2022
Revised: 18 December 2022
Accepted: 23 December 2022
Published: 4 January 2023

The second study on migratory properties by Toedebusch and colleagues [5] is about the involvement of microglia-derived olfactomedin-like 3 (Olfml3) in glioma progression. Toedebusch et al. performed CRISPR-Cas9-mediated Olfml3 gene editing in N9 cells and clearly demonstrated that this extracellular matrix protein promoted glioma cell migration and invasion. Other essential mechanisms examined were stemness and self-renewal, as reported by Wu and coauthors [6]. In their study, the authors reported that the expression of BIRC3, a member of the IAP family of proteins that inhibit apoptosis, promotes glioblastoma stemness and tumorigenicity of glioma stem cells (GSCs) through inactivation of BMP4 signaling pathway. Elucidating the mechanisms of GBM stemness reprogramming and adaptation, which is believed to be a primary cause of therapeutic failures, is very important for advancing our understanding of glioblastoma resistance to therapy (Figure 1).

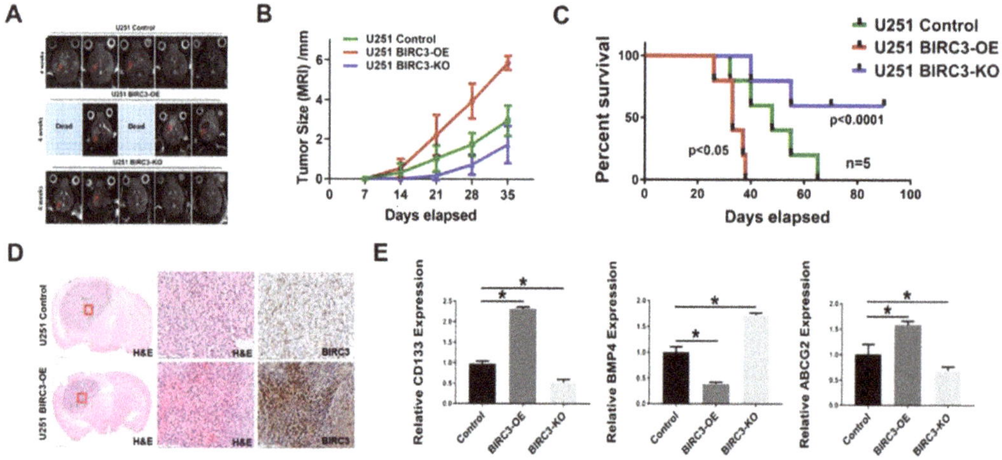

Figure 1. BIRC3 influences tumor initiation and progression in GBM orthotopic xenograft model. From paper by Wu et al., [6]. (**A**). Horizontal axial MRI scan of mouse brain tumors 4 weeks after implantation. Two of BIRC3-OE mice were already dead at 4 weeks. (**B**). Tumor size calculation from MRI scan. $n = 5$. (**C**). Kaplan-Meier survival curve of U251 control BIRC3-OE and BIRC3-KO intracranial injection mice. $n = 5$ mice/group. (**D**). Mice were sacrificed at different timepoints and brain tissues of U251 control and BIRC3-OE groups were fixed in 10% neutral formalin. H&E staining and BIRC3 immunohistochemistry was performed as described in the Material and Methods Section 4. Five mice were included in this histological study and similar results were observed in each animal. (**E**). When mice were sacrificed, part of tumor tissues were isolated. mRNA from tumor tissues were extracted. BMP4, CD133 and ABCG2 mRNA expression analyzed by real-time PCR in extracted tumor tissues. $n = 3$, * $p < 0.05$.

Moreover, studies also reported tumorigenesis mechanisms controlled by the circadian gene cry [7]. Circadian rhythm regulation [8] has only recently been introduced in cancer research. However, the deregulation of circadian genes in tumors has been recognized as important. Using an in-house *Drosophila* model, Jarabo et al. reported a novel role of the light-regulated protein Cry that acts as a core component of the circadian clock. The model uses the system of co-activation of EGFR and PI3K signaling pathways in Drosophila glial cells.

The mechanisms affected by temozolomide in pediatric glioblastoma was investigated by Damanskienė et al. [9]. The authors tested the differences in efficacy of temozolomide doses between PBT24 and SF8628 cell lines of high-grade pediatric glioblastoma xenografts in a chicken chorioallantoic membrane (CAM) model. The study stressed the importance of personalized therapy for glioblastoma, emphasizing that it should be specifically tailored to the pediatric population.

The importance of metabolic dysregulation has long been recognized as a driving force of cancer, particularly glioblastoma. In their contribution, Franceschi et al. addressed a very important aspect of glioblastoma progression: the involvement of a metabolic pathway parallel to glycolysis, specifically the pentose phosphate pathway (PPP), which plays a critical role in sustaining cancer-cell survival and growth [10]. The authors evaluated the role of sedoheptulose kinase (SHPK), an enzyme involved in the nonoxidative arm of the PPP, by conducting a functional enrichment analysis using microarray data on SHPK expression in glioblastoma patients. Further, they evaluated the effects of SHPK overexpression in three different glioblastoma cell lines. Their results showed that the increased SHPK expression was significantly correlated with a worse glioblastoma prognosis.

Moreover, two studies tackled RNA interference in glioma. Namely, the need to identify miRNAs as specific non-invasive biomarkers for the prognosis of glioma is highlighted in the work of Levallet et al. [11]. The authors reported that the expression level of a panel of seven pro-angiogenic and/or pro-hypoxic miRNAs (has-miR-200b-3p, -200c-3p, -210-3p, -126-5p, -221-3p, -424-5p, and -451-5p) was affected in patients with glioma and related to the glioma histology grade. The results suggest that pro-angiogenic and/or pro-hypoxic miRNAs can be used as tools for monitoring patients, specifically with IDH-mutated low-grade tumors, since they are easily measurable in plasma. In a different study, Clausing et al. [12] examined the effect of $IDH1^{R132H}$ mutation on the redox system in a CRISPR/Cas edited glioblastoma model and compared them with IDH1 wild-type (IDH1wt) cells. This model is suitable for portrait $IDH1^{R132H}$-dependent alterations in tumor cell metabolism. In the study, they also showed an increase in NAD+ in $IDH1^{R132H}$ glioblastoma cells compared to $IDH1^{wt}$. Their findings underline the therapeutic potential of targeting the NAD+ synthesis pathway, but authors recommend caution for small-molecule inhibitors.

In the Special Issue, we also included one study about the importance of FET PET/CT scans [13]. With the purpose of treatment improvement, Skoblar Vidmar et al. tested the performance of O-(2-[^{18}F] fluoroethyl-)-L-tyrosine (^{18}F-FET) PET for the differentiation between glioma patients based on IDH mutational status. The enzyme isocitrate dehidrogenase (IDH) is a biomarker that improves diagnostic accuracy, but also influences the response and the course of treatment, and thus, overall survival [14]. This is the first reported study that assessed the diagnostic performance of different ^{18}F-FET PET segmentation approaches for differentiation between treatment-related changes (TRC) and true progression (TP). The neuro-oncological therapy can lead to the development of TRC that mimic TP and distinguishing TRCs from TP in treated patients remains a challenge in glioma cases, since both share similar clinical symptoms and imaging characteristics [14]. In their contribution, Skoblar Vidmar and co-authors stressed the importance of molecular biomarkers that have clinicopathologic utilities.

One study focused on a different group of most common primary brain tumors i.e., meningiomas [15]. Bukovac and coauthors investigated the role of DVL1 that is the central mediator of Wnt signaling pathway. The results revealed that the central PDZ region of DVL1 gene harbored frequent mutations. The study further showed that the samples containing mutations in the PDZ domain expressed significantly less DVL1 protein and that the nuclear expression of DVL1 was significantly correlated with a higher expression of active β-catenin ($p = 0.029$) and a higher meningioma grades ($p = 0.030$). Their genomic instability, sequencing and immunohistochemistry results indicate that Wnt signaling is activated in meningioma and that DVL1 could potentially represent a good biomarker for meningioma progression.

To obtain a broader view of the field, we published three review papers that provided comprehensive and thoroughly updated critical standpoints on several topics such as targeted therapies for vestibular schwannomas [16], as well as molecular biomarkers [17] and precision oncology for complex diseases such as glioblastoma [18]. Tamua and Toda prepared an in-depth update of the currently available knowledge on the molecular biology of vestibular schwannomas and its relevance to treatment. The importance of the tumor microenvironment, inflammation, and stress reaction in the development and progression of vestibular schwannomas is also critically surveyed. The review discusses a range of therapeutic approaches, from surgery and radiation therapy to gene therapy. Finally, the authors indicate that tumor-microenvironment-targeted therapy may also be supportive and recommend multimodal therapy for patients with refractory vestibular schwannomas. Next, Sareen et al. [17] conducted a systematic review and performed meta-analysis of key molecular biomarkers that have been investigated for their predictive value in recent glioblastoma clinical trials. Analysis of the prognostic significance of IDH1 mutation showed significantly better overall survival in patients with IDH1 mutation. Meta-analysis including 575 glioblastoma patients presenting with either amplification or high expression of EGFR gene did not reveal prognostic significance, which the authors contributed to limited patient numbers; they recommended more homogeneous studies on larger patient cohorts. Lastly, Panovska and De Smet [18] surveyed the current knowledge on therapy approaches for complex and heterogeneous diseases such as glioblastoma that display formidable inter- and intra-tumoral heterogeneity. The authors appeal on personalized and distinct therapeutic approaches in order to achieve clinical benefits since the clinical trials of the past 20 years have failed to improve the outcome for the vast majority of glioblastoma patients. The review highlights the need for procedures that can precisely select the appropriate patients who could benefit from the given therapy, but also to the drug sensitivity of specific tumor cells of a particular patient. The authors discuss the current state of the art of transforming technologies, tools and challenges for functional precision oncology, and conclude that that the personalization of cancer medicine is the way to tackle this disease.

In conclusion, in order to better understand the biology of brain tumors and ultimately improve diagnostic and therapeutic approaches, this Special Issue intends to elucidate the diversity of problems behind brain tumors and their potential solutions. This is also nicely illustrated with the diversity of methodologies used in the papers published in our Special Issue. As brain tumor incidence increases with age, with the growing numbers of elderly population, the number of patients is expected to rise. Many cases of brain tumors are characterized with discouraging prognosis. The treatment resistance and the possibility of recurrence is the cause of poor survival, especially for the glioma branch [19].

To improve patient outcome, numerous novel ideas and alternative approaches are constantly being explored by the research community. Molecular testing in modern cancer diagnostics, combined with the development of personalized therapies, is the main avenue for the successful outcomes and progress. However, there is still room for improvement, which demands that we search for novel crucial molecular players and development of new concepts about the initiation and progression of brain tumors. As reported in this Special Issue, brain tumors are investigated on various cellular levels with the ultimate goal of greater patient benefit. We hope that the themes and fields covered in this Special Issue will attract readers from the broader scientific community, contribute to expanding our knowledge about the biology of brain tumors, and inspire further studies that will improve the diagnosis and clinical management of brain tumors.

Lastly, as Guest Editors, we would like to gratefully acknowledge the input of all the authors of both original research articles and review papers, and their contribution to this relevant and valuable research topic.

Author Contributions: Conceptualization, writing—original draft preparation, writing—review and editing N.P.-Š. and I.J. All authors have read and agreed to the published version of the manuscript.

Conflicts of Interest: The authors declare no conflict of interest.

References

1. Yang, Y.; Schubert, M.C.; Kuner, T.; Wick, W.; Winkler, F.; Venkataramani, V. Brain Tumor Networks in Diffuse Glioma. *Neurotherapeutics* **2022**, *19*, 1832–1843. [CrossRef] [PubMed]
2. Pellerino, A.; Bruno, F.; Palmiero, R.; Pronello, E.; Bertero, L.; Soffietti, R.; Ruda, R. Clinical Significance of Molecular Alterations and Systemic Therapy for Meningiomas: Where Do We Stand? *Cancers* **2022**, *14*, 2256. [CrossRef] [PubMed]

3. Majercikova, Z.; Dibdiakova, K.; Gala, M.; Horvath, D.; Murin, R.; Zoldak, G.; Hatok, J. Different Approaches for the Profiling of Cancer Pathway-Related Genes in Glioblastoma Cells. *Int. J. Mol. Sci.* **2022**, *23*, 10883. [CrossRef] [PubMed]
4. Odrzywolski, A.; Jarosz, B.; Kielbus, M.; Telejko, I.; Ziemianek, D.; Knaga, S.; Rola, R. Profiling Glioblastoma Cases with an Expression of DCX, OLIG2 and NES. *Int. J. Mol. Sci.* **2021**, *22*, 13217. [CrossRef] [PubMed]
5. Toedebusch, R.G.; Lucchesi, C.A.; Debebe, E.T.; Wittenburg, L.A.; Chen, X.; Toedebusch, C.M. Microglia-Derived Olfactomedin-like 3 Promotes Pro-Tumorigenic Microglial Function and Malignant Features of Glioma Cells. *Int. J. Mol. Sci.* **2021**, *22*, 13052. [CrossRef] [PubMed]
6. Wu, Q.; Berglund, A.E.; MacAulay, R.J.; Etame, A.B. A Novel Role of BIRC3 in Stemness Reprogramming of Glioblastoma. *Int. J. Mol. Sci.* **2021**, *23*, 297. [CrossRef] [PubMed]
7. Jarabo, P.; de Pablo, C.; Gonzalez-Blanco, A.; Casas-Tinto, S. Circadian Gene cry Controls Tumorigenesis through Modulation of Myc Accumulation in Glioblastoma Cells. *Int. J. Mol. Sci.* **2022**, *23*, 2043. [CrossRef] [PubMed]
8. Smyllie, N.J.; Bagnall, J.; Koch, A.A.; Niranjan, D.; Polidarova, L.; Chesham, J.E.; Chin, J.W.; Partch, C.L.; Loudon, A.S.I.; Hastings, M.H. Cryptochrome proteins regulate the circadian intracellular behavior and localization of PER2 in mouse suprachiasmatic nucleus neurons. *Proc. Natl. Acad. Sci. USA* **2022**, *119*, e2113845119. [CrossRef] [PubMed]
9. Damanskiene, E.; Balnyte, I.; Valanciute, A.; Lesauskaite, V.; Alonso, M.M.; Stakisaitis, D. The Comparative Experimental Study of Sodium and Magnesium Dichloroacetate Effects on Pediatric PBT24 and SF8628 Cell Glioblastoma Tumors Using a Chicken Embryo Chorioallantoic Membrane Model and on Cells In Vitro. *Int. J. Mol. Sci.* **2022**, *23*, 10455. [CrossRef] [PubMed]
10. Franceschi, S.; Lessi, F.; Morelli, M.; Menicagli, M.; Pasqualetti, F.; Aretini, P.; Mazzanti, C.M. Sedoheptulose Kinase SHPK Expression in Glioblastoma: Emerging Role of the Nonoxidative Pentose Phosphate Pathway in Tumor Proliferation. *Int. J. Mol. Sci.* **2022**, *23*, 5978. [CrossRef] [PubMed]
11. Levallet, G.; Dubois, F.; Leclerc, A.; Petit, E.; Bekaert, L.; Faisant, M.; Creveuil, C.; Emery, E.; Zalcman, G.; Lechapt-Zalcman, E. The Use of Pro-Angiogenic and/or Pro-Hypoxic miRNAs as Tools to Monitor Patients with Diffuse Gliomas. *Int. J. Mol. Sci.* **2022**, *23*, 6042. [CrossRef] [PubMed]
12. Clausing, M.; William, D.; Preussler, M.; Biedermann, J.; Grutzmann, K.; Richter, S.; Buchholz, F.; Temme, A.; Schrock, E.; Klink, B. Different Effects of RNAi-Mediated Downregulation or Chemical Inhibition of NAMPT in an Isogenic IDH Mutant and Wild-Type Glioma Cell Model. *Int. J. Mol. Sci.* **2022**, *23*, 5787. [CrossRef]
13. Skoblar Vidmar, M.; Doma, A.; Smrdel, U.; Zevnik, K.; Studen, A. The Value of FET PET/CT in Recurrent Glioma with a Different IDH Mutation Status: The Relationship between Imaging and Molecular Biomarkers. *Int. J. Mol. Sci.* **2022**, *23*, 6787. [CrossRef] [PubMed]
14. Qin, D.; Yang, G.; Jing, H.; Tan, Y.; Zhao, B.; Zhang, H. Tumor Progression and Treatment-Related Changes: Radiological Diagnosis Challenges for the Evaluation of Post Treated Glioma. *Cancers* **2022**, *14*, 3771. [CrossRef]
15. Bukovac, A.; Dragicevic, K.; Kafka, A.; Oreskovic, D.; Cesarec-Augustinovic, S.; Pecina-Slaus, N. Decoding the Role of DVL1 in Intracranial Meningioma. *Int. J. Mol. Sci.* **2021**, *22*, 11996. [CrossRef]
16. Tamura, R.; Toda, M. A Critical Overview of Targeted Therapies for Vestibular Schwannoma. *Int. J. Mol. Sci.* **2022**, *23*, 5462. [CrossRef] [PubMed]
17. Sareen, H.; Ma, Y.; Becker, T.M.; Roberts, T.L.; de Souza, P.; Powter, B. Molecular Biomarkers in Glioblastoma: A Systematic Review and Meta-Analysis. *Int. J. Mol. Sci.* **2022**, *23*, 8835. [CrossRef] [PubMed]
18. Panovska, D.; De Smet, F. Functional Precision Oncology: The Next Frontier to Improve Glioblastoma Outcome? *Int. J. Mol. Sci.* **2022**, *23*, 8637. [CrossRef] [PubMed]
19. Le Rhun, E.; Preusser, M.; Roth, P.; Reardon, D.A.; van den Bent, M.; Wen, P.; Reifenberger, G.; Weller, M. Molecular targeted therapy of glioblastoma. *Cancer Treat Rev.* **2019**, *80*, 101896. [CrossRef] [PubMed]

International Journal of *Molecular Sciences*

MDPI

Article

Different Approaches for the Profiling of Cancer Pathway-Related Genes in Glioblastoma Cells

Zuzana Majercikova [1], Katarina Dibdiakova [1], Michal Gala [2], Denis Horvath [3], Radovan Murin [1], Gabriel Zoldak [3] and Jozef Hatok [1,*]

1 Department of Medical Biochemistry, Jessenius Faculty of Medicine in Martin, Comenius University in Bratislava, Mala Hora 11161/4D, SK-03601 Martin, Slovakia
2 Department of Biophysics, Faculty of Science, University of P.J. Safarik, Jesenna 5, SK-04001 Kosice, Slovakia
3 Center for Interdisciplinary Biosciences, Technology and Innovation Park, University of P.J. Safarik, Trieda SNP 1, SK-04011 Kosice, Slovakia
* Correspondence: jozef.hatok@uniba.sk; Tel.: +421-43-2633-469

Abstract: Deregulation of signalling pathways that regulate cell growth, survival, metabolism, and migration can frequently lead to the progression of cancer. Brain tumours are a large group of malignancies characterised by inter- and intratumoral heterogeneity, with glioblastoma (GBM) being the most aggressive and fatal. The present study aimed to characterise the expression of cancer pathway-related genes ($n = 84$) in glial tumour cell lines (A172, SW1088, and T98G). The transcriptomic data obtained by the qRT-PCR method were compared to different control groups, and the most appropriate control for subsequent interpretation of the obtained results was chosen. We analysed three widely used control groups (non-glioma cells) in glioblastoma research: Human Dermal Fibroblasts (HDFa), Normal Human Astrocytes (NHA), and commercially available mRNAs extracted from healthy human brain tissues (hRNA). The gene expression profiles of individual glioblastoma cell lines may vary due to the selection of a different control group to correlate with. Moreover, we present the original multicriterial decision making (MCDM) for the possible characterization of gene expression profiles. We observed deregulation of 75 genes out of 78 tested in the A172 cell line, while T98G and SW1088 cells exhibited changes in 72 genes. By comparing the delta cycle threshold value of the tumour groups to the mean value of the three controls, only changes in the expression of 26 genes belonging to the following pathways were identified: angiogenesis *FGF2*; apoptosis *APAF1, CFLAR, XIAP*; cellular senescence *BM1, ETS2, IGFBP5, IGFBP7, SOD1, TBX2*; DNA damage and repair *ERCC5, PPP1R15A*; epithelial to mesenchymal transition *SNAI3, SOX10*; hypoxia *ADM, ARNT, LDHA*; metabolism *ATP5A1, COX5A, CPT2, PFKL, UQCRFS1*; telomeres and telomerase *PINX1, TINF2, TNKS*, and *TNKS2*. We identified a human astrocyte cell line and normal human brain tissue as the appropriate control group for an in vitro model, despite the small sample size. A different method of assessing gene expression levels produced the same disparities, highlighting the need for caution when interpreting the accuracy of tumorigenesis markers.

Keywords: glioblastoma; cancer pathway; mRNA; multicriterial analysis

Citation: Majercikova, Z.; Dibdiakova, K.; Gala, M.; Horvath, D.; Murin, R.; Zoldak, G.; Hatok, J. Different Approaches for the Profiling of Cancer Pathway-Related Genes in Glioblastoma Cells. *Int. J. Mol. Sci.* **2022**, *23*, 10883. https://doi.org/10.3390/ijms231810883

Academic Editors: Ivana Jovčevska and Nives Pećina-Šlaus

Received: 5 August 2022
Accepted: 13 September 2022
Published: 17 September 2022

1. Introduction

Glioblastoma (GBM) is one of the most prevalent primary malignant brain tumours in adults, as classified by the World Health Organization (WHO). According to histopathological and molecular characteristics, glioblastoma is classified into four grades (I-IV) [1]. Despite aggressive multimodal therapy consisting of surgical resection, radiation, and chemotherapy with the alkylating agent temozolomide, the prognosis remains dismal, with a median overall survival of 12–15 months after diagnosis [2,3]. The intensive study of malignant gliomas over the past three decades has determined several molecular hallmarks that have enhanced classification and therapeutical strategies. As with all other types of brain tumours, glioblastomas have a heterogeneous character. The result was the introduction of the most recent WHO classification of central nervous system tumours in 2021 [4,5]. In addition to the histological approach and somatic mutations, gene expression signatures contribute significantly to the overall classification of glioblastoma subtypes [6]. Currently, successful treatment response and outcome predictions for patients with GBM are made using next generation sequencing analyses that account for intratumoral heterogeneity [7]. The determination of *O^6-methylguanine-DNA methyltransferase* (MGMT), *isocitrate dehydrogenase* (IDH), *tumour protein p53* (TP53), *platelet-derived growth factor receptor alpha* (PDGFRA), or *epidermal growth factor receptor* (EGFR) gene expression levels, as well as the presence of their mutations and methylation status, are important prognostic factors [8–11]. One of the most commonly used methods in the aforementioned molecular analysis is real-time PCR, indicating a clear application with minimal errors. The all real-time PCR techniques (classical method based on dyes, TaqMan probes assay, microarray, RNA sequencing analysis, etc.) are still highly quantitative and sensitive methods for the detection of gene expression levels but are generally best for examining a relatively small number of transcripts in a large set of samples. One of the few disadvantages is the necessity of involving a biostatistics expert in the evaluation process. However, with proper optimisation, we may encounter erroneous analysis and, subsequently, variable interpretations of results. Properly chosen biostatistical methods also clearly contribute to successful analysis. Cell models have also been used to characterise the mechanisms underlying glioblastoma formation [12].

In the present study, we examined the transcriptomic profiles of glial cancer cell lines: A172, T98G, and SW1088. We focused on the detection of representative cancer genes (Human Cancer PathwayFinder^{TM} PCR Array), which were divided into nine pathways: angiogenesis, apoptosis, cell cycle, cellular senescence, DNA damage and repair, epithelial-to-mesenchymal transition, hypoxia signalling, metabolism, and telomeres and telomerase (Table 1). This study's primary objective is based on a different approach for selecting the control group and, consequently, for interpreting the results. We analysed three different control groups (non-glioma cells) widely used in glioblastoma research: Human Dermal Fibroblasts (HDFa), Normal Human Astrocytes (NHA), and commercially available mRNAs acquired from healthy human brain tissues (hRNA). Using real-time PCR analysis, the gene expression profiles of glioblastoma cell lines differ when compared to a control group based on correlational differences. Our comprehensive approach, which incorporates numerous statistical analyses, contributes to the most accurate interpretation of the results.

Table 1. List of analysed genes and their assignment to cellular pathways.

Pathway	Symbol	Gene Name
Angiogenesis	*ANGPT1*	Angiopoietin 1
	ANGPT2	Angiopoietin 2
	CCL2	Chemokine (C-C motif) ligand 2
	FGF2	Fibroblast growth factor 2
	FLT1	Fms-related tyrosine kinase 1
	KDR	Kinase insert domain receptor
	PGF	Placental growth factor
	SERPINF1	Serpin peptidase inhibitor, clade F member 1
	TEK	TEK tyrosine kinase, endothelial
	VEGFC	Vascular endothelial growth factor C

Table 1. *Cont.*

Pathway	Symbol	Gene Name
Apoptosis	*APAF1*	Apoptotic peptidase activating factor 1
	BCL2L11	BCL2-like 11 (apoptosis facilitator)
	BIRC3	Baculoviral IAP repeat containing 3
	CASP2	Caspase 2
	CASP7	Caspase 7
	CASP9	Caspase 9
	CFLAR	CASP8 and FADD-like apoptosis regulator
	FASLG	Fas ligand (TNF superfamily, member 6)
	NOL3	Nucleolar protein 3 (apoptosis repressor)
	XIAP	X-linked inhibitor of apoptosis
Cell cycle	*AURKA*	Aurora kinase A
	CCND2	Cyclin D2
	CCND3	Cyclin D3
	CDC20	Cell division cycle 20 homolog (S. cerevisiae)
	E2F4	E2F transcription factor 4, p107/p130-binding
	MCM2	Minichromosome maintenance complex component 2
	MKI67	Antigen identified by monoclonal antibody Ki-67
	SKP2	S-phase kinase-associated protein 2 (p45)
	STMN1	Stathmin 1
	WEE1	WEE1 homolog (S. pombe)
Cellular senescence	*BMI1*	BMI1 polycomb ring finger oncogene
	ETS2	V-Ets erythroblastosis virus E26 oncogene homolog 2
	IGFBP3	Insulin-like growth factor binding protein 3
	IGFBP5	Insulin-like growth factor binding protein 5
	IGFBP7	Insulin-like growth factor binding protein 7
	MAP2K1	Mitogen-activated protein kinase kinase 1
	MAP2K3	Mitogen-activated protein kinase kinase 3
	MAPK14	Mitogen-activated protein kinase 14
	SERPINB2	Serpin peptidase inhibitor, clade B, member 2
	SOD1	Superoxide dismutase 1, soluble
	TBX2	T-box 2
DNA damage and repair	*DDB2*	Damage-specific DNA binding protein 2, 48kDa
	DDIT3	DNA-damage-inducible transcript 3
	ERCC3	Excision repair cross-complementing rodent repair deficiency, complementation group 3
	ERCC5	Excision repair cross-complementing rodent repair deficiency, complementation group 5
	GADD45G	Growth arrest and DNA-damage-inducible, γ
	LIG4	DNA Ligase 4, ATP-dependent
	POLB	DNA Polymerase beta
	PPP1R15A	Protein phosphatase 1, regulatory subunit 15A

Table 1. *Cont.*

Pathway	Symbol	Gene Name
Epithelial-to-mesenchymal transition (EMT)	*CDH2*	Cadherin 2, type 1, N-cadherin (neuronal)
	DSP	Desmoplakin
	FOXC2	Forkhead box C2
	GSC	Goosecoid homeobox
	KRT14	Keratin 14
	OCLN	Occludin
	SNAI1	Snail homolog 1 (Drosophila)
	SNAI2	Snail homolog 2 (Drosophila)
	SNAI3	Snail homolog 3 (Drosophila)
	SOX10	SRY (sex determining region Y)-box 10
Hypoxiasignalling	*ADM*	Adrenomedullin
	ARNT	Aryl hydrocarbon receptor nuclear translocator
	CA9	Carbonic anhydrase 9
	EPO	Erythropoietin
	HMOX1	Heme oxygenase 1
	LDHA	Lactate dehydrogenase A
	SLC2A1	Solute carrier family 2, member 1
Metabolism	*ACLY*	ATP citrate lyase
	ACSL4	Acyl-CoA synthetase long-chain family member 4
	ATP5A1	Mitochondrial ATP synthase alpha subunit 1
	COX5A	Cytochrome c oxidase subunit 5A
	CPT2	Carnitine palmitoyltransferase 2
	G6PD	Glucose-6-phosphate dehydrogenase
	GPD2	Glycerol-3-phosphate dehydrogenase 2 (mitochondrial)
	LPL	Lipoprotein lipase
	PFKL	Phosphofructokinase, liver
	UQCRFS1	Ubiquinol-cytochrome c reductase Rieske iron-sulfur polypeptide 1
Telomeres and telomerase	*DKC1*	Dyskerin
	PINX1	PIN2/TERF1 interacting, telomerase inhibitor 1
	TEP1	Telomerase-associated protein 1
	TERF1	Telomeric repeat binding factor (NIMA-interacting) 1
	TERF2IP	Telomeric repeat binding factor 2, interacting protein
	TINF2	TERF1-interacting nuclear factor 2
	TNKS	Tankyrase
	TNKS2	Tankyrase 2

2. Results

We achieved the results on human RNA isolated from cell lines (tumour and non-tumour) and normal brain tissues. The expression of cancer pathway-related genes at the mRNA level was compared among tumour and control group of samples by the real-time qPCR. Out of the total number of monitored genes (n = 84) in the Human Cancer

PathwayFinder^{TM} PCR Array, we were able to detect all of them, and only in the control group was there not a gene for the *Fas ligand* in any of the samples. In the next sections, we describe the relationship between the groups of samples for individual genes as well as the unique multicriteria analysis used for the correct interpretation of the results.

2.1. Relative mRNA Ratio of Control Cells

The selection of an appropriate control sample is a crucial initial step for group qRT-PCR analyses. On three separate controls (NHA, hRNA, and HDFa) relative mRNA gene levels associated with cancer development were detected. The Normal Human Astrocyte cell line was characterised by a significantly higher Ct ratio in the following genes: *CDH2*, *MKI67*, *LDH2*, *IGFBP7*, and *IGFBP5* (Figure 1). The *Keratin 14* gene was amplified in the NHA line exclusively. Only *Angiopoietin 1* exhibited a significantly lower mRNA ratio in NHA compared to both control groups (hRNA and HDFa). We demonstrated that there is no amplification of *AURKA*, *ANGPT2*, *FASLG*, or *GSC* products (Figure 1). *Goosecoid homeobox* gene amplification was only detected in Human Dermal Fibroblasts among the control samples. Compared to NHA and hRNA samples, the HDFa cell line exhibited overexpression of a single gene, namely *Adrenomedullin*. In contrast to the other controls, most genes were found to have lower levels. As undetected mRNA, *AURKA*, *FASLG*, *KRT14*, and *SNAI3* were considered (Figure 1a).

The total human RNA (hRNA) control sample consists of total RNAs from healthy brain donors that are commercially available. We found only hRNA control to contain the *AURKA* gene. *ANGPT2*, *CASP9*, *E2F4*, *ETS2*, *FLT1*, *GPD2*, *LPL*, *MAP2K1*, *SERPINF1*, *SNAI3*, *SOD1*, *SOX10*, *STMN1*, *TEP1*, *TERF1*, *TERF2IP*, and *TINF2* exhibited significant overexpression. In contrast, mRNA levels for *IGFBP3*, *SNAI2*, and *WEE1* were significantly higher compared to NHA and HDFa. In addition, the amplification of mRNA for Desmoplakin, Fas ligand, Goosecoid homeobox, Keratin 14, and vascular endothelial growth factor C was not detectable (Figure 1a).

2.2. Relative mRNA Ratio in Tumour Cell Lines

Results of the relative mRNA expression ratio comparison between A172, SW1088, and T98G cell lines are shown in Figure 1b. Similar cancer-related mRNA ratios were found in all tumour cell lines for 60 out of 84 genes (71.4%). The genes *CDH2*, *DDB2*, *DSP*, *EPO*, *FGF2*, and *TEK* were overexpressed in an astrocytoma cell line. In addition, the expression of *Cyclin D2*, *kinase insert domain receptor*, and *Keratin 14* genes was present exclusively in the A172 line. In contrast, we found a significantly lower mRNA ratio for *ACSL4*, *ADM*, *ANGPT1*, and *SNAI2*. Only the mRNA for the *Fas ligand* was not detected. Except for *CCND2*, *FLT1*, *FOXC2*, *KDR*, *KRT14*, and *TEK*, the mRNA expression ratio of cancer pathway-associated genes in the astrocytoma grade III cell line (SW1088) was comparable to that of other tumour cell lines (Figure 1b).

In the glioblastoma cell line (T98G), we detected the overexpression of the Cyclin D3 transcript. Significantly lower mRNA levels were detected for *ANGPT2*, *DSP*, *KDR*, *KRT14*, and *LPL*. Four genes (*CCND2*, *DSP*, *FASLG*, and *PGF*) lacked fluorescent signals.

2.3. Data Preparation before Determination of Gene Expression Level

The selection of the appropriate control is crucial when comparing mRNA levels between groups using the ΔCt method and qRT-PCR analysis. Therefore, with respect to the control group, the genes of the tested group (tumour) sought may appear to be inaccurate. Here, we used the ΔCt method by comparing the relative expression of tumour genes to that of three control genes separately to identify the validity of the results. We used a combination of several mathematical and statistical methods to prepare gene expression data.

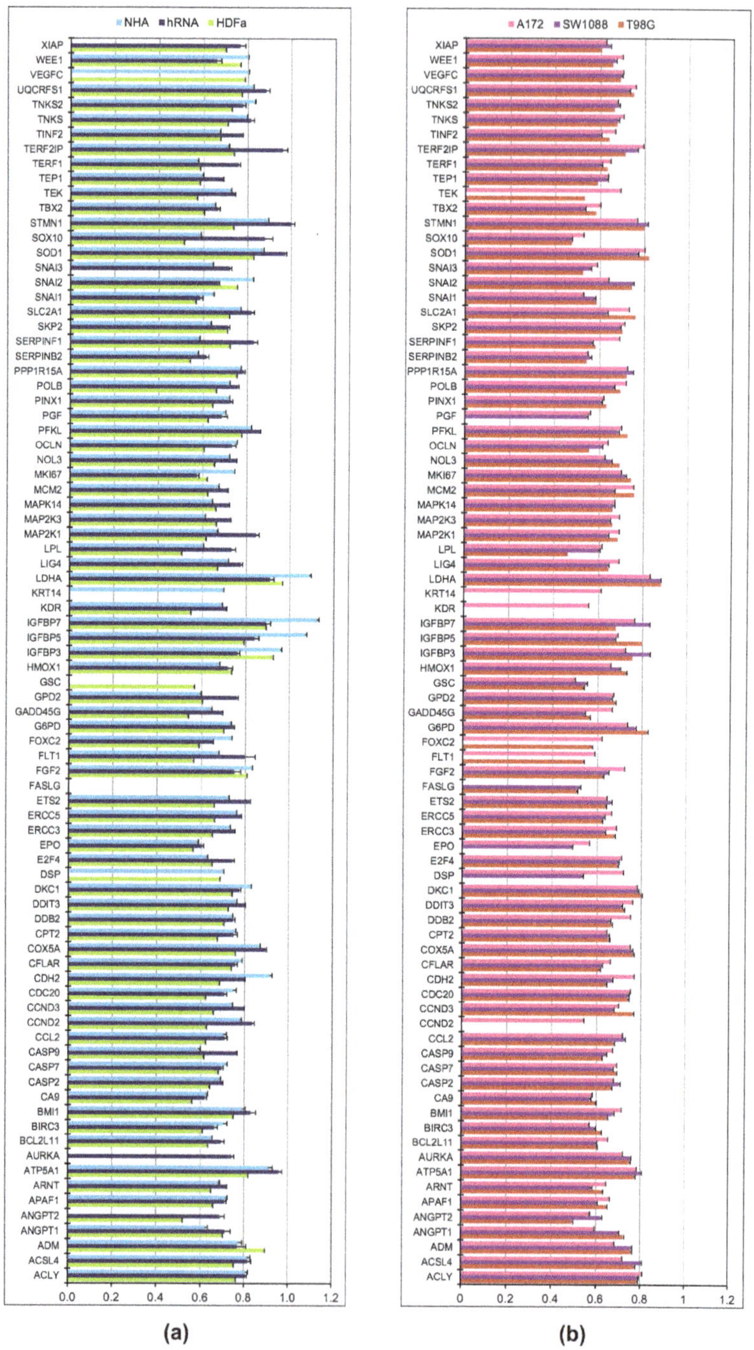

Figure 1. The mean of Ct ratio of mRNA levels. (**a**) The mRNA expression levels in control group. (**b**) The mRNA expression levels in group of tumour cell lines. Ct ratio, gene of interest/housekeeping genes. Each mRNA level included three replicates.

2.3.1. Principal Component Analysis

Principal component analysis (PCA) was utilised to examine the variability of gene expression profiles among various samples (see Material and Methods). Figure 2 shows the first two components, PC1 vs. PC2 as a scatter plot, with a variance of 41.2% for PC1 and 25.8% for PC2, respectively. As seen in the scatter plot, the gene profiles corresponding to tumour tissues (red circles) are closer together than the gene profiles of controls. From an overall perspective, tumour lines can be defined by a putative cluster, whereas other data are more dispersed. Expression profiles of controls demonstrate a wide variation, indicating their distinct and cell-line-specific expression. Regarding variability within the same sample–experimental replicas, only small variations are visible, and all replicas can be attributed to a given cell line.

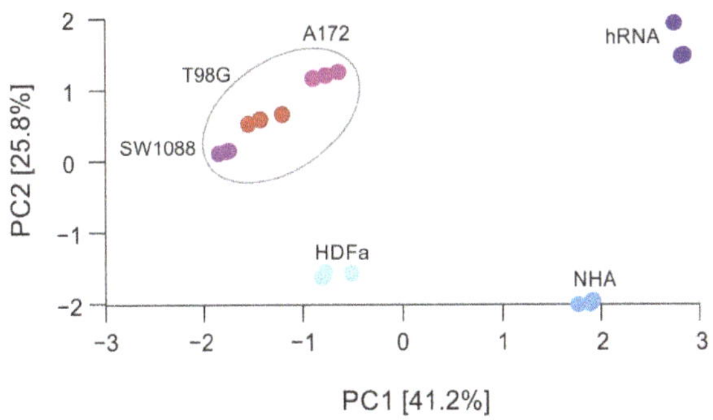

Figure 2. Scatter plot of principal component analysis. The grey elliptical line represents data of tumour cell lines, dots outside the line are control samples. All values of each sample were measured in triplicates.

2.3.2. Correlations between Gene Expressions

In the previous analysis, we discovered that LDA can identify and pinpoint the differences in individual genes and distinguish between the classes of control cell lines and tumour samples. Subsequently, we have decided to analyse the relationships between gene expressions in greater detail. By calculating the Pearson correlation coefficient, we focused on the relationship between the expressions of individual genes. In control cells, a general gene regulation network controls gene expression. Therefore, significant positive and negative correlations between gene expressions will result from the regulation network. Gene correlations can be lost if the regulatory network is dysregulated, such as in tumour cells. Figure 3a depicts a correlation between the genes in the control samples and the genes in the tumour samples. In control samples, the average correlation is low (0.33), but there are substantial differences between the individual genes. For example, *TERF1* and *TERF2IPa* genes had the highest correlation of 0.999, and *SNAI2* and *CASP9* genes had the highest anti-correlation of −0.996. In tumour lines, the average correlation is even lower, 0.031, indicating a significant loss of coordinated gene expressions. Given that the overall expression profiles of individual genes in tumour cells are more similar (dots in red elliptical line, Figure 2) than in control cells, which appeared more heterogeneously in the PCA plot, this may be a surprising result. The difference between PCA and gene–gene correlation is that PCA describes the variance between full profiles, whereas gene–gene correlation provides information about possible causal relations between expression values. In tumour cell lines, the relationship between the *SNAI1* and *SNAI2* genes is the highest at 0.9999, while the *LPL* and *IGFBP5* genes have the highest anti-correlation (−0.997). The descending arrangement of the gene correlation (Figure 3b) indicates that tumour samples have lower overall values than control samples.

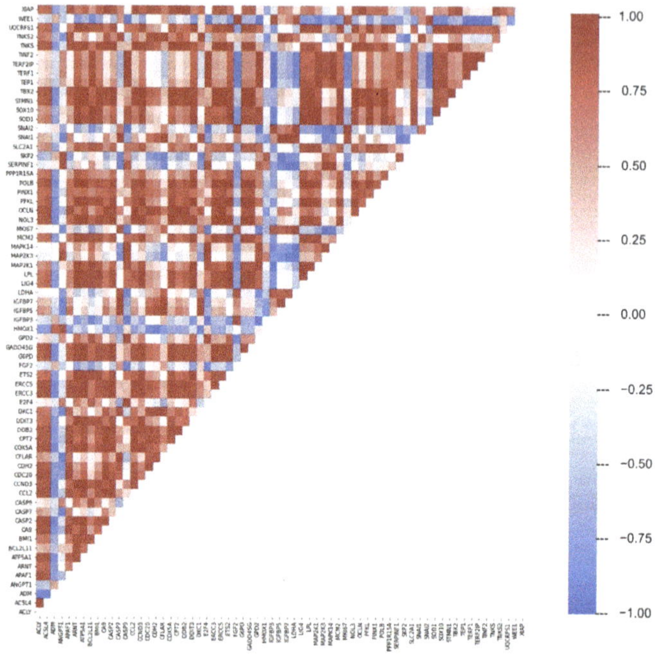

(a)

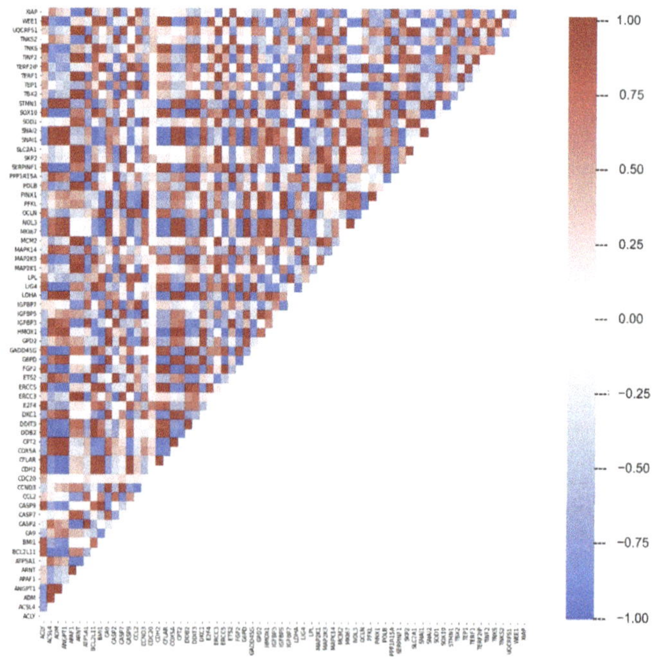

(b)

Figure 3. Pearson correlation of individual genes expressions of control (**a**) and tumour (**b**) cell lines.

Kernel density estimation revealed a different relationship between gene expressions (Figure 4). The distribution of gene expression correlations deviates from the Gaussian distribution for both control and tumour cell lines. In contrast to tumours, the peak density of control samples is centred around the value of 0.95. Maximum density in the positive correlation range is roughly in the same place as in control cell lines, but the amplitude is slightly lower, indicating that some gene–gene correlations have changed. The missing positive gene–gene correlations have become negative in tumour cell lines. The density analysis of the control vs. tumour correlation values approaches 0, as shown in Figure 4.

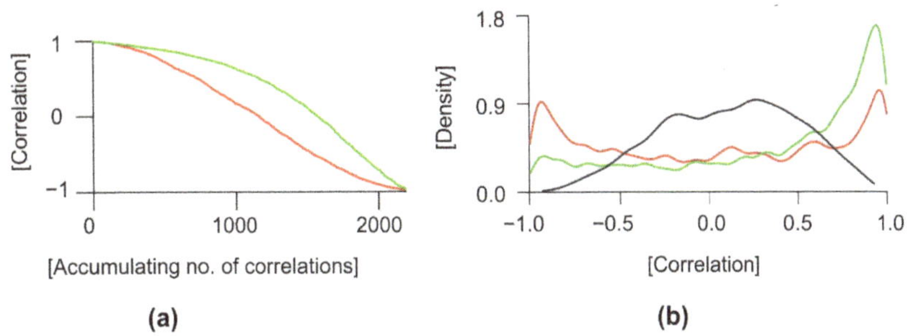

Figure 4. Statistical analysis of gene expression correlations. (**a**) Descending order of correlations. (**b**) Kernel density of correlations. The green line represents data from control, red line from tumour and grey line is the mix of both groups.

In the next step, we calculated Pearson correlations between gene expression pro-files of analysed cell lines (Figure 5). First, the correlation of gene expression profiles within the control group and within the tumour group was calculated. We found that the average Pearson correlation coefficient among the control group is 0.71, which is slightly lower than the average correlation between tumour cells at 0.84. When we pairwise correlated the expression profiles of control and tumour cells, we arrived at the average correlation coefficient of 0.56. Some of the gene expression profiles of control vs. tumour cells showed higher correlation coefficients than profiles of controls-only or tumour-only. For example, the profiles of HDFa_3 versus the expression profile of hRNA_1 control shows a correlation coefficient of 0.47. Conversely, the expression profile of HDFa_3 versus expression profile of SW1088_3 from the tumour group shows the correlation coefficient of 0.72.

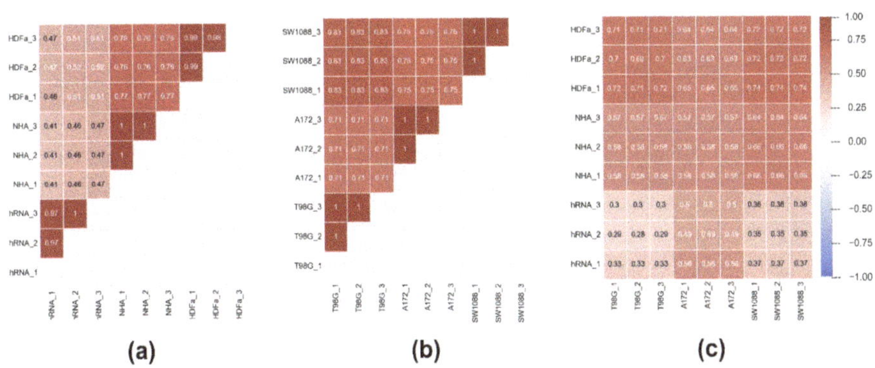

Figure 5. Pearson correlations between gene expression profiles of analysed cell lines. (**a**) Correlation of gene expression profiles within control group, (**b**) within tumour group and (**c**) between samples of control and tumour group.

2.4. Fold Change of Cancer Pathway-Related Genes

Figure 6 shows the fold regulation data obtained by correlating cancer cell lines (A172, SW1088, and T98G) with three distinct control groups (HDFa, NHA, and hRNA). Fold downregulation is represented by blue values, while fold upregulation is represented by red values. Changes in regulation that do not reach significance are highlighted in grey. Not available values (N/A) representing changes in expression are marked in white.

When comparing the expression of genes associated with angiogenesis, we obtained heterogeneous results (Figure 6). The significantly higher expression of *ANGPT1* was detected in the T98G glioblastoma cell line (2.20 compared to HDFa; 2.48 compared to hRNA; 27.98 compared to NHA; and 5.48 compared to all controls) and the SW1088 cell line compared to the NHA control (16.96). In contrast, data analysis of Angiopoietin 1 in A172 revealed a statistically significant decrease in expression levels relative to HDFa and hRNA controls. At the same time, the expression of the Angiopoietin 1 antagonist coding gene, *ANGPT2*, was found to be opposite in comparisons between SW1088 and a combination of controls (5.35). The fold regulation of *ANGPT2* was significantly lower in all tested cell lines compared to control RNAs isolated from the whole brain. In contrast, A172 and SW1088 lines exhibited elevated expression levels (7.57 and 56.2) compared to HDFa control. Compared to control samples, the expression of *CCL2* appeared to be stably increased across tested cell lines, whereas the expression of *FGF2* appeared to be stably decreased. We discovered the reduced expression of *FLT1* and *PGF*. *FLT1* expression was statistically significantly decreased in both glioblastoma cell lines, whereas *PGF* was not detected in glioblastoma cell line T98G (Figure 6).

Among the investigated apoptosis genes, lower expression of *APAF1*, *BIRC3*, *CFLAR*, and *XIAP* was detected in all test groups, with statistical significance and fold regulation varying according to divergent control groups (Figure 6). For instance, in the A172 group, expression of *BIRC3* was three times lower in correlation with NHA than in the hRNA control group. The same pattern was observed when fold regulation values were correlated with HDFa and hRNA. In the astrocytoma cell line, we found that the gene encoding Caspase 2 was upregulated in relation to HDFa control (3.15). *CASP2* deregulation was not significant in either glioblastoma cell line, but it was constantly increasing. In contrast to *CASP2*, the expression of another caspase family protease, *CASP9*, was inconsistently deregulated (Figure 6). CASP9 expression was only found to be negatively regulated in glioma cancer cell lines when compared to the hRNA control group. However, we found that all lines correlated to NHA had relatively high positive fold changes (Figure 6).

Our results in cell cycle-related genes indicate, with a few exceptions, a significant increase in the expression of genes involved in cell division (*CDC20*, *E2F4*, *MCM2*, *MKI67*, and *SKP2*). Notable is the similarity between the expression fold changes of the *CDC20* and *E2F4* genes in the control group (Figure 6). Decreased expression of *CCND3* was prominently identified in SW1088 in terms of hRNA and NHA control groups (−7.52 and −2.45). The expression of *Stathmin* was found to be decreased in all test groups. However, there are differences in fold regulation relative to controls. *WEE1* expression was found to be statistically significant in all glioma cell lines, but only when compared to hRNA controls (4.75, 2.79, 2.24).

We observed decreased expression of *BMI1*, *ETS2*, *IGFBP5*, *IGFBP7*, *SOD1*, and *TBX2* from all investigated genes associated with cellular ageing molecular pathways, but differences in negative fold regulation with respect to control groups were remarkable (Figure 6). For instance, among all glioma cancer cell lines, the fold change of *IGFBP7* expression was decreased approximately 16.5 times more in comparison to the NHA control group than in comparison to the hRNA control and approximately 2.5 times more in comparison to human dermal fibroblasts. To mention the regulation of mitogen-activated kinase's expression in glioblastoma cell lines, *MAP2K1*, *MAP2K3*, and *MAPK14* genes were only upregulated when compared to HDFa and NHA. All three MAP kinases within the tested cell lines were downregulated with varying statistical significance in terms of hRNA control.

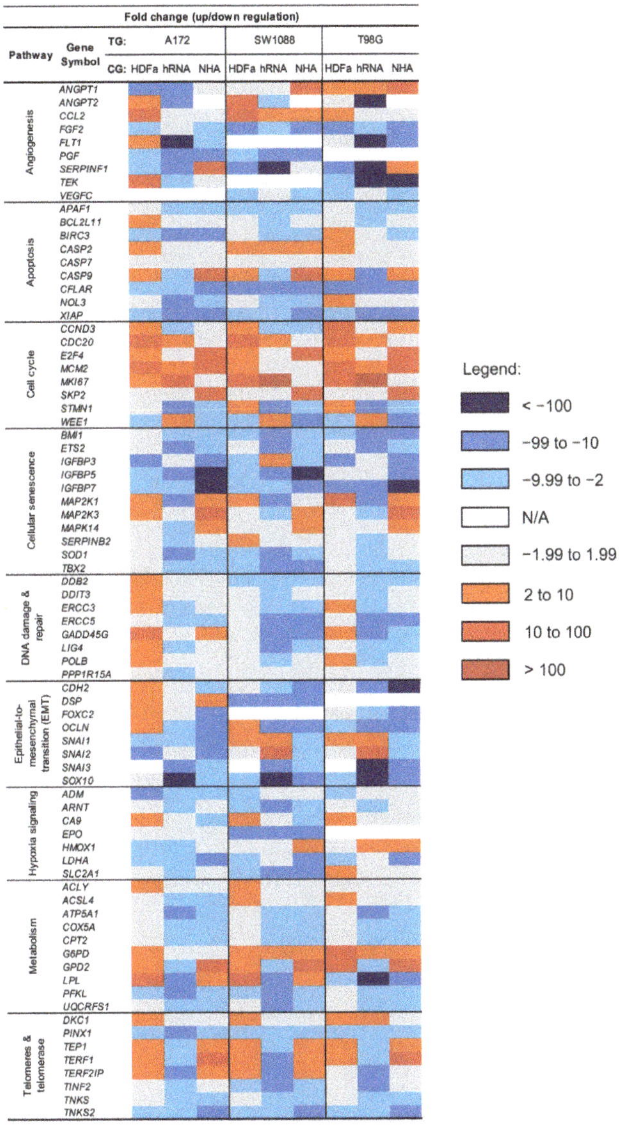

Figure 6. Heat map of fold change regulation between test and control group. Test group (TG): A172, SW1088 and T98G; Control group (CG): HDFa, hRNA and NHA.

Among the eight investigated genes involved in the response to DNA damage in the A172 group, *DDB2*, *GADD45G*, and *POLB* were significantly upregulated when compared to the HDFa control group (Figure 6). There was downregulation of *LIG4* in glioma cancer cell lines. In A172, the expression of this gene was approximately 4.2 times lower when compared to *LIG4* expression in hRNA control. SW1088 and T98G cell lines exhibit a similar pattern. Despite differences in fold change values due to control group selection, other statistically significant results from Figure 6 were broadly in line with expectations. *SNAI1* and *SNAI2* were downregulated genes related to the EMT pathway that we discovered in all cell lines, but only with regard to NHA control (Figure 6). When compared to HDFa and hRNA control groups, the expression of *SNAI1* and *SNAI2* in SW1088 and T98G lines was increased. In terms of hRNA, all glioma cell lines showed a significant decrease in *SOX10* expression (−4502; −26,634; −19,800), indicating that *SOX10* is highly expressed in our normal human brain control. In comparison to HDFa and NHA controls, *SOX10* appeared to be downregulated. Other detected downregulated genes were *CDH2*, *FOXC2* and *OCLN*. Downregulation of *FOXC2* was present only in the glioblastoma cell line. Fold regulation in SW1088 cell lines was not determined.

Analysis of hypoxia-signalling factors revealed a decrease in expression in the majority of genes, including *SLC2A1*, the gene- encoding Glucose transporter type 1 (Figure 6). This result is statistically significant only in SW1088 (−7.31; −32.3; −12.8) when compared to the hRNA control in the A172 cell line (−3.89). The Erythropoietin-encoding gene *EPO* appears to be exclusively downregulated in SW1088. In contrast, fold change was detected without statistical significance in the other two cell lines (A172) or was not detected at all (T98G).

Most cellular metabolism genes investigated in glioma cell lines (*ATP5A1*, *COX5A*, *CPT2*, *PFKL*, and *UQCRFS1*) were found to be statistically downregulated. *ACSL4* was found to be downregulated in only one of the glioma cell lines, A172. Although *GPD2* and *LPL* were found to be upregulated in A172 and SW1088, with similar fold changes and statistical significance when compared to HDFa, NHA, and a combination of controls, we discovered that both genes were downregulated when compared to the hRNA control group. In contrast, *LPL* expression in T98G was significantly reduced, with the lowest achieved value in comparison to the hRNA control group. *G6PD* upregulation was revealed to be statistically significant only in the SW1088 (5.46; 2.67; 3.34) and T98G (13.7; 6.67; 8.35) cell lines (Figure 6). We also observed increased expression of *ACLY*, but none of these differences was statistically significant.

Of the eight investigated genes involved in telomere maintenance and telomerase regulation, three (*PINX1*, *TNKS*, and *TNKS2*) were found to be downregulated in all glioma cell lines when compared to individual controls (Figure 6). No significant differences in fold changes of *PINX1* expression were found in A172 or SW1088 in comparison to all three negative controls (Figure 6). The fold change in expression in T98G was approximately two times lower (−2.31; −4.74; −3.98). We identified an increase in expression of *TEP1*, *TERF1*, and *TERF2IP* in glioblastoma and astrocytoma cell lines compared to HDFa and NHA control cell lines. The study found no statistically significant changes in *TERF2IP* expression in the T98G cell line. In contrast, we found a statistically significant decrease in *TEP1*, *TERF1*, and *TERF2IP* expression when compared to hRNA control.

2.5. Multi-Criteria Decision of Gene Expression in Sample Correlation

To determine the power of individual genes in groups, a multi-criteria decision support system was used. As a result, we were able to identify a group of genes in the correlation between controls and tumours.

The alternative method for assessing gene expression is based on separate, two-value quantification of the consequences of over- or underexpression. Not only is knowledge of the data carriers and gene expression important here, but so is the emphasis on the form of the set ordering in sets of triplicates. The results in the plots are represented by a normalised (dimensionless) two-factor form called R^+, R^-. Each R represents a proportion of the numerator's dissimilarity tendencies. Both measures preferentially account for when there is either an increase (superscript +) or decrease (superscript −) in gene expression in the test subject relative to the control. The denominators in the measures only serve for normalisation because they track intra-group differences in expression. For a robust linear combinatorial representation of stochastic measures, we proposed a data-driven approach inspired by the weighting theory of measures derived from multi-criteria decision support systems theory [11]. The details of our procedures for deriving R^+, R^-, as well as the methods of calculation, are difficult to explain concisely; therefore, we provide a detailed description of them in the Supplementary Material.

Figure 7 depicts the resultant correlation of genes (up or down) for individual pathways. As mentioned above, the individual points represent triplicate measurements of the logarithmic values of delta Ct for all samples, correlating the control and tumour groups. Correlation 2D plots of in-plane gene expression (R^+, R^-) constructed for tumour cell lines include comparison with controls. A 45-degree line separates the regions of overexpression and underexpression; at this line, overexpression and underexpression are therefore balanced, the scenarios when noise predominates are covered by the (0.1) × (0.1) square. This square region is hence statistically less important. In addition to the shown genes, it is necessary to mention points that far exceeded our proposed trend scale and were therefore deemed irrelevant. We identified the following genes as having decreased expression in tumour lines compared to controls: *FGF2, APAF1, CFLAR, XIAP, STMN1, WEE1, ERCC5, LIG4, PPP1R15A, OCLN, SOX10, ARNT, LDHA, ATP5A1, COX5A, CPT2, PFKL, UQCRFS1, ETS2, IGFBP3, IGFBP5, IGFBP7, SOD1, TBX2, PINX1, TINF2, TNKS*, and *TNKS2* (out of range). *CCL2, CASP2, CDC20* (out of range), *E2F4, MCM2, MKI67, SKP2, ACLY, G6PD, GPD2, MAPK14, DKC1*, and *TERF1* were the genes with increased levels in the tumour lines.

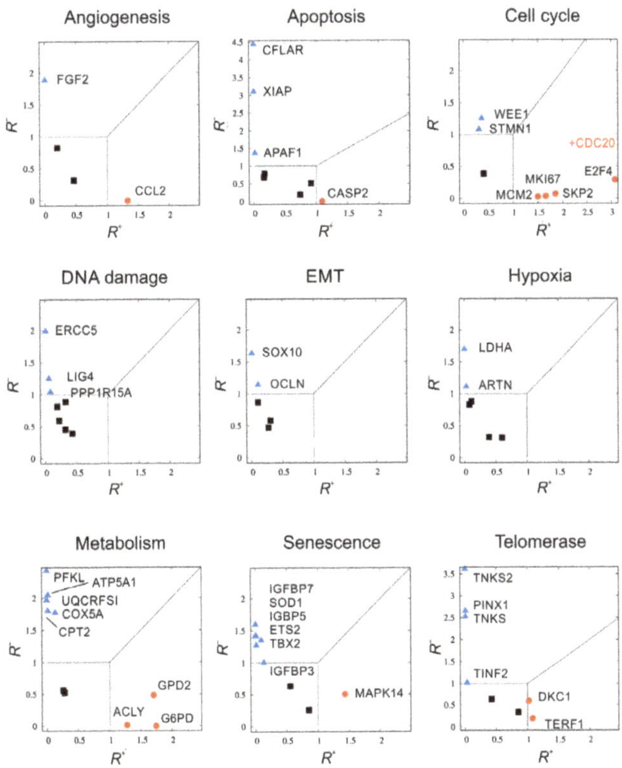

Figure 7. Multi-criterial correlation plots of gene expression in tumour cell lines relative to the controls. R^+ represents a negative proportion of the numerator's dissimilarity tendencies, R^- positive. A 45-degree line separates the regions of overexpression and underexpression; at this line, overexpression and underexpression are therefore balanced. The scenarios when noise predominates are covered by the (0.1) × (0.1) square. There are three main regions in the graph to categorize according to the level of gene expression with three ways of marking the corresponding points: statistically significant and overexpressed with $R^+ > 1$ (red circles), statistically significant underexpressed with $R^- > 1$ (blue triangles), finally statistically less significant (black squares) bordered by $0 \leq R^+ \leq 1, 0 \leq R^- \leq 1$.

3. Discussion

Glial tumours are biologically aggressive neoplasms with an abnormally high proliferative capacity and a diffuse invasion pattern. Glioblastoma (grade IV astrocytoma), composed of poorly differentiated neoplastic astrocytes, is the most malignant astrocytic tumour. Based on histopathological and molecular criteria, the WHO grading system categorises gliomas into grades I through IV [4]. Although the majority of neurological tumours derive from the glial lineage, it is unknown whether tumour cells arise from the transformation of an immature precursor or the dedifferentiation of a mature glial cell. Several genetic pathways are involved in the initiation and progression of these neoplasms, especially in the emergence of secondary GBMs.

In our study, we focused on the transcriptomic analysis of genes associated with the cancer pathways in glial tumour cells. As the experimental models, we chose the human glioma cell lines A172 (glioblastoma), SW1088 (astrocytoma), and T98G (glioblastoma). A172 and T98G cell lines are currently the most commonly used glioma cell lines for gene expression analysis. Weller's team performed the first large-scale analysis of 12 glioma cell lines, estimating the profile of 5800 genes. Their cluster and gene expression correlation analysis identified subsets of genes whose expression levels exhibited significant associations with drug sensitivity profiles [13]. Kiseleva et al. identified morphological, surface markers, and several growth factor genes or extracellular matrix genes in the characterisation of both glioblastomas, A172 and T98G [14]. Among the nine tested genes, the expression of *Alpha actin 2* was notably high in both cell lines. In addition, the data revealed a high level of activity of genes encoding major angiogenesis inducers (*VEGF, FGF2, TGFb1*) and *Thrombospondin-1*. The transcriptomic analysis of SW1088 cells was associated with individual genes or various inhibitory effects [15–17]. In our previous study, we determined the effect of ABT-737 and MIM-1 inhibitors on the mRNA level of apoptosis-associated genes in the T98G cell line [18]. As a control group, human astrocyte (HA) cells were used. Significant changes in apoptotic gene expression were obtained in both cell lines, with the greatest number of altered genes ($n = 42$) occurring in the HA line following MIM-1 treatment. Regardless of the genes involved in determining fold regulation between groups of samples, the choice of control samples will always be decisive. In addition, based on our final multivariate criterion, it is evident that results vary not only according to the choice of control but also according to the evaluation method employed. Therefore, the discussion will centre solely on the genes selected using the MCDA method.

Angiogenesis, as one of the hallmarks of cancer [19], plays a crucial role in glioblastoma growth through oncogene activation and/or downregulation of tumour suppressor genes, resulting in the upregulation of angiogenic pathways [20]. The initial step in the induction of angiogenesis in GBM is the overexpression and secretion of angiogenic growth factors, such as vascular endothelial growth factor (VEGF), followed by their binding to the receptors on epithelial cells [21]. Fibroblast growth factor 2 encoded by the *FGF2* gene is a crucial positive regulator of glioblastoma cell proliferation and survival [22]. However, our results showed a decrease in *FGF2* expression. Although the loss of the *FGF2 receptor* gene is associated with a poor prognosis in glioma patients [23], *FGF2* seems to be persistently expressed because it has been identified as an oncogenic factor in GBM [24], and its expression has been confirmed in other gliomas and meningiomas [25].

Apoptosis refers to a programmed cell death characterised by non-inflammatory cellular fragmentation [26]. It is an essential regulatory mechanism for cell proliferation and death. Intrinsic or extrinsic pathways can initiate apoptosis, with both leading to proteolytic activation of caspases and controlled cell death. Cancer cells have evolved mechanisms to sustain proliferative signalling, thereby sustaining cell growth and avoiding cell death. In the current study, only nine apoptotic genes were included in the array, with *caspase-2* and *caspase-7* being the most expressed in tumour lines. The remaining genes involved in apoptosis regulation were downregulated, including *CFLAR*, *XIAP* and *APAF1*. X-linked inhibitor of apoptosis protein is the most potent and best-defined anti-apoptotic IAP family member that directly counteracts apoptosis by binding to caspase-9 and the effector caspases-3 and -7 [27]. The X-linked inhibitor of apoptosis protein is abnormally expressed in a variety of human cancers [28,29]. Although this is not evident in brain tissue, Murphy et al. investigated the low levels of XIAP in GBM patients and brain cell lines [30]. The protein encoded by the antiapoptotic gene *CFLAR* is a Caspase-8 and FADD-like apoptosis regulator. By binding to the death receptor, it protects cells from cell death signalling and inhibits receptor-mediated apoptosis [31]. Induction of hypoxia in the A172 glioblastoma cell line results in the expression of *CFLAR* [32]. *CFLAR* expression was detected in glioblastoma tissue samples [33]. Despite these findings, our analysis showed a significant reduction in *CFLAR* expression. Apoptotic peptidase activating factor 1 (APAF1) is a proapoptotic protein that participates in the formation of apoptosomes in response to cell death signals [34]. Overexpression of *APAF1* induced apoptosis in U-373MG human glioma cells [35]. Our previous study on apoptotic gene expression revealed a slight decrease in the expression of *APAF1* in glioblastoma patient samples [36]. In this study, we confirmed our previous findings regarding glioma cell lines. A decrease in *APAF1* expression may lead to apoptosis reduction, thereby favouring cancer cell survival [36].

The main goal of the cell cycle is to ensure accurate DNA replication in the S phase and the final formation of two identical daughter cells in the mitotic phase. The cells use various checkpoints to maintain the optimal progression of the cell cycle, which will slow down or stop the event if necessary [37]. Ki-67, a prognostic and proliferative marker expressed by the *MKI67* gene in cell nuclei during the active phases of the cell cycle (G1-M) with maximum expression at the G2/M phase interface, is used to control the malignant nature of cells [38]. In gliomas, its elevated expression, which increases with malignancy grade, has been well characterised [39]. The absolute highest expression of all genes was observed in *CDC20*. The protein of the same name is responsible for regulating the mitotic phase of the cell cycle. Jeremy Rich's team identified an increased expression of *CDC20* in GBM compared to lower grade gliomas and healthy brain tissue. Their results also indicate the importance of *CDC20* proto-oncogene expression in glioblastoma stem cells, as it plays an essential role in the regulation of proliferation, self-renewal, and survival of these cells [40]. It even contributes to glioma chemoresistance. In accordance with the aforementioned studies, we also identified upregulation of the *CDC20* gene in all monitored groups.

Stathmin is an oncoprotein "18" that is distributed throughout the cytoplasm of cells and regulates microtubule kinetics, thereby affecting cell cycle proliferation and differentiation. Many studies indicate that *STMN1* expression is elevated in glioblastomas [41] and a variety of human cancers [42,43]. Our analyses identified a significant downregulation of *STMN1* in all tumour lines relative to the average control, suggesting reduced cell proliferation and tumour cell migration. The essential cell cycle regulator *WEE1* kinase was similarly underexpressed in tumour cell lines. Its primary function is to stop the progression of the cell cycle at the transition from G2 to the mitotic phase in cells with defectively replicated or damaged DNA [44]. In addition, glioblastoma patients whose *WEE1* expression is upregulated have a shorter survival rate [45,46]. We identified elevated levels of the *WEE1* gene in all tumour lines when compared to human RNA from healthy brain tissues.

The main role of DNA repair mechanisms is to respond to environmental factors that cause DNA damage [47]. These gene mutations can result in a diminished or impaired capacity to repair DNA and an accumulation of damaged DNA, which ultimately increases the risk of cancer. Furthermore, tumour cells overexpress the genes encoding DNA repair mechanisms, increasing repair capacity and treatment resistance [48]. DNA ligase IV joins single-strand breaks in a double-stranded polydeoxynucleotide in an ATP-dependent reaction, and its low expression results in inefficient function of the repair system. Compared to normal astrocytes, brain tumour lines had lower levels of the *LIG4* coding gene, and these findings correlated with transcriptomic and genomic analyses [49]. In all tumour cell lines, decreased *LIG4* expression was observed. We also observed a significant reduction in expression of the *ERCC5* gene, whose product is part of the nucleotide excision repair system [50]. Borderline low levels of the gene encoding Protein Phosphatase 1 Regulatory Subunit 15A have been confirmed.

Epithelial to mesenchymal transition (EMT) is the process by which epithelial cells lose their epithelial characteristics and acquire a mesenchymal phenotype, resulting in increased mobility and chemoresistance [51]. Although glioma cells are not of epithelial origin, an EMT-like process in GBM can be induced [52]. Overexpression of various growth factors, such as transforming growth factor (TGF), epidermal growth factor (EGF), fibroblast growth factor (FGF), and HIF-1, cause EMT in cancer cells [53]. As a result of growth factor-mediated signalling, transcription factors (Snail, Slug, dEF1, SIP1, Twist1, and FOXC2) are activated and induce an EMT-like phenotype [54]. Multiple signalling pathways participate in these processes. The phosphatidylinositol-3-kinase (PI3K)/Akt signalling pathway plays an important role in regulating cell growth and maintaining cancer biology. Cooperation with other signalling pathways such as transforming growth factor β (TGF-β), nuclear factor (NF)-κB, and Ras and Wnt signalling pathways leads to direct or indirect induction of the EMT process, resulting in enhanced invasiveness, aggression, chemoresistance, and apoptosis resistance of the tumour mass [55]. The transcription factor *SOX10* is one of the key determinants of oligodendroglial differentiation. Therefore, Bannykh and colleagues decided to compare the presence of *SOX10* in oligodendrogliomas and astrocytomas to determine its specificity. Although at lower levels [56], the majority of oligodendrogliomas and a significant proportion of astrocytomas, including glioblastomas, produced *SOX10*. Consistent with previous research, multivariable analysis confirmed a decrease in the expression of *SOX10* and *OCLN*, which belongs to the EMT group [57,58].

Cancer cells surrounding the necrotic nucleus lack nutrients and oxygen. Hypoxia is the primary physiological trigger of angiogenesis [59], which is activated by Hypoxia-inducible factor 1 [60]. *HMOX1* is one of the many genes expressed during hypoxia induced by HIF-1. Due to its antioxidant and antiapoptotic effects, Hemoxigenase 1 plays a crucial role in tumour growth [61]. Because *HMOX1* activity stimulates angiogenesis, this enzyme is a suitable indicator of glioma neovascularization [62]. Only the T98G glioblastoma cell line was found to have elevated levels of the *HMOX1* gene. In contrast, only *LDHA* and *ARNT* were downregulated relative to controls. Lactate dehydrogenase A (LDHA) is a key enzyme in the anaerobic glycolytic pathway [63]. In addition to promoting acidification of the microenvironment, lactate production promotes the metastatic nature of the tumour [64]. Several authors have reported on the significance of glioblastoma *LDHA* expression [65–67]. Chesnelong and colleagues found low expression and high methylation of *LDHA* in IDHmt glioblastomas [68]. Kathagen-Buhmann et al. identified a decline in LDHA production in non-migrated cells [67]. Since *LDHA* expression is promoted by hypoxia, low levels of *LDHA* in gliomas that were cultivated under standard conditions in the presence of oxygen may be attributable to an oxygenated environment.

The largest number of changes in gene expression (n = 8) between tumour and non-malignant groups were identified in metabolic genes (*PFKL*, *ATP5A1*, *UQCRFS1*, *CPT2*, *COX5A*, *ACLY*, *GPD2*, and *G6PD*). Even under aerobic conditions, tumour cells are known for their high glycolytic activity [69]. Along with an increase in glucose consumption and lactate production, this promotes rapid cell proliferation and GBM growth, which is correlated with the elevated activity of glycolytic enzymes [70]. Phosphofructokinase-1 is a regulatory glycolytic enzyme catalysing the phosphorylation of fructose-6-phosphate to fructose-1,6-bisphosphate. The presence of its liver isoform (PFKL) in gliomas was analysed by Stanke et al. However, they did not observe any statistically significant changes compared to healthy tissue [71]. In our samples, we identified a statistically significant decrease in *PFKL* expression, confirming that the prevalent isoform in brain tissue is the platelet isoform (PFKP), not PFKL [72]. A high expression of *Glucose-6-phosphate dehydrogenase* was spotted. Glucose-6-phosphate dehydrogenase (G6PD) is one of the pentose phosphate pathway (PPP) enzymes that catalyses the production of NADPH [73]. During normoxia, glioma and non-neoplastic brain cells both produce an abundance of these enzymes. A negative association between *G6PD* expression and survival in patients with low-grade glioma was discovered [74]. COX5A is one of the three subunits of cytochrome c oxidase, a respiratory chain complex IV encoded by mitochondrial DNA [75]. UQCRFS1 is a respiratory chain complex III subunit. Both *COX5A* and the gene encoding another subunit of complex III, *UQCRB*, are downregulated in glioblastoma patients compared to healthy individuals. In contrast, when compared to expression in gliomas of lower grade malignancy, *COX5A* expression is significantly increased in GBM, and *UQCRB* expression is at approximately the same level. The reduced expression has also been linked to a poor prognosis [71]. In line with previous findings, we were able to identify the downregulation of *COX5A* and *UQCRFS1*.

Cellular senescence is an irreversible process of cell cycle arrest [76]. During this process, senescent cells undergo morphological changes that include flattening, increased cytoplasmic volume, or increased granularity. Only the *MAPK14* gene was found to be overexpressed in our sample cohort, while six other genes were found to be underexpressed (*IGFBP3, IGFBP5, IGFBP7, SOD1, TBX2*, and *ETS2*). *ETS2* is a transcription factor that regulates apoptotic and angiogenic genes, as well as genes involved in proliferation and differentiation [77]. Cam et al. identified *ETS2* expression in glioblastomas and, in association with ΔNp73, confirmed its role in tumour progression, angiogenesis, and improved tumour cell survival [78]. On the contrary, bioinformatic analysis of transcriptomic data from glioma patients revealed a decrease in gene expression of *ETS2* regardless of the degree of malignancy [79]. In glioma cell lines, we found a statistically significant decrease in *ETS2* expression. Superoxide dismutase 1, encoded by the *SOD1* gene, is an enzyme that converts free superoxide radicals into less harmful hydrogen peroxide and oxygen [80]. A decrease in *SOD1* expression has been identified in glioblastomas and is associated with improved response to radiotherapy and a better prognosis for patients [81,82]. *SOD1* expression was also reduced in glioma cell lines when compared to non-malignant cell controls. The only overexpressed gene involved in the regulation of senescence is *MAPK14*. Mitogen-activated protein kinase 14, a protein product of *MAPK14*, is an essential component of the MAP kinase signal transduction pathway that influences the direct activation of transcription factors in response to cell stress stimuli [83]. *MAPK14* expression was found to be elevated in glioma cells, which is in contrast to the findings of other studies, which indicated that the expression of this gene was decreased in glioblastoma samples [84].

Human telomeres, located at the ends of chromatids, are tandem nucleotide repeats of a short DNA sequence associated with various telomere-binding proteins with a predominantly protective function [85]. The primary function of telomeres is to compensate for incomplete DNA replication at chromosome ends, thereby maintaining intact genetic information [86]. However, as a result of cell division, telomeres become progressively shorter, resulting in cellular senescence and apoptosis induction [87]. PINX1 was identified as a potent telomerase inhibitor that interacts directly with the catalytic activity of telomerase [88]. Our analysis revealed a decrease in *PINX1* expression. Previous studies have shown a correlation between a decrease in *PINX1* expression and the metastatic nature and poor prognosis of cancer patients [89]. In glioblastoma cell lines with induced overexpression of *PINX1*, there was a reduction in cell migration and proliferation due to cell cycle arrest at the G1 phase [90]. In contrast, there is evidence that *PINX1* expression is associated with poor survival in glioma patients because it promotes cell proliferation [75,91]. The *DKC1* gene encodes Dyskerin, an additional protein that regulates telomerase activity [92]. Glioma is one of several human cancers in which *DKC1* is upregulated [93,94]. Consistent with previous findings, elevated expression of *DKC1* in glioma cell lines was also identified. Tankyrases (TNKS, TNKS2) are proteins involved in telomere length maintenance [95], which, together with regulation of the Wnt/β-catenin pathway, is important for cancer cell renewal and survival [96]. Expression of *TNKS* and *TNKS2* was decreased in glioma cells compared to non-malignant cells and normal brain tissue. Additionally, the expression of *TINF2* was reduced. We identified an increase in *TERF* gene expression.

Glial tumours are biologically aggressive neoplasms with an elevated, often aberrant, and diffusely invading proliferative capacity. Composed of poorly differentiated neoplastic astrocytes, glioblastoma (grade IV astrocytoma) is the most malignant astrocytic tumour. According to histopathological and molecular criteria, the WHO grading system categorises gliomas into grades I through IV, based on their degree of malignancy. Although the majority of neurological tumours derive from the glial lineage, it is unclear whether tumour cells result from the transformation of an immature precursor or the dedifferentiation of a mature glial cell. Several genetic pathways are involved in the initiation and progression of these neoplasms, particularly during the manifestation of secondary GBMs.

4. Materials and Methods

4.1. Cell Culturing

The glioma tumour cell panel (T98G, A172 and SW1088) was purchased from American Type Culture Collection (ATCC) under catalogue numbers: CRL-1690™, CRL-1620™, HTB-12™; respectively). Cell cultures were maintained as monolayer in Dulbecco's modified Eagle's media with 25 mM glucose, and supplemented with foetal bovine serum (10%, v/v), and penicillin/streptomycin (1×; PAA). Normal Human Astrocytes (NHA) were provided from ATCC and cultured in Dulbecco's modified Eagle's media high glucose/F12 (1:1; Merck KGaA, Darmstadt, Germany) supplemented with foetal bovine serum (10%, v/v), and penicillin/streptomycin (1×; PAA). The Human Dermal Fibroblasts (HDFa; Gibco—Thermo Fisher Scientific, Waltham, MA, USA) was used as non-specific tissue control and cultured in HAM´s Nutrient Mixture F12 (Merck KGaA, Darmstadt, Germany) supplemented with foetal bovine serum (10% v/v; Gibco—Thermo Fisher Scientific, Waltham, MA, USA) and penicillin/streptomycin (1×; PAA Laboratories GmbH, Austria). Cells were cultured at 37 °C in an atmosphere of 5% CO_2. Before each experiment, single-cell suspension was prepared using 0.05% trypsin/EDTA solution, and cells were counted using Countess™ automated cell counter (Thermo Fisher Scientific, Waltham, MA, USA).

4.2. Control Brain RNA

Commercially available total RNA from human brain tissue of single healthy normal donor was used as a control group (HR-201, Human Brain Total RNA—Amsbio, Abingdon, UK) for quantitative PCR. For quantitative PCR analysis, we used three independent transcripts into cDNA.

4.3. RNA Extraction and cDNA Synthesis

Total RNA was isolated using AllPrep® DNA/RNA Mini Kit (Qiagen Inc., Germantown, MD, USA). Concentration of isolated RNA was measured in Implen P300 NanoPhotometer (Implen GmbH, München, Germany). Two micrograms of purified cellular RNA was converted to single-stranded cDNA using RT^2 First Strand Kit (330,401; Qiagen Inc., Germantown, MD, USA) according to the protocol supplied by the manufacturer.

4.4. Real-Time PCR Array

Real-time PCR (quantitative PCR) was carried out using RT^2 SYBR® Green Rox™ qPCR Mastermix (330,502; Qiagen Inc., Germantown, MD, USA) in 96-well plate format of the Human Cancer PathwayFinder™ PCR Array (PAHS-033ZC; Qiagen Inc., USA). The PCR reaction mix (SYBR® Green Rox™ qPCR Mastermix (1340 µL), PCR water (1290 µL) and cDNA (50 µL) was distributed into the 96-well plate to a final volume of 25 µL per well. The sealed plate was briefly centrifuged at 1000× g for 1 min. Amplification was performed in the ViiA7 Real-Time PCR system (Thermo Fisher Scientific, Waltham, MA, USA). After denaturation at 95 °C for 10 min, fluorescence was detected over 40 cycles (95 °C for 15 s, 60 °C for 1 min).

4.5. Statistical Analysis

Samples of cDNA were measured in triplicate, and the levels of the genes of interest were normalized to the three endogenous controls (β-actin, *ACTB*; Ribosomal protein large unit P0, *RPLP0* and Glyceraldehyde-3-phosphate dehydrogenase, *GAPDH*), determined using the ΔΔCt method. The expression data from the separate control group:

- Human Dermal Fibroblasts (HDFa);
- Normal Human Astrocytes (NHA);
- Human Brain Total RNA (hRNA).

These were used as a reference in the ΔΔCt method calculation for each glial cell line (T98G, A172 and SW1088) individually. The relative expression of 84 genes in tumour cell lines and non-neoplastic samples was calculated using the RT^2 Profiler PCR Array Data Analysis Web Portal (Qiagen) based on $2^{-\Delta\Delta Ct}$ method [97], where $\Delta\Delta Ct = (Ct_{GOI} - Ct_{HKG})_{TESTING\ GROUP} - (Ct_{GOI} - Ct_{HKG})_{CONTROL\ GROUP}$. Fold-change calculations were performed using Qiagen data analysis software (https://dataanalysis2.qiagen.com/pcr, accessed on 1 January 2022). The genes with a significant difference in expression were those with an average fold-change of ≤ -2.0 or ≥ 2.0, and statistically significant differences were those with a corresponding p value of <0.05.

Gene expression values were normalised to a 0–1 scale for both control and tumour cell lines. Using the KNIME Analytics Tool, sample normalisation, principal component analysis, and linear discriminant analysis were calculated.

For the statistical analyses mentioned above, only genes with detectable signals in all samples were selected from the raw data set. The Euclidean distance was used to calculate the distance between the samples. Using Pearson correlation, gene–gene expressions of control and tumour samples were correlated, respectively. The Pearson correlation coefficient for the two populations (X, and Y) is calculated as follows:

$$\rho_{X,Y} = \frac{cov(X,Y)}{\sigma_X \sigma_Y}$$

where $cov(X,Y)$ is the covariance; σ_X is the standard deviation of X; and σ_Y is the standard deviation of Y. The analyses were calculated in Python using Anaconda Navigator and JupyterLab. The Pandas, NumPy, and SciPy libraries were used. Kernel density and bandwidth optimisation were calculated using the Shimazaki and Shinomoto web application (https://www.neuralengine.org/res/kernel.html, accessed on 8 March 2022). We used the Matplotlib Python library to visualise the heatmap of gene correlations.

5. Conclusions

We focused on the transcriptomic analysis of genes associated with cancer pathways in glial tumour cells. As the experimental models, we selected the human glioblastoma cell lines A172 and T98G and the astrocytoma cell line SW1088. Sixty genes were deregulated in glioblastoma cell line A172 in comparison to the HDFa control group; 57 genes in comparison to the human RNA control group; and 54 genes in comparison to the human astrocytes control group, according to transcriptomic data. In the astrocytoma cell line SW1088, we found differences in the expression levels of 57, 60, and 59 genes related to HDFa, hRNA, and NHA control groups, respectively. In correlation with T98G and HDFa, 47 significantly deregulated genes were discovered. With hRNA, 57 genes, and the NHA control group, 52 genes with varying expression levels were identified. By combining the PCA method and multi-criteria decision in the analysis of gene expression, we were able to identify altered genes involved in cancer pathways in heterogeneous sample groups. We managed to reduce the selection of significant genes based on a combined mathematical analysis. In tumour cells, we finally identified 26 genes that showed a deregulated state compared to the average expression value of three different controls. The most changed genes represented pathways involved in cellular senescence (*BM1, ETS2, IGFBP5, IGFBP7, SOD1* and *TBX2*) and then metabolism (*ATP5A1, COX5A, CPT2, PFKL, UQCRFS1*).

Supplementary Materials: The following supporting information can be downloaded at: https://www.mdpi.com/article/10.3390/ijms231810883/s1. References [98–100] are cited in the supplementary materials.

Author Contributions: Conceptualization, J.H. and G.Z.; methodology, Z.M., M.G. and D.H.; formal analysis, R.M. and G.Z.; investigation, M.G. and G.Z.; resources, J.H. and G.Z.; data curation, Z.M., K.D., M.G. and D.H.; writing—original draft preparation, Z.M., M.G. and D.H.; writing—review and editing, J.H., D.H. and R.M.; visualization, J.H.; supervision, J.H. and G.Z.; project administration, J.H.; funding acquisition, J.H., M.G. and G.Z. All authors have read and agreed to the published version of the manuscript.

Funding: This work was supported by the Slovak Research and Development Agency under the contract no. APVV-18-0088 and by the project implementation: Open scientific community for modern interdisciplinary research in medicine (Acronym: OPENMED), ITMS2014+: 313011V455 supported by the Operational Programme Integrated Infrastructure. M.G. was funded by the ERDF and Internal Scientific Grants System (VVGS) of Faculty of Science UPJŠ (no. vvgs-pf-2022-2116).

Institutional Review Board Statement: Not applicable.

Informed Consent Statement: Not applicable.

Data Availability Statement: The data generated and analysed during the current study are available from the corresponding author upon reasonable request.

Conflicts of Interest: The authors declare no conflict of interest.

References

1. Louis, D.N.; Perry, A.; Reifenberger, G.; von Deimling, A.; Figarella-Branger, D.; Cavenee, W.K.; Ohgaki, H.; Wiestler, O.D.; Kleihues, P.; Ellison, D.W. The 2016 World Health Organization Classification of Tumors of the Central Nervous System: A summary. *Acta Neuropathol.* **2016**, *131*, 803–820. [CrossRef]
2. Stupp, R.; Mason, W.P.; Van den Bent, M.J.; Weller, M.; Fisher, B.; Taphoorn, M.J.; Belanger, K.; Brandes, A.A.; Marosi, C.; Bogdahn, U.; et al. Radiotherapy plus concomitant and adjuvant temozolomide for glioblastoma. *N. Engl. J. Med.* **2005**, *352*, 987–996. [CrossRef]
3. Wen, P.Y.; Kesari, S. Malignant gliomas in adults. *N. Engl. J. Med.* **2008**, *359*, 492–507. [CrossRef]
4. Louis, D.N.; Perry, A.; Wesseling, P.; Brat, D.J.; Cree, I.A.; Figarella-Branger, D.; Hawkins, C.; Ng, H.K.; Pfister, S.M.; Reifenberger, G.; et al. The 2021 WHO Classification of Tumors of the Central Nervous System: A summary. *Neuro-Oncology* **2021**, *23*, 1231–1251. [CrossRef]
5. Gritsch, S.; Batchelor, T.T.; Gonzalez Castro, L.N. Diagnostic, therapeutic, and prognostic implications of the 2021 World Health Organization classification of tumors of the central nervous system. *Cancer* **2022**, *128*, 47–58. [CrossRef]
6. Soomro, S.H.; Ting, L.R.; Qing, Y.Y.; Ren, M. Molecular biology of glioblastoma: Classification and mutational locations. *J. Pak. Med. Assoc.* **2017**, *67*, 1410–1414. [PubMed]
7. Yang, P.H.; Tao, Y.; Luo, J.; Paturu, M.; Lu, H.C.; Ramkissoon, S.; Heusel, J.W.; Leuthardt, E.C.; Chicoine, M.R.; Dowling, J.L.; et al. Multivariate analysis of associations between clinical sequencing and outcome in glioblastoma. *Neurooncol. Adv.* **2022**, *4*, vdac002. [CrossRef]
8. Aum, D.J.; Kim, D.H.; Beaumont, T.L.; Leuthardt, E.C.; Dunn, G.P.; Kim, A.H. Molecular and cellular heterogeneity: The hallmark of glioblastoma. *Neurosurg. Focus* **2014**, *37*, E11. [CrossRef]
9. Richterová, R.; Jurečeková, J.; Evinová, A.; Kolarovszki, B.; Benčo, M.; De Riggo, J.; Sutovský, J.; Mahmood, S.; Račay, P.; Dobrota, D. Most frequent molecular and immunohistochemical markers present in selected types of brain tumors. *Gen. Physiol. Biophys.* **2014**, *33*, 259–279. [CrossRef]
10. Mansouri, A.; Hachem, L.D.; Mansouri, S.; Nassiri, F.; Laperriere, N.J.; Xia, D.; Lindeman, N.I.; Wen, P.Y.; Chakravarti, A.; Mehta, M.P.; et al. MGMT promoter methylation status testing to guide therapy for glioblastoma: Refining the approach based on emerging evidence and current challenges. *Neuro-Oncology* **2019**, *21*, 167–178. [CrossRef]
11. Zhou, W.; Wahl, D.R. Metabolic Abnormalities in Glioblastoma and Metabolic Strategies to Overcome Treatment Resistance. *Cancers* **2019**, *11*, 1231. [CrossRef]
12. Lee, Y.J.; Seo, H.W.; Baek, J.H.; Lim, S.H.; Hwang, S.G.; Kim, E.H. Gene expression profiling of glioblastoma cell lines depending on TP53 status after tumor-treating fields (TTFields) treatment. *Sci. Rep.* **2020**, *10*, 12272. [CrossRef]
13. Weller, M.; Rieger, J.; Grimmel, C.; Van Meir, E.G.; De Tribolet, N.; Krajewski, S.; Reed, J.C.; Von Deimling, A.; Dichgans, J. Predicting chemoresistance in human malignant glioma cells: The role of molecular genetic analyses. *Int. J. Cancer* **1998**, *18*, 640–644. [CrossRef]
14. Kiseleva, L.N.; Kartashev, A.V.; Vartanyan, N.L.; Pinevich, A.A.; Samoilovich, M.P. Characteristics of A172 and T98G cell lines. *Tsitologiia* **2016**, *58*, 349–355. [CrossRef]
15. García-Claver, A.; Lorente, M.; Mur, P.; Campos-Martín, Y.; Mollejo, M.; Velasco, G.; Meléndez, B. Gene expression changes associated with erlotinib response in glioma cell lines. *Eur. J. Cancer* **2013**, *49*, 1641–1653. [CrossRef]
16. Lee, S.Y.; Slagle-Webb, B.; Rizk, E.; Patel, A.; Miller, P.A.; Sung, S.S.; Connor, J.R. Characterization of a novel anti-cancer compound for astrocytomas. *PLoS ONE* **2014**, *9*, e108166. [CrossRef]

17. Menezes, W.P.; Silva, V.; Gomes, I.; Rosa, M.N.; Spina, M.; Carloni, A.C.; Alves, A.; Melendez, M.; Almeida, G.C.; Silva, L.; et al. Loss of 5′-Methylthioadenosine Phosphorylase (MTAP) is Frequent in High-Grade Gliomas; Nevertheless, It Is Not Associated with Higher Tumor Aggressiveness. *Cells* **2020**, *9*, 492. [CrossRef]
18. Vidomanova, E.; Racay, P.; Pilchova, I.; Halasova, E.; Hatok, J. Microfluidic profiling of apoptosis-related genes after treatment with BH3-mimetic agents in astrocyte and glioblastoma cell lines. *Oncol. Rep.* **2016**, *36*, 3188–3196. [CrossRef]
19. Hanahan, D.; Weinberg, R.A. The hallmarks of cancer. *Cell* **2000**, *100*, 57–70. [CrossRef]
20. Takano, S.; Yamashita, T.; Ohneda, O. Molecular therapeutic targets for glioma angiogenesis. *J. Oncol.* **2010**, *2010*, 351908. [CrossRef]
21. Plate, K.H.; Risau, W. Angiogenesis in malignant gliomas. *Glia* **1995**, *15*, 339–347. [CrossRef]
22. Wang, F.; Yang, L.; Shi, L.; Li, Q.; Zhang, G.; Wu, J.; Zheng, J.; Jiao, B. Nuclear translocation of fibroblast growth factor-2 (FGF2) is regulated by Karyopherin-β2 and Ran GTPase in human glioblastoma cells. *Oncotarget* **2015**, *6*, 21468–21478. [CrossRef]
23. Ohashi, R.; Matsuda, Y.; Ishiwata, T.; Naito, Z. Downregulation of fibroblast growth factor receptor 2 and its isoforms correlates with a high proliferation rate and poor prognosis in high-grade glioma. *Oncol. Rep.* **2014**, *32*, 1163–1169. [CrossRef]
24. Haley, E.M.; Kim, Y. The role of basic fibroblast growth factor in glioblastoma multiforme and glioblastoma stem cells and in their in vitro culture. *Cancer Lett.* **2014**, *346*, 1–5. [CrossRef]
25. Takahashi, J.A.; Mori, H.; Fukumoto, M.; Igarashi, K.; Jaye, M.; Oda, Y.; Kikuchi, H.; Hatanaka, M. Gene expression of fibroblast growth factors in human gliomas and meningiomas: Demonstration of cellular source of basic fibroblast growth factor mRNA and peptide in tumor tissues. *Proc. Natl. Acad. Sci. USA* **1990**, *87*, 5710–5714. [CrossRef]
26. Wong, R.S. Apoptosis in cancer: From pathogenesis to treatment. *J. Exp. Clin. Cancer Res.* **2011**, *30*, 87. [CrossRef]
27. Obexer, P.; Ausserlechner, M.J. X-linked inhibitor of apoptosis protein—A critical death resistance regulator and therapeutic target for personalized cancer therapy. *Front. Oncol.* **2014**, *4*, 197. [CrossRef]
28. Gao, X.; Zhang, L.; Wei, Y.; Yang, Y.; Li, J.; Wu, H.; Yin, Y. Prognostic Value of XIAP Level in Patients with Various Cancers: A Systematic Review and Meta-Analysis. *J. Cancer* **2019**, *10*, 1528–1537. [CrossRef]
29. Mizutani, Y.; Nakanishi, H.; Li, Y.N.; Matsubara, H.; Yamamoto, K.; Sato, N.; Shiraishi, T.; Nakamura, T.; Mikami, K.; Okihara, K.; et al. Overexpression of XIAP expression in renal cell carcinoma predicts a worse prognosis. *Int. J. Oncol.* **2007**, *30*, 919–925. [CrossRef]
30. Murphy, Á.C.; Weyhenmeyer, B.; Schmid, J.; Kilbride, S.M.; Rehm, M.; Huber, H.J.; Senft, C.; Weissenberger, J.; Seifert, V.; Dunst, M.; et al. Activation of executioner caspases is a predictor of progression-free survival in glioblastoma patients: A systems medicine approach. *Cell Death Dis.* **2013**, *4*, e629. [CrossRef]
31. Chandrasekaran, Y.; McKee, C.M.; Ye, Y.; Richburg, J.H. Influence of TRP53 status on FAS membrane localization, CFLAR (c-FLIP) ubiquitinylation, and sensitivity of GC-2spd (ts) cells to undergo FAS-mediated apoptosis. *Biol. Reprod.* **2006**, *74*, 560–568. [CrossRef] [PubMed]
32. Baghbani, F.; Raoofian, R.; Hasanzadeh Nazarabadi, M.; Hamzehloei, T.; Soukhtanloo, M.; Heidari, M.; Afsharzadeh, S.M.; Shekouhi, S.; Moradi, F.; Sarli, A.A.; et al. Identification of novel hypoxia response genes in human glioma cell line a172. *Iran. J. Basic Med. Sci.* **2013**, *16*, 675–682.
33. Brahm, C.G.; Abdul, U.K.; Houweling, M.; van Linde, M.E.; Lagerweij, T.; Verheul, H.; Westerman, B.A.; Walenkamp, A.; Fehrmann, R. Data-driven prioritization and preclinical evaluation of therapeutic targets in glioblastoma. *Neurooncol. Adv.* **2020**, *2*, vdaa151. [CrossRef] [PubMed]
34. Fullstone, G.; Bauer, T.L.; Guttà, C.; Salvucci, M.; Prehn, J.H.M.; Rehm, M. The apoptosome molecular timer synergises with XIAP to suppress apoptosis execution and contributes to prognosticating survival in colorectal cancer. *Cell Death Differ.* **2020**, *27*, 2828–2842. [CrossRef] [PubMed]
35. Shinoura, N.; Sakurai, S.; Asai, A.; Kirino, T.; Hamada, H. Over-expression of APAF-1 and caspase-9 augments radiation-induced apoptosis in U-373MG glioma cells. *Int. J. Cancer* **2001**, *93*, 252–261. [CrossRef]
36. Blahovcova, E.; Richterova, R.; Kolarovszki, B.; Dobrota, D.; Racay, P.; Hatok, J. Apoptosis-related gene expression in tumor tissue samples obtained from patients diagnosed with glioblastoma multiforme. *Int. J. Mol. Med.* **2015**, *36*, 1677–1684. [CrossRef] [PubMed]
37. Vitovcova, B.; Skarkova, V.; Rudolf, K.; Rudolf, E. Biology of Glioblastoma Multiforme-Exploration of Mitotic Catastrophe as a Potential Treatment Modality. *Int. J. Mol. Sci.* **2020**, *21*, 5324. [CrossRef]
38. Mateoiu, C.; Pirici, A.; Bogdan, F. Immunohistochemical nuclear staining for p53, PCNA, Ki-67 and bcl-2 in different histologic variants of basal cell carcinoma. *Rom. J. Morphol. Embryol.* **2011**, *52* (Suppl. S1), 315–319.
39. Virga, J.; Szivos, L.; Hortobágyi, T.; Chalsaraei, M.K.; Zahuczky, G.; Steiner, L.; Tóth, J.; Reményi-Puskár, J.; Bognár, L.; Klekner, A. Extracellular matrix differences in glioblastoma patients with different prognoses. *Oncol. Lett.* **2019**, *17*, 797–806. [CrossRef]
40. Xie, Q.; Wu, Q.; Mack, S.C.; Yang, K.; Kim, L.; Hubert, C.G.; Flavahan, W.A.; Chu, C.; Bao, S.; Rich, J.N. CDC20 maintains tumor initiating cells. *Oncotarget* **2015**, *6*, 13241–13254. [CrossRef]
41. Dong, B.; Mu, L.; Qin, X.; Qiao, W.; Liu, X.; Yang, L.; Xue, L.; Rainov, N.G.; Liu, X. Stathmin expression in glioma-derived microvascular endothelial cells: A novel therapeutic target. *Oncol. Rep.* **2012**, *27*, 714–718. [CrossRef] [PubMed]
42. Hemdan, T.; Lindén, M.; Lind, S.B.; Namuduri, A.V.; Sjöstedt, E.; de Ståhl, T.D.; Asplund, A.; Malmström, P.U.; Segersten, U. The prognostic value and therapeutic target role of stathmin-1 in urinary bladder cancer. *Br. J. Cancer* **2014**, *111*, 1180–1187. [CrossRef] [PubMed]

43. Sun, R.; Liu, Z.; Wang, L.; Lv, W.; Liu, J.; Ding, C.; Yuan, Y.; Lei, G.; Xu, C. Overexpression of stathmin is resistant to paclitaxel treatment in patients with non-small cell lung cancer. *Tumour Biol.* **2015**, *36*, 7195–7204. [CrossRef]
44. Ghelli Luserna di Rorà, A.; Cerchione, C.; Martinelli, G.; Simonetti, G. A WEE1 family business: Regulation of mitosis, cancer progression, and therapeutic target. *J. Hematol. Oncol.* **2020**, *13*, 126. [CrossRef] [PubMed]
45. Lescarbeau, R.S.; Lei, L.; Bakken, K.K.; Sims, P.A.; Sarkaria, J.N.; Canoll, P.; White, F.M. Quantitative Phosphoproteomics Reveals Wee1 Kinase as a Therapeutic Target in a Model of Proneural Glioblastoma. *Mol. Cancer. Ther.* **2016**, *15*, 1332–1343. [CrossRef] [PubMed]
46. Tashnizi, A.H.; Jaberipour, M.; Razmkhah, M.; Rahnama, S.; Habibagahi, M. Tumour suppressive effects of WEE1 gene silencing in neuroblastomas. *J. Cancer Res. Ther.* **2016**, *12*, 221–227. [CrossRef]
47. Wood, R.D.; Mitchell, M.; Sgouros, J.; Lindahl, T. Human DNA repair genes. *Science* **2001**, *291*, 1284–1289. [CrossRef] [PubMed]
48. Dizdaroglu, M. Oxidatively induced DNA damage and its repair in cancer. *Mutat. Res. Rev. Mutat. Res.* **2015**, *763*, 212–245. [CrossRef]
49. Toma, M.; Witusik-Perkowska, M.; Szwed, M.; Stawski, R.; Szemraj, J.; Drzewiecka, M.; Nieborowska-Skorska, M.; Radek, M.; Kolasa, P.; Matlawska-Wasowska, K.; et al. Eradication of LIG4-deficient glioblastoma cells by the combination of PARP inhibitor and alkylating agent. *Oncotarget* **2018**, *9*, 36867–36877. [CrossRef]
50. Qian, T.; Zhang, B.; Qian, C.; He, Y.; Li, Y. Association between common polymorphisms in ERCC gene and glioma risk: A meta-analysis of 15 studies. *Medicine* **2017**, *96*, e6832. [CrossRef]
51. Armento, A.; Ehlers, J.; Schötterl, S.; Naumann, U. Molecular Mechanisms of Glioma Cell Motility. In *Glioblastoma*; De Vleeschouwer, S., Ed.; Codon Publication: Brisbane, Australia, 2017; Chapter 5; pp. 73–93, ISBN 978-0-9944381-2-6.
52. Iwadate, Y. Epithelial-mesenchymal transition in glioblastoma progression. *Oncol. Lett.* **2016**, *11*, 1615–1620. [CrossRef] [PubMed]
53. Lindsey, S.; Langhans, S.A. Crosstalk of oncogenic signaling pathways during epithelial-mesenchymal transition. *Front. Oncol.* **2014**, *4*, 358. [CrossRef] [PubMed]
54. Yang, J.; Weinberg, R.A. Epithelial-mesenchymal transition: At the crossroads of development and tumor metastasis. *Dev. Cell* **2008**, *14*, 818–829. [CrossRef] [PubMed]
55. Xu, W.; Yang, Z.; Lu, N. A new role for the PI3K/Akt signaling pathway in the epithelial-mesenchymal transition. *Cell Adh. Migr.* **2015**, *9*, 317–324. [CrossRef] [PubMed]
56. Bannykh, S.I.; Stolt, C.C.; Kim, J.; Perry, A.; Wegner, M. Oligodendroglial-specific transcriptional factor SOX10 is ubiquitously expressed in human gliomas. *J. Neurooncol.* **2006**, *76*, 115–127. [CrossRef] [PubMed]
57. Schaffenrath, J.; Wyss, T.; He, L.; Rushing, E.J.; Delorenzi, M.; Vasella, F.; Regli, L.; Neidert, M.C.; Keller, A. Blood-brain barrier alterations in human brain tumors revealed by genome-wide transcriptomic profiling. *Neuro. Oncol.* **2021**, *23*, 2095–2106. [CrossRef]
58. Ferletta, M.; Uhrbom, L.; Olofsson, T.; Pontén, F.; Westermark, B. Sox10 has a broad expression pattern in gliomas and enhances platelet-derived growth factor-B-induced gliomagenesis. *Mol. Cancer Res.* **2007**, *5*, 891–897. [CrossRef]
59. Liao, D.; Johnson, R.S. Hypoxia: A key regulator of angiogenesis in cancer. *Cancer Metastasis Rev.* **2007**, *26*, 281–290. [CrossRef]
60. Wykoff, C.C.; Beasley, N.J.; Watson, P.H.; Turner, K.J.; Pastorek, J.; Sibtain, A.; Wilson, G.D.; Turley, H.; Talks, K.L.; Maxwell, P.H.; et al. Hypoxia-inducible expression of tumor-associated carbonic anhydrases. *Cancer Res.* **2000**, *60*, 7075–7083.
61. Tanaka, S.; Akaike, T.; Fang, J.; Beppu, T.; Ogawa, M.; Tamura, F.; Miyamoto, Y.; Maeda, H. Antiapoptotic effect of haem oxygenase-1 induced by nitric oxide in experimental solid tumour. *Br. J. Cancer* **2003**, *88*, 902–909. [CrossRef]
62. El Andaloussi, A.; Lesniak, M.S. CD4+ CD25+ FoxP3+ T-cell infiltration and heme oxygenase-1 expression correlate with tumor grade in human gliomas. *J. Neurooncol.* **2007**, *83*, 145–152. [CrossRef] [PubMed]
63. Ward, P.S.; Thompson, C.B. Metabolic reprogramming: A cancer hallmark even warburg did not anticipate. *Cancer Cell* **2012**, *21*, 297–308. [CrossRef]
64. Han, T.; Kang, D.; Ji, D.; Wang, X.; Zhan, W.; Fu, M.; Xin, H.B.; Wang, J.B. How does cancer cell metabolism affect tumor migration and invasion? *Cell Adh. Migr.* **2013**, *7*, 395–403. [CrossRef] [PubMed]
65. Li, J.; Zhu, S.; Tong, J.; Hao, H.; Yang, J.; Liu, Z.; Wang, Y. Suppression of lactate dehydrogenase A compromises tumor progression by downregulation of the Warburg effect in glioblastoma. *Neuroreport* **2016**, *27*, 110–115. [CrossRef] [PubMed]
66. Crane, C.A.; Austgen, K.; Haberthur, K.; Hofmann, C.; Moyes, K.W.; Avanesyan, L.; Fong, L.; Campbell, M.J.; Cooper, S.; Oakes, S.A.; et al. Immune evasion mediated by tumor-derived lactate dehydrogenase induction of NKG2D ligands on myeloid cells in glioblastoma patients. *Proc. Natl. Acad. Sci. USA* **2014**, *111*, 12823–12828. [CrossRef]
67. Kathagen-Buhmann, A.; Schulte, A.; Weller, J.; Holz, M.; Herold-Mende, C.; Glass, R.; Lamszus, K. Glycolysis and the pentose phosphate pathway are differentially associated with the dichotomous regulation of glioblastoma cell migration versus proliferation. *Neuro Oncol.* **2016**, *18*, 1219–1229. [CrossRef]
68. Chesnelong, C.; Chaumeil, M.M.; Blough, M.D.; Al-Najjar, M.; Stechishin, O.D.; Chan, J.A.; Pieper, R.O.; Ronen, S.M.; Weiss, S.; Luchman, H.A.; et al. Lactate dehydrogenase A silencing in IDH mutant gliomas. *Neuro-Oncology* **2014**, *16*, 686–695. [CrossRef]
69. Hanahan, D.; Weinberg, R.A. Hallmarks of cancer: The next generation. *Cell* **2011**, *144*, 646–674. [CrossRef]
70. Sanzey, M.; Abdul Rahim, S.A.; Oudin, A.; Dirkse, A.; Kaoma, T.; Vallar, L.; Herold-Mende, C.; Bjerkvig, R.; Golebiewska, A.; Niclou, S.P. Comprehensive analysis of glycolytic enzymes as therapeutic targets in the treatment of glioblastoma. *PLoS ONE* **2015**, *10*, e0123544. [CrossRef]

71. Stanke, K.M.; Wilson, C.; Kidambi, S. High Expression of Glycolytic Genes in Clinical Glioblastoma Patients Correlates with Lower Survival. *Front. Mol. Biosci.* **2021**, *8*, 752404. [CrossRef]
72. Lee, J.H.; Liu, R.; Li, J.; Zhang, C.; Wang, Y.; Cai, Q.; Qian, X.; Xia, Y.; Zheng, Y.; Piao, Y.; et al. Stabilization of phosphofructokinase 1 platelet isoform by AKT promotes tumorigenesis. *Nat. Commun.* **2017**, *8*, 949. [CrossRef] [PubMed]
73. Kathagen, A.; Schulte, A.; Balcke, G.; Phillips, H.S.; Martens, T.; Matschke, J.; Günther, H.S.; Soriano, R.; Modrusan, Z.; Sandmann, T.; et al. Hypoxia and oxygenation induce a metabolic switch between pentose phosphate pathway and glycolysis in glioma stem-like cells. *Acta Neuropathol.* **2013**, *126*, 763–780. [CrossRef] [PubMed]
74. Yang, C.A.; Huang, H.Y.; Lin, C.L.; Chang, J.G. G6PD as a predictive marker for glioma risk, prognosis and chemosensitivity. *J. Neurooncol.* **2018**, *139*, 661–670. [CrossRef] [PubMed]
75. Bai, J.; Chen, Y.S.; Mei, P.J.; Liu, Q.H.; Du, Y.; Zheng, J.N. PinX1 is up-regulated and associated with poor patients' survival in gliomas. *Int. J. Clin. Exp. Pathol.* **2015**, *8*, 6952–6959.
76. Sharpless, N.E.; Sherr, C.J. Forging a signature of in vivo senescence. *Nat. Rev. Cancer* **2015**, *15*, 397–408. [CrossRef]
77. Fry, E.A.; Inoue, K. Aberrant expression of ETS1 and ETS2 proteins in cancer. *Cancer Rep. Rev.* **2018**, *2*, 1000151. [CrossRef]
78. Cam, M.; Charan, M.; Welker, A.M.; Dravid, P.; Studebaker, A.W.; Leonard, J.R.; Pierson, C.R.; Nakano, I.; Beattie, C.E.; Hwang, E.I.; et al. ΔNp73/ETS2 complex drives glioblastoma pathogenesis—Targeting downstream mediators by rebastinib prolongs survival in preclinical models of glioblastoma. *Neuro-Oncology* **2020**, *22*, 345–356. [CrossRef]
79. Babal, Y.K.; Kandemir, B.; Kurnaz, I.A. Gene Regulatory Network of ETS Domain Transcription Factors in Different Stages of Glioma. *J. Pers. Med.* **2021**, *11*, 138. [CrossRef]
80. Noor, R.; Mittal, S.; Iqbal, J. Superoxide dismutase—Applications and relevance to human diseases. *Med. Sci. Monit.* **2002**, *8*, RA210-5.
81. Gao, Z.; Sarsour, E.H.; Kalen, A.L.; Li, L.; Kumar, M.G.; Goswami, P.C. Late ROS accumulation and radiosensitivity in SOD1-overexpressing human glioma cells. *Free Radic. Biol. Med.* **2008**, *45*, 1501–1509. [CrossRef]
82. Lu, W.C.; Xie, H.; Yuan, C.; Li, J.J.; Li, Z.Y.; Wu, A.H. Identification of potential biomarkers and candidate small molecule drugs in glioblastoma. *Cancer Cell Int.* **2020**, *20*, 419. [CrossRef] [PubMed]
83. Igea, A.; Nebreda, A.R. The stress kinase p38a as a target for cancer therapy. *Cancer Res.* **2015**, *75*, 3997–4002. [CrossRef] [PubMed]
84. Ariey-Bonnet, J.; Carrasco, K.; Le Grand, M.; Hoffer, L.; Betzi, S.; Feracci, M.; Tsvetkov, P.; Devred, F.; Collette, Y.; Morelli, X.; et al. In silico molecular target prediction unveils mebendazole as a potent MAPK14 inhibitor. *Mol. Oncol.* **2020**, *14*, 3083–3099. [CrossRef] [PubMed]
85. Griffith, J.D.; Comeau, L.; Rosenfield, S.; Stansel, R.M.; Bianchi, A.; Moss, H.; de Lange, T. Mammalian telomeres end in a large duplex loop. *Cell* **1999**, *97*, 503–514. [CrossRef]
86. Shore, D.; Bianchi, A. Telomere length regulation: Coupling DNA end processing to feedback regulation of telomerase. *EMBO J.* **2009**, *28*, 2309–2322. [CrossRef]
87. Schmitt, C.A. Cellular senescence and cancer treatment. *Biochim. Biophys. Acta* **2007**, *1775*, 5–20. [CrossRef]
88. Zhou, X.Z.; Lu, K.P. The Pin2/TRF1-interacting protein PinX1 is a potent telomerase inhibitor. *Cell* **2001**, *107*, 347–359. [CrossRef]
89. Shi, R.; Zhao, Z.; Zhou, H.; Wei, M.; Ma, W.L.; Zhou, J.Y.; Tan, W.L. Reduced expression of PinX1 correlates to progressive features in patients with prostate cancer. *Cancer Cell Int.* **2014**, *14*, 46. [CrossRef]
90. Mei, P.J.; Chen, Y.S.; Du, Y.; Bai, J.; Zheng, J.N. PinX1 inhibits cell proliferation, migration and invasion in glioma cells. *Med. Oncol.* **2015**, *32*, 73. [CrossRef]
91. Yongjie, L.; Naiquan, D.; Shibo, D. Amplification of PINX1 in glioblastoma promotes cell proliferation and is targeted by miR-627. *Int. J. Clin. Exp. Med.* **2017**, *10*, 6720–6727.
92. Heiss, N.S.; Knight, S.W.; Vulliamy, T.J.; Klauck, S.M.; Wiemann, S.; Mason, P.J.; Poustka, A.; Dokal, I. X-linked dyskeratosis congenita is caused by mutations in a highly conserved gene with putative nucleolar functions. *Nat. Genet.* **1998**, *19*, 32–38. [CrossRef] [PubMed]
93. Montanaro, L.; Brigotti, M.; Clohessy, J.; Barbieri, S.; Ceccarelli, C.; Santini, D.; Taffurelli, M.; Calienni, M.; Teruya-Feldstein, J.; Trerè, D.; et al. Dyskerin expression influences the level of ribosomal RNA pseudo-uridylation and telomerase RNA component in human breast cancer. *J. Pathol.* **2006**, *210*, 10–18. [CrossRef]
94. Miao, F.A.; Chu, K.; Chen, H.R.; Zhang, M.; Shi, P.C.; Bai, J.; You, Y.P. Increased DKC1 expression in glioma and its significance in tumor cell proliferation, migration and invasion. *Investig. New Drugs* **2019**, *37*, 1177–1186. [CrossRef] [PubMed]
95. Canudas, S.; Houghtaling, B.R.; Kim, J.Y.; Dynek, J.N.; Chang, W.G.; Smith, S. Protein requirements for sister telomere association in human cells. *EMBO J.* **2007**, *28*, 4867–4878. [CrossRef]
96. Clevers, H.; Loh, K.M.; Nusse, R. Stem cell signaling. An integral program for tissue renewal and regeneration: Wnt signaling and stem cell control. *Science* **2014**, *346*, 1248012. [CrossRef]
97. Schmittgen, T.D.; Livak, K.J. Analyzing real-time PCR data by the comparative C(T) method. *Nat. Protoc.* **2008**, *3*, 1101–1108. [CrossRef] [PubMed]
98. Paradowski, B.; Shekhovtsov, A.; Baczkiewicz, A.; Kizielewicz, B.; Salabun, W. Similarity Analysis of Methods for Objective Determination of Weights in Multi-Criteria Decision Support Systems. *Symmetry* **2021**, *13*, 1874. [CrossRef]
99. Weisstein, E.W. Lehmer Mean. From MathWorld, A Wolfram Web Resource. Available online: https://mathworld.wolfram.com/LehmerMean.html (accessed on 2 February 2022).
100. Bullen, P.S. *Handbook of Means and Their Inequalities*; Kluwer: Dordrecht, The Netherlands, 2003; pp. 175–177.

International Journal of *Molecular Sciences*

MDPI

Communication

The Value of FET PET/CT in Recurrent Glioma with a Different IDH Mutation Status: The Relationship between Imaging and Molecular Biomarkers

Marija Skoblar Vidmar [1,2,*], Andrej Doma [2,3,*], Uroš Smrdel [1,3], Katarina Zevnik [2,3] and Andrej Studen [4,5]

1 Division of Radiotherapy, Institute of Oncology Ljubljana, 1000 Ljubljana, Slovenia; usmrdel@onko-i.si
2 Department of Nuclear Medicine, Institute of Oncology Ljubljana, 1000 Ljubljana, Slovenia; kzevnik@onko-i.si
3 Faculty of Medicine, University of Ljubljana, 1000 Ljubljana, Slovenia
4 Experimental Particle Physics Department, Jožef Stefan Institute, 1000 Ljubljana, Slovenia; andrej.studen@ijs.si
5 Faculty of Mathematics and Physics, University of Ljubljana, 1000 Ljubljana, Slovenia
* Correspondence: mskoblar@onko-i.si (M.S.V.); adoma@onko-i.si (A.D.)

Abstract: The evaluation of treatment response remains a challenge in glioma cases because the neuro oncological therapy can lead to the development of treatment-related changes (TRC) that mimic true progression (TP). Positron emission tomography (PET) using O-(2-[^{18}F] fluoroethyl-)-L-tyrosine (^{18}F-FET) has been shown to be a useful tool for detecting TRC and TP. We assessed the diagnostic performance of different ^{18}F-FET PET segmentation approaches and different imaging biomarkers for differentiation between late TRC and TP in glioma patients. Isocitrate dehydrogenase (IDH) status was evaluated as a predictor of disease outcome. In our study, the proportion of TRC in IDH wild type (IDHwt) and IDH mutant (IDHm) subgroups was without significant difference. We found that the diagnostic value of static and dynamic biomarkers of ^{18}F-FET PET for discrimination between TRC and TP depends on the IDH mutation status of the tumor. Dynamic ^{18}F-FET PET acquisition proved helpful in the IDH wild type (IDHwt) subgroup, as opposed to the IDH mutant (IDHm) subgroup, providing an early indication to discontinue dynamic imaging in the IDHm subgroup.

Keywords: glioma; treatment-related changes; true progression; pseudoprogression; radiation necrosis; biomarkers; ^{18}F-FET PET; IDH mutation

Citation: Skoblar Vidmar, M.; Doma, A.; Smrdel, U.; Zevnik, K.; Studen, A. The Value of FET PET/CT in Recurrent Glioma with a Different IDH Mutation Status: The Relationship between Imaging and Molecular Biomarkers. *Int. J. Mol. Sci.* **2022**, *23*, 6787. https://doi.org/10.3390/ijms23126787

Academic Editors: Nives Pećina-Šlaus and Ivana Jovčevska

Received: 13 May 2022
Accepted: 14 June 2022
Published: 17 June 2022

Publisher's Note: MDPI stays neutral with regard to jurisdictional claims in published maps and institutional affiliations.

1. Introduction

Molecular biomarkers have fundamentally changed the understanding of glioma over the last decade. Accordingly, the fifth edition of the World Health Organization Classification of Tumors of the Central Nervous System (WHO CNS5) incorporates numerous molecular biomarkers with clinicopathologic utility that are important for more accurate classification of CNS neoplasms [1]. Molecular biomarkers also improve diagnostic accuracy and influence the course of treatment by changing treatment recommendations [2]. A marker of particular importance is isocitrate dehidrogenase (IDH). Mutations in genes encoding IDH are known to play a crucial role in the classification of gliomas. IDHm glioma generally exhibits a better disease outcome than IDHwt. The IDHm is an independent predictor of prolonged survival and its prevalence is inversely correlated with tumor grades [3,4]. In adults, diffuse gliomas have been divided into three types according to the new classification: (1) astrocytoma, IDHm; (2) oligodendroglioma, IDHm and 1p/19q-codeleted; and (3) glioblastoma, IDHwt. Prior to that, glioblastomas were diagnosed based on the histologic findings including both IDHm and IDHwt tumors with very different biological features and prognoses. In WHO CNS5, glioblastomas comprise only IDHwt tumors. In addition, IDHwt diffuse astrocytic tumors in adults without the histologic features of glioblastoma but having one or more of three genetic parameters (TERT promoter mutation, EGFR gene amplification, combined gain of entire chromosome 7 and loss of entire chromosome 10) are classified as glioblastomas. In the new classification, all IDHm diffuse astrocytic tumors are considered a single type astrocytoma, IDHm and are graded as gradus 2, 3, or 4. Grading of gliomas also takes into account some other molecular findings such as the presence of CDKN2A/B homozygous deletion, which results in a worse prognosis and is subsequently graded as a WHO grade 4. The separation into IDH wild type and mutant tumors is an important advancement and a key factor in the treatment, follow-up, and understanding of glial tumors [5].

The treatment of gliomas includes maximal surgical resection, possibly followed by radiotherapy (RT) and chemotherapy with either procarbazine/lomustine/vincristine (PCV) or temozolomide (TMZ). Due to the proliferative, radioresistant, and chemoresistant nature of the gliomas and high levels of intratumoral heterogeneity, the disease often recurs, and the possibilities of additional treatment are very limited [6].

In the regular clinical work, the glioma treatment response assessment is based on imaging diagnostics, primarily MRI. MRI is the mainstay of imaging gliomas to monitor both treatment and response. T1-weighted MRI without and with contrast medium, T2-weighted as well as fluid-attenuated inversion recovery (FLAIR) MRI sequences are used for anatomic imaging [7–9].

Irradiation of brain tumors causes damage to the blood-brain barrier, which can lead to extravascular leakage of the contrast medium, which may have the same appearance on magnetic resonance imaging (MRI) images as a vital residual tumor or TP. This side effect of oncology treatment is defined as pseudoprogression and begins to occur approximately three months after irradiation, with the incidence of up to 50% in combined oncology treatment. Another radiation-induced side effect is radionecrosis (RN), possibly due to vascular injury and glial cell damage, usually occurring six months after the irradiation. RN can have the appearance of tumour residue or tumor recurrence on postcontrast MRI. The incidence of RN is estimated at up to 30% and increases with the length of the time from irradiation [10,11].

TRC, such as pseudoprogression and RN, overlaps with TP. This makes the differentiation challenging, and can consequently complicate the treatment course and compromise care. Therefore, the correct differentiation between TRC and actual TP continues to be a crucial issue [12,13]. For these reasons, additional imaging methods such as perfusion MRI or MR spectroscopy and functional methods such as ^{18}F-FET PET are used. ^{18}F-FET PET CT is based on the evaluation of transport of ^{18}F labeled tyrosine in tissues. In gliomas, ^{18}F-FET uptake significantly correlates with tumor cell density and neoangiogenesis, all biological hallmarks of highly malignant glial tumors. [14–16]. The aim of this study was to assess the diagnostic performance of different ^{18}F-FET PET segmentation approaches for differentiation between late TRC and TP in glioma patients with different IDH mutation statuses.

Since we investigated the late effects of radiochemotherapy, TRC was associated with radiation necrosis in our study.

2. Materials and Methods

2.1. Subjects

This retrospective study included 47 patients who were treated at the Division of Radiotherapy, Institute of Oncology in Ljubljana and, on the recommendation of the multidisciplinary tumor board, were referred to our Nuclear Medicine Department for ^{18}F -FET PET imaging between April 2019 and October 2021 in order to distinguish between TP and TRC. All patients who had undergone a standard MRI were able to understand the reason for additional ^{18}F-FET PET imaging. All patients had previously been diagnosed with adult diffuse gliomas and had a prior biopsy and radiochemotherapy according to EANO guidelines. All patients had a prior MRI suspicious of TP, as determined by the Response Assessment in Neuro-Oncology (RANO) working group criteria.

The study was approved by the institutional review board committee (approval number ERIDNPVO-0073/2021). All involved persons gave their written informed consent prior to study inclusion. The study conformed to the ethical norms and standards in the Declaration of Helsinki. All biological material was administered according to international guidelines.

2.2. Determination of IDH Genotype

The IDH mutation status was assessed by the IDH1R132H protein expression level evaluated by immunohistochemistry until early 2017 (15 pts), and after that using Next Generation Sequencing of a Glioma-Tailored Gene Panel (29 pts). For ^{18}F-FET PET analysis, patients were split into IDHm and IDHwt groups. Next-generation sequencing (NGS) is being increasingly used in routine clinical practice, including for the diagnostics of rare entities like gliomas because it can replace multiple single-gene genomic testing technologies while requiring only one test. Gene-targeted NGS offers a cost-effective approach to simultaneous detection of multiple genetic alterations with a minimal amount of sampled DNA while achieving high sensitivity. This makes this method highly attractive for use in gliomas. Specifically designed panels for gliomas are needed for the routine diagnosis of these tumors. We use the isolation of DNA from FFPE tissue using the Maxwell RSC FFPE Plus DNA Purification kit (Promega). The NGS panel assesses mutations in specific target regions ("Hotspots") in 9 genes: BRAF, H3F3A, HIST1H3B, HIST1H3C, IDH1, IDH2, KRAS, NRAS, pTERT, analysis of the entire coding region of 11 genes: ACVR1, ATRX, CIC, FUBP1, EGFR, FGFR1, PIK3CA, PIK3R1, PTEN, SETD2, TP53,12 gene copy number (CNV) analysis: CDKN2A, CDKN2B, EGFR, FGFR1, MDM2, MDM4, MET, MYCN, PDGFRA, PIK3CA, PIK3R1, PTEN and chromosome level analysis. Analysis of results is performed with the IonReporter software package (Thermo Fisher Scientific) (reference genome hg19). Border detection (default filter) is set to 5.0% (mutation rate relative to unmutated DNA). The sensitivity of the method is 99.21% (hotspot), 96.88% (indel), 97.10% (de novoSNV), 85.71% (de novo indel), and 95.35% (fusion). Negative results (no mutations) do not exclude the presence of mutations, amplifications, or deletions below the limit of detection [17].

2.3. ^{18}F-FET PET Imaging

The synthesis of ^{18}F-FET was performed by IASON GmbH (Graz, Austria). We used an integrated PET/CT system (Biograph mCT 64; Siemens, Erlangen, Germany) for acquisition of dynamic PET images over 40 min, starting immediately after injection of 3MBq of ^{18}F-FET per kg of body weight. All patients fasted for at least 6 h prior to PET acquisition. Dynamic 40-min scans were acquired using 35 sequences (200 $\times$ 200 matrix; 12 $\times$ 5 s; 6 $\times$ 10 s; 6 $\times$ 30 s; 5 $\times$ 60 s; 6 $\times$ 5 min). PET images were reconstructed with ordered-subset expectation maximization (OSEM) algorithm using 2 iterations with 21 subsets and gauss filtering to a full width at a half maximum (FWHM) of 5mm. High-resolution static images (400 $\times$ 400 matrix) were reconstructed from 20–40 min post-injection scans with OSEM algorithm consisting of two iterations with 21 subsets and Gaussian filtering to a FWHM of 3 mm.

2.4. ^{18}F-FET PET Image Analysis

PET scans were interpreted by two experienced nuclear medicine physicians, who were blinded to the histological and clinical data. The assessment of the tumor maximal standardized uptake value (SUV_{max}) was performed for each lesion by placing a spherical Volume-of-Interest (VOI) over the area of maximal amino acid uptake in the tumor on summed 20–40 min post-injection PET images. The mean standardized uptake value of the normal background brain tissue (SUV_{mean_bg}) was determined by placing a crescent-shaped VOI over the contralateral, unaffected hemisphere including white and grey matter. A tumor volume segmentation using a 3-dimensional auto contouring process with a tumor-to-background ratio (TBR) cutoff of at least 1.6 g/mL was used to determine the mean standardized uptake value of the tumor (SUV_{mean_tumor}). This cutoff was based on the results of a biopsy-controlled study in which a lesion-to-brain ratio of 1.6 g/mL resulted in the best separation between tumor and non-tumor tissue [16]. Respective ratios of the TBR_{mean} and TBR_{max} were calculated by dividing SUV_{mean_tumor} and SUV_{max}, respectively, by the SUV_{mean_bg}. For Time Activity Curve (TAC) evaluation, a spherical VOI was centered over the area of the highest tumor uptake to the entire dynamic datasets. Time-to-Peak (TTP) was determined as a time (in minutes) from the beginning of the dynamic PET acquisition up to the peak activity in the lesion. In lesions with constantly increasing TAC without an identifiable peak, TTP was defined as the end of the dynamic PET acquisition. We identified a cut-off point at 22.5 min post-injection as a dip in a distinct two-peak distribution of patients with respect to TTP with an early group reaching a distinct maximum before the prescribed threshold and a late group reaching it later or not exhibiting a peak at all. Using the cut-off, we identified the following curve categories, in order of increasing shape score:

- TAC score of −1: lesions with an early peak in SUV, followed by a constant descent of activity;
- TAC score of 0: lesions with ascending SUV reaching an early peak before 22.5 min, followed by a plateau or small descent of less than 5%;
- TAC score of 1: lesions with constantly increasing SUV without an identifiable peak.

2.5. Diagnosis of TP

Diagnosis of TP was based on histopathologic analysis following surgery, by clinical deterioration, and/or further radiological progression in a follow-up MRI at least four weeks after the initial assessment. In contrast, the diagnosis of TRC was applied in cases of negative histopathology, stable clinical conditions (with no treatment changes within the follow-up time), or stabilization/regression of the contrast-enhancing lesions at follow-up MRI (at least four weeks following initial assessment), respectively. The diagnosis for some patients was confirmed by more than one modality. Thus, the classification criteria in our study were similar to those of previous investigations [18,19].

2.6. Statistical Analysis

Descriptive statistics were used to calculate typical measures in patients' demographic and clinical characteristics. Data were expressed as median with a range, and categorical data were expressed as counts and frequencies. Statistical analyses were carried out using IBM SPSS Statistics software version 26 (Statistical package for the Social Sciences Statistical Software; SPSS Inc, IBM Corporation, Armonk, NY, USA).

We analyzed the FET outcome data using R statistical software (R version 3.1.1 (2014-07-10), R Foundation for Statistical Computing, Vienna, Austria, https://www.R-project.org/ (accessed on 16 June 2022). To statistically assess significant differences in ROC curves we performed a non-parametric ROC analysis [20] We used the optimal operating point on the ROC curve using the Youden index, and uncertainty in cut-off values was modelled using a large number approach [21]. We compared patient groups using the Mann-Whitney-Wilcoxon rank-sum U test [22]. We combined TAC score, IDH mutation status, and TBR_{mean} predictions using the logistic regression (LR) model. When used on patients grouped by IDH mutation status, we only used TAC score and TBR_{mean} variables in the LR model. Results with a p-value below 0.05 were deemed statistically significant, and 95% confidence intervals ($CI_{95\%}$) were used to quantify uncertainty in statistically derived values.

3. Results

We analyzed the data of 47 patients with glial tumors who underwent ^{18}F-FET PET for differentiation between TP and TRC. The interval between the end of radiation therapy and subsequent PET imaging was no less than 12 weeks in all cases.

Forty-four patients were eligible for analysis, and their median age was 44 (17 to 72, SD 14 years). Twenty-seven (61.4%) were male and 17 (38.6%) were female.

IDHm and IDHwt were present in 26 (59.1%) and 18 (40.9%) patients, respectively.

TP and TRC were confirmed in 32 (72.7%) and 12 (17.7%), respectively. The proportion of TRC in IDHwt 5/18 (27.8%) and IDHm 7/26 (26.9%) groups, was without significant difference (p = 0.61).

Diagnosis of TP/TRC was confirmed through histopathological analysis following surgery, through MRI, or based on clinical deterioration.

Regarding the verification of the diagnosis, 11 patients had surgery, 38 patients got a confirmed diagnosis after repeated MRI and 34 patients experienced clinical deterioration. Two out of 12 patients with TRC (2/12) had surgery, as opposed to nine patients with true progression (9/32). The proportion of those operated on is not significantly different between these groups.

The time from diagnosis to ^{18}F-FET PET was 104 weeks (84 weeks in IDHwt and 130 weeks in IDHm; the difference in median time to evaluation was without significance regarding the TRC or TP (p = 0.5), as well as IDH status (p = 0.9).

The proportion of patients with TRC according to IDH status was not significantly different (χ^2 p = 0.9).

In this group of patients, the overall survival was excellent, with median survival exceeding 500 weeks. The median survival according to IDH status was not different, though the median survival was not reached in the TRC group, the analysis is underpowered to detect significance.

^{18}F-FET-PET parameters were then analyzed for the whole group and selectively according to different IDH mutation statuses. For the whole group, the SUV_{max} value had a mean of 4.04 and a median of 3.78, with the SD of 1.83 (IDHm: 4.10, 3.82, and 1.73, IDHwt: 4.02, 3.76, and 2.05, respectively). The difference between groups was not statistically significant.

When comparing the mean values for SUV parameters according to the IDH mutation status, we found that while in IDHwt patients there are no significant differences in SUV values according to radio necrosis and progression, in the IDHm, the mean values of SUV_{max} (TRC/TP (p): 4.7/2.49 (0.001)), TBR_{max} and TBR_{mean} differed significantly (Table 1).

Figure 1 shows a comparison of ROC curves for different FET- derived variables split by IDH mutation status. For all patients, TBR_{max} and TBR_{mean} together with logReg show a statistically significantly better performance compared to TTP. In groups split by IDH status, no statistically significant differences in ROC curves could be determined. Relapse was associated with larger values of TBR_{max} and TBR_{mean}, while the opposite association to lower values was identified for TTP. In IDHm, the TTP ROC is below the diagonal, indicating an inverse relationship between higher TTP and relapse, which is not statistically confirmed. In IDHwt cases, TTP is the best predictor, its difference to either TBR_{max} or TBR_{mean} is, however, not significant. We found a single significant variable, TBR_{mean}, with a coefficient of 1.6 (p = 0.03) and equivalent odds ratio of 4.9 ($CI_{95\%}$: 2.1–8.5) per unit change in TBR_{mean} in the LR model for the full patient group. We identified no significant variables in IDH-specific LR models.

Table 1 shows the ability of FET in predicting the tumor status. For each IDH group we show the predictive quality of FET variables -TBR_{max}, TBR_{mean}, TTP and LR. For each variable we show its mean and range in the TRC group, mean and range in TP group, optimum cut-off based on the Youden index, sensitivity and specificity with associated 95% confidence intervals at cut-off point and the p-value associated with the MWU test for patients grouped by tumor outcome. The variables that can identify tumor outcome with statistical significance are shown in bold. Two regimes can be identified: in the full and IDHm group, TBR_{max} and TBR_{mean} are significant and TTP is irrelevant. Conversely, TTP becomes significant and TBR_{mean} and TBR_{max} irrelevant in the IDHwt group. In all cases the LR model which combines TBRmax and TAC score is a significant predictor of the tumor outcome.

Table 1. Analysis of sensitivities, specificities, thresholds and diagnostic accuracy of ^{18}F-FET PET biomarkers in differentiation between late TP and TRC in glioma patients.

All (N = 42)				IDHm (N = 23)				IDHwt (N = 18)			
TBR		TTP		TBR		TTP		TBR		TTP	
Max	Mean	(min)	LR	Max	Mean	(min)	LR	Max	Mean	(min)	LR
				TP (count, mean, median, range)							
N = 31				N = 17				N = 14			
4.1	2.2	26	0.8	4.2	2.2	30	0.8	4.0	2.1	22	0.8
4.0	2.1	32		4.1	2.1	40		3.8	2.2	14.5	
1.1–8.0	0–3.2	5–40	0.1–1	2.1–6.4	1.7–3.2	7–40	0.3–1	1.1–8.0	0–3.1	7–40	0.4–1
				TRC (mean, median, range)							
N = 11				N = 6				N = 4			
2.6	1.5	35	0.5	2.6	1.6	30	0.5	2.7	1.4	40	0.6
2.3	1.9	40		2.2	1.8	32		2.6	1.9	40	
1.6–4.2	0–2.2	12–40	0.1–0.8	1.9–4.1	0–2.2	12–40	0–0.9	1.6–4.2	0–2.0	40–40	0.4–0.6
				Threshold (optimum, $CI_{95\%}$)							
3.03	2.04	32	0.79	3.03	1.96	32	0.66	2.9	2.09	40	0.65
2.6–3.4	1.8–2.3	28–36	0.7–0.9	2.6–3.4	1.7–2.3	27–37	0.6–0.8	2.2–3.6	1.6–2.6	36–40	0.6–0.8
				Sensitivity (%, value at optimum, $CI_{95\%}$)							
77	71	48	58	94	88	83	88	64	64	79	79
60–89	53–84	32–65	41–74	73–99	66–97	54–97	66–97	39–84	39–84	52–92	52–92
				Specificity (%, value at optimum, $CI_{95\%}$)							
82	91	91	100	83	83	53	83	75	100	100	100
52–92	62–98	62–98	74–100	44–97	44–97	31–74	44–97	30–95	51–100	51–100	51–100
				Accuracy (%, value at optimum, $CI_{95\%}$)							
79	76	60	69	91	87	61	87	67	72	83	83
64–88	61–87	44–73	54–81	73–98	68–95	41–78	68–95	44–84	49–88	61–94	61–94
				***p*-value**							
0.001	0.001	0.18	0.002	0.004	0.01	0.61	0.01	0.33	0.14	0.05	0.05

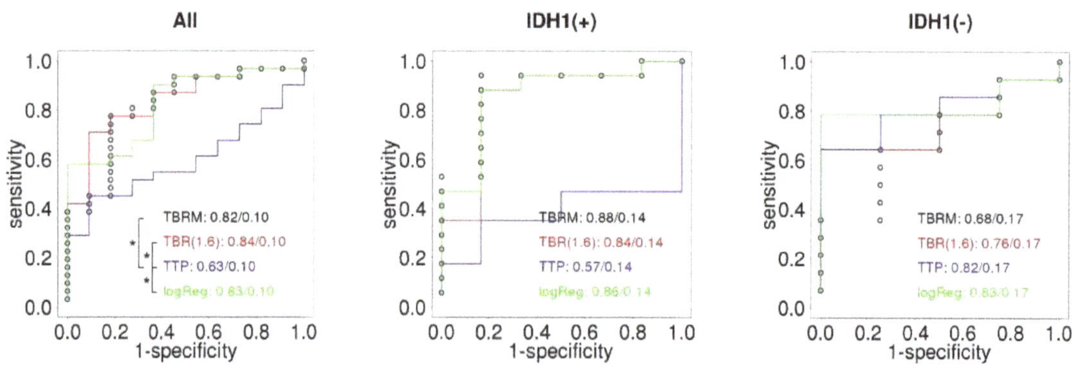

Figure 1. Receiver–operator characteristic (ROC) curves in classifying tumor outcome based on FET-derived parameters for different IDH mutation status groups. All: all patients in the study. IDH1(m): patients with mutated IDH gene. IDH1(wt): patients with IDH wild type. TBRM is TBR_{max}, TBR(1.6) is TBR_{mean} at SUV 1.6 g/mL cutoff, TTP is time to peak and log Reg is the logistic regression model. The image shows AUC values with associated standard deviation. The asterisk associated brackets identify variables with statistically significant ROC curves as evaluated by a non-parametric significance test.

4. Discussion

This is the first reported study of performance of [18]F -FET PET for the differentiation between TP and TRC in glioma patients based on IDH mutation status. We evaluated the diagnostic potential of static and dynamic [18]F -FET PET parameters for differentiation between late TRC and TP based on different IDH mutation subgroups in mixed diffuse glioma patients. The differentiation between TP and TRC represents one of the most frequent indications for the use of amino acid PET in clinical practice. PET using radiolabeled amino acids is gaining increasing interest for the diagnostics of brain tumors because a conventional MRI is limited in differentiating tumor tissue from nonspecific tissue changes following neuro-oncological treatment. Recently, the RANO working group has recommended the additional use of amino acid PET imaging for brain tumor management [23].

There is a growing body of evidence that IDH mutations play a role in the formation of brain tumors and influence the response to neuro oncological treatment and overall survival [24,25]. *IDH* mutations in glioma are associated with significantly prolonged progression-free and overall survival compared with IDHwt tumors. There is now a greater appreciation that the biology of IDHm glioma is quite different from that of IDHwt tumors and that tumorigenic processes most likely are different as well. While the specific mechanism of IDH mutation that results in the oncogenic switch in gliomas remains unknown, potential mechanisms have been identified, including the inhibition of hypoxia-related proline hydroxylases, inhibition of DNA demethylases, inhibition of histone demethylases, and alterations in glutamate metabolism. Further work is needed to elucidate the specific role of IDH mutation and the pathological consequences that clearly affect tumor evolution and prognosis. Given the complex role of the IDH mutation in the progression, aggressive biological behavior, and response to the treatment of diffuse gliomas, it seems reasonable to observe and analyze these two molecular groups separately [26–30].

In our study, the static ^{18}F-FET PET measures TBRmax and TBRmean outperformed the dynamic parameters in the IDH (all) group (accuracy 79 and 76% respectively, $p = 0.001$ both) and the IDHm group (accuracy 91 and 87%, p of 0.004 and 0.01, respectively); dynamic measure TTP only achieved an accuracy of 60 and 61% in IDH (all) and the IDHm group, respectively, and the p value of the MWU test was not significant. The situation was reversed in the IDHwt group, where TTP was found to be a significant predictor (accuracy of 83%, $p = 0.05$) while static measures showed a mediocre performance (accuracy of 67 and 72% respectively, non-significant p values of MWU test).

The results of our study contradict several previous reports, where authors described improved diagnostic sensitivity when dynamic metrics were added to static ^{18}F-FET metrics alone.

The difference in the results reported could be due to distinctly different patient populations: while previous studies [31–35] did not take into account the IDH mutation status of the participants, IDH mutation was present in 59.1% of our patients, this being a high proportion compared to usual rates in glioma patients. IDH mutation rates vary substantially between different types of glioma, nonetheless they are not common [28,29].

Therefore, the significant diagnostic performance of dynamic PET acquisition in a non-divided group of glioma patients, the majority of which would typically be IDHwt, was impaired by a large IDHm subgroup included in our study.

Due to the limited availability of the radiotracer at our institution we consider performing ^{18}F-FET PET/CT investigation particularly when MRI yields inconclusive results between TRC and TP. Important for our study is that ^{18}F-FET PET/CT imaging was considered appropriate only if it resulted in therapeutic consequences. The patients with poor performance status and without further treatment options, typically being IDHwt, were therefore following the standard of care not assigned to receive ^{18}F-FET PET imaging, thus further reducing the number of IDHwt subjects in our study. Therefore, a higher proportion of IDH mutations in our study is a consequence of more frequent equivocal decisions at a multidisciplinary tumor board in IDHm cases, and represents a selection of notably difficult cases [31]. A high rate of IDHm in our patients also enabled this IDH mutation status-based study.

In comparison to the static acquisition, the use of dynamic acquisition of ^{18}F-FET PET is very time-consuming and can therefore be challenging in an otherwise busy imaging department. Dynamic imaging incurs additional costs as a consequence of the prolonged decay of the tracer intended for consecutive patients, and fewer scans are performed within normal working hours, hence shorter imaging times are preferred. In our study, the shape of the dynamic curve, in general, has not been statistically significantly associated with the TRC or TP, although there was a trend that associated the plateau-shaped curve with the progressive disease. Our results suggest that the exclusion of dynamic acquisition and the performance of only static acquisition in IDHm patients could be a cost-effective strategy without the diagnostic potential of the investigation being hampered. Nonetheless, since dynamic imaging metrics proved useful in the IDHwt subgroup, these patients should be provided with the dynamic acquisition. [32,33].

While the cutoff values of TBR_{mean} in the all, IDHm and IDHwt subgroups (2.04, $CI_{95\%}$:1.8–2.3; 1.96, $CI_{95\%}$:1.7–2.3; and 2.09, $CI_{95\%}$:1.6–2.6, respectively) as determined in our study, within CL agree with previously reported cutoff values of 1.9–2.0 for the differentiation of both early and late TRC from true progression, the cutoff values of TBR_{max} for both all and IDHm (3.03, $CI_{95\%}$:2.6–3.4), and IDHwt (2.9, $CI_{95\%}$: 2.2–3.6) subgroups, as determined in our study, were above the values, reported by the majority of authors [31–34]. An optimal TBR_{max} cutoff value of about 1.9 with an accuracy of 85% was determined in late TRC glioblastoma patients, while a TBR_{max} cutoff of 2.3 (accuracy 96%) was determined in the early glioblastoma pseudoprogression by Galldiks et al. [18,34,35]. However, Kartels et al. estimated the optimal TBR_{max} cutoff value to be 3.52 in a late glioblastoma multiforme group of patients, indicating non-uniform outcomes in different patient group settings [36]. The reason for only moderate accuracy in differentiating TP from TRC might be due to a non-homogeneous group of patients because we did not limit our study to high-grade gliomas or specific treatment regimens. As reported in a meta-analysis by Cui et al., the accuracy of FET is known to be higher in high-grade glioma than in the mixed glioma patients group [37].

5. Conclusions

Differentiating TRC from TP is of critical importance for patient management and prognosis and it can often be challenging. In our study, the proportion of TRC in IDHwt and IDHm subgroups was without significant difference. We found that the diagnostic value of static and dynamic biomarkers of ^{18}F-FET PET for discrimination between TRC and TP depends on the IDH mutation status of the tumor. Dynamic biomarkers play an important role in the IDHwt subgroup, and as opposed to the case of IDHm, the dynamic acquisition of ^{18}F-FET PET might eventually be discontinued. Further prospective research in large sample sizes is needed to determine the value of ^{18}F-FET PET in different molecular biomarker settings and to confirm our findings.

Author Contributions: Conceptualization, M.S.V. and A.D.; methodology, M.S.V., A.D. and A.S.; validation, M.S.V., A.D. and K.Z.; formal analysis, M.S.V., A.D., K.Z., U.S. and A.S.; investigation, M.S.V., A.D.; data curation M.S.V., A.D. and K.Z.; writing—original draft preparation, M.S.V., A.D., U.S. and A.S.; writing—review and editing, M.S.V. and A.D.; visualization, A.S.; project administration, M.S.V. All authors have read and agreed to the published version of the manuscript.

Funding: This research received no external funding.

Institutional Review Board Statement: The study was conducted in accordance with the Declaration of Helsinki and approved by the Institutional Review Board of the Oncology Institute Ljubljana, approval number ERIDNPVO-0073/2021.

Informed Consent Statement: Informed consent was obtained from all subjects involved in the study.

Data Availability Statement: The data presented in this study are available on request from the corresponding author. The patient data are not publicly available due to ethical restrictions.

Conflicts of Interest: The authors declare no conflict of interest.

References

1. Louis, D.N.; Perry, A.; Wesseling, P.; Brat, D.J.; Cree, I.A.; Figarella-Branger, D.; Hawkins, C.; Ng, H.K.; Pfister, S.M.; Reifenberger, G.; et al. 2021 WHO Classification of Tumors of the Central Nervous System: A summary. *Neuro Oncol.* **2021**, *23*, 1231–1251. [CrossRef] [PubMed]
2. Burke, H.B. Predicting Clinical Outcomes Using Molecular Biomarkers. *Biomark. Cancer* **2016**, *8*, 89–99. [CrossRef]
3. Huang, J.; Yu, J.; Tu, L.; Huang, N.; Li, H.; Luo, Y. Isocitrate Dehydrogenase Mutations in Glioma: From Basic Discovery to Therapeutics Development. *Front. Oncol.* **2019**, *9*, 506. [CrossRef] [PubMed]
4. Kaminska, B.; Czapski, B.; Guzik, R.; Król, S.K.; Gielniewski, B. Consequences of IDH1/2 Mutations in Gliomas and an Assessment of Inhibitors Targeting Mutated IDH Proteins. *Molecules* **2019**, *24*, 968. [CrossRef]
5. Wen, P.Y.; Packer, R.J. The 2021 WHO Classification of Tumors of the Central Nervous System: Clinical implications. *Neuro Oncol.* **2021**, *23*, 1215–1217. [CrossRef] [PubMed]
6. Weller, M.; van den Bent, M.; Preusser, M.; Le Rhun, E.; Tonn, J.C.; Minniti, G.; Bendszus, M.; Balana, C.; Chinot, O.; Dirven, L.; et al. EANO guidelines on the diagnosis and treatment of diffuse gliomas of adulthood. *Nat. Rev. Clin. Oncol.* **2021**, *18*, 170–186. [CrossRef] [PubMed]
7. Poon, M.T.C.; Sudlow, C.L.M.; Figueroa, J.D.; Brennan, P.M. Longer-term (≥2 years) survival in patients with glioblastoma in population-based studies pre- and post-2005: A systematic review and meta-analysis. *Sci. Rep.* **2020**, *10*, 11622. [CrossRef]
8. Villanueva-Meyer, J.E.; Mabray, M.C.; Cha, S. Current Clinical Brain Tumor Imaging. *Neurosurgery* **2017**, *81*, 397–415. [CrossRef]
9. Chourmouzi, D.; Papadopoulou, E.; Marias, K.; Drevelegas, A. Imaging of brain tumors. *Surg. Oncol. Clin. N. Am.* **2014**, *23*, 629–684. [CrossRef]
10. Verma, N.; Cowperthwaite, M.C.; Burnett, M.G.; Markey, M.K. Differentiating tumor recurrence from treatment necrosis: A review of neuro-oncologic imaging strategies. *Neuro Oncol.* **2013**, *15*, 515–534. [CrossRef]
11. Brandsma, D.; Stalpers, L.; Taal, W.; Sminia, P.; van den Bent, M.J. Clinical features, mechanisms, and management of pseudoprogression in malignant glioma. *Lancet Oncol.* **2008**, *9*, 453–461. [CrossRef]
12. Vellayappan, B.; Tan, C.L.; Yong, C.; Khor, L.K.; Koh, W.Y.; Yeo, T.T.; Detsky, J.; Lo, S.; Sahgal, A. Diagnosis and Management of Radiation Necrosis in Patients with Brain Metastases. *Front. Oncol.* **2018**, *8*, 395. [CrossRef] [PubMed]
13. Munier, S.; Ginalis, E.E.; Patel, N.V.; Danish, S.; Hanft, S. Radiation Necrosis in Intracranial Lesions. *Cureus* **2020**, *12*, e7603. [CrossRef]
14. Galldiks, N.; Stoffels, G.; Filss, C.; Rapp, M.; Blau, T.; Tscherpel, C.; Ceccon, G.; Dunkl, V.; Weinzierl, M.; Stoffel, M.; et al. The use of dynamic O-(2-18F-fluoroethyl)-l-tyrosine PET in the diagnosis of patients with progressive and recurrent glioma. *Neuro Oncol.* **2015**, *17*, 1293–1300. [CrossRef] [PubMed]
15. Santo, G.; Laudicella, R.; Linguanti, F.; Nappi, A.G.; Abenavoli, E.; Vergura, V.; Rubini, G.; Sciagrà, R.; Arnone, G.; Schillaci, O.; et al. The Utility of Conventional Amino Acid PET Radiotracers in the Evaluation of Glioma Recurrence also in Comparison with MRI. *Diagnostics* **2022**, *12*, 844. [CrossRef] [PubMed]
16. Pauleit, D.; Floeth, F.; Hamacher, K.; Riemenschneider, M.J.; Reifenberger, G.; Müller, H.-W.; Zilles, K.; Coenen, H.H.; Langen, K.-J. O-(2-[18F]fluoroethyl)-L-tyrosine PET combined with MRI improves the diagnostic assessment of cerebral gliomas. *Brain* **2005**, *128*, 678–687. [CrossRef]
17. Zacher, A.; Kaulich, K.; Stepanow, S.; Wolter, M.; Köhrer, K.; Felsberg, J.; Malzkorn, B.; Reifenberger, G. Molecular diagnostics of gliomas using next generation sequencing of a glioma-tailored gene panel. *Brain Pathol.* **2017**, *27*, 146–159. [CrossRef]
18. Werner, J.M.; Stoffels, G.; Lichtenstein, T.; Borggrefe, J.; Lohmann, P.; Ceccon, G.; Shah, N.J.; Fink, G.R.; Langen, K.J.; Kabbasch, C.; et al. Differentiation of treatment-related changes from tumor progression: A direct comparison between dynamic FET PET and ADC values obtained from DWI MRI. *Eur. J. Nucl. Med. Mol. Imaging* **2019**, *46*, 1889–1901. [CrossRef] [PubMed]
19. Young, R.J.; Gupta, A.; Shah, A.D.; Graber, J.J.; Zhang, Z.; Shi, W.; Holodny, A.I.; Omuro, A.M. Potential utility of conventional MRI signs in diagnosing pseudoprogression in glioblastoma. *Neurology* **2011**, *76*, 1918–1924. [CrossRef]
20. DeLong, E.R.; DeLong, D.M.; Clarke-Pearson, D.L. Comparing the areas under two or more correlated receiver operating characteristic curves: A nonparametric approach. *Biometrics* **1988**, *44*, 837–845. [CrossRef]
21. Lai, C.Y.; Tian, L.; Schisterman, E.F. Exact confidence interval estimation for the Youden index and its corresponding optimal cut-point. *Comput. Stat. Data Anal.* **2012**, *56*, 1103–1114. [CrossRef] [PubMed]
22. Wu, P.; Han, Y.; Chen, T.; Tu, X.M. Causal inference for Mann-Whitney-Wilcoxon rank sum and other nonparametric statistics. *Stat. Med.* **2014**, *33*, 1261–1271. [CrossRef] [PubMed]
23. Albert, N.L.; Weller, M.; Suchorska, B.; Galldiks, N.; Soffietti, R.; Kim, M.M.; la Fougère, C.; Pope, W.; Law, I.; Arbizu, J.; et al. Response Assessment in Neuro-Oncology working group and European Association for Neuro-Oncology recommendations for the clinical use of PET imaging in gliomas. *Neuro Oncol.* **2016**, *18*, 1199–1208. [CrossRef] [PubMed]
24. Appin, C.L.; Brat, D.J. Molecular pathways in gliomagenesis and their relevance to neuropathologic diagnosis. *Adv. Anat. Pathol.* **2015**, *22*, 50–58. [CrossRef] [PubMed]
25. Dimitrov, L.; Hong, C.S.; Yang, C.; Zhuang, Z.; Heiss, J.D. New developments in the pathogenesis and therapeutic targeting of the IDH1 mutation in glioma. *Int. J. Med. Sci.* **2015**, *12*, 201–213. [CrossRef] [PubMed]
26. Han, S.; Liu, Y.; Cai, S.J.; Qian, M.; Ding, J.; Larion, M.; Gilbert, M.R.; Yang, C. IDH mutation in glioma: Molecular mechanisms and potential therapeutic targets. *Br. J. Cancer* **2020**, *122*, 1580–1589. [CrossRef]

27. Turkalp, Z.; Karamchandani, J.; Das, S. IDH Mutation in Glioma: New Insights and Promises for the Future. *JAMA Neurol.* **2014**, *71*, 1319–1325. [CrossRef]
28. Bai, H.; Harmanci, A.S.; Erson-Omay, E.Z.; Li, J.; Coskun, S.; Simon, M.; Krischek, B.; Özduman, K.; Omay, S.B.; Sorensen, E.A.; et al. Integrated genomic characterization of IDH1-mutant glioma malignant progression. *Nat. Genet.* **2016**, *48*, 59–66. [CrossRef]
29. Nicholson, J.G.; Fine, H.A. Diffuse Glioma Heterogeneity and Its Therapeutic Implications. *Cancer Discov.* **2021**, *11*, 575–590. [CrossRef]
30. Kloosterhof, N.K.; Bralten, L.B.; Dubbink, H.J.; French, P.J.; van den Bent, M.J. Isocitrate dehydrogenase-1 mutations: A fundamentally new understanding of diffuse glioma? *Lancet Oncol.* **2011**, *12*, 83–91. [CrossRef]
31. El Saghir, N.S.; Keating, N.L.; Carlson, R.W.; Khoury, K.E.; Fallowfield, L. Tumor boards: Optimizing the structure and improving efficiency of multidisciplinary management of patients with cancer worldwide. *Am. Soc. Clin. Oncol. Educ. Book* **2014**, *34*, e461–e466. [CrossRef] [PubMed]
32. Maurer, G.D.; Brucker, D.P.; Stoffels, G.; Filipski, K.; Filss, C.P.; Mottaghy, F.M.; Galldiks, N.; Steinbach, J.P.; Hattingen, E.; Langen, K.J. 18F-FET PET Imaging in Differentiating Glioma Progression from Treatment-Related Changes: A Single-Center Experience. *J. Nucl. Med.* **2020**, *61*, 505–511. [CrossRef] [PubMed]
33. Mihovilovic, M.I.; Kertels, O.; Hänscheid, H.; Löhr, M.; Monoranu, C.M.; Kleinlein, I.; Samnick, S.; Kessler, A.F.; Linsenmann, T.; Ernestus, R.I.; et al. O-(2-(18F)fluoroethyl)-L-tyrosine PET for the differentiation of tumour recurrence from late pseudoprogression in glioblastoma. *J. Neurol. Neurosurg. Psychiatry* **2019**, *90*, 238–239. [CrossRef] [PubMed]
34. Galldiks, N.; Dunkl, V.; Stoffels, G.; Hutterer, M.; Rapp, M.; Sabel, M.; Reifenberger, G.; Kebir, S.; Dorn, F.; Blau, T.; et al. Diagnosis of pseudoprogression in patients with glioblastoma using O-(2-[18F]fluoroethyl)-L-tyrosine PET. *Eur. J. Nucl. Med. Mol. Imaging* **2015**, *42*, 685–695. [CrossRef]
35. Kebir, S.; Fimmers, R.; Galldiks, N.; Schäfer, N.; Mack, F.; Schaub, C.; Stuplich, M.; Niessen, M.; Tzaridis, T.; Simon, M.; et al. Late Pseudoprogression in Glioblastoma: Diagnostic Value of Dynamic O-(2-[18F]fluoroethyl)-L-Tyrosine PET. *Clin. Cancer Res.* **2016**, *22*, 2190–2196. [CrossRef]
36. Kertels, O.; Mihovilovic, M.I.; Linsenmann, T.; Kessler, A.F.; Tran-Gia, J.; Kircher, M.; Brumberg, J.; Monoranu, C.M.; Samnick, S.; Ernestus, R.I.; et al. Clinical Utility of Different Approaches for Detection of Late Pseudoprogression in Glioblastoma With O-(2-[18F]Fluoroethyl)-L-Tyrosine PET. *Clin. Nucl. Med.* **2019**, *44*, 695–701. [CrossRef]
37. Cui, M.; Zorrilla-Veloz, R.I.; Hu, J.; Guan, B.; Ma, X. Diagnostic Accuracy of PET for Differentiating True Glioma Progression From Post Treatment-Related Changes: A Systematic Review and Meta-Analysis. *Front. Neurol.* **2021**, *12*, 671867. [CrossRef]

International Journal of
Molecular Sciences

MDPI

Article

The Use of Pro-Angiogenic and/or Pro-Hypoxic miRNAs as Tools to Monitor Patients with Diffuse Gliomas

Guénaëlle Levallet [1,2,*], Fatéméh Dubois [1,2], Arthur Leclerc [3], Edwige Petit [1], Lien Bekaert [3], Maxime Faisant [2], Christian Creveuil [4], Evelyne Emery [3], Gérard Zalcman [5,6] and Emmanuèle Lechapt-Zalcman [1,2,7,8]

1 Normandie University, UNICAEN, CNRS, ISTCT Unit, GIP CYCERON, F-14074 Caen, France; fatemeh.dubois@unicaen.fr (F.D.); epetit@cyceron.fr (E.P.); emmanuele.lechapt@aphp.fr (E.L.-Z.)
2 Department of Pathology, CHU de Caen, F-14033 Caen, France; faisant-m@chu-caen.fr
3 Department of Neurosurgery, CHU de Caen, F-14033 Caen, France; arthur.leclerc@neurochirurgie.fr (A.L.); lien.bekaert@gmail.com (L.B.); emery-e@chu-caen.fr (E.E.)
4 Biomedical Research Unit, CHU de Caen, F-14033 Caen, France; creveuil-cr@chu-caen.fr
5 Department of Thoracic Oncology & CIC1425, Paris Cité University, Hôpital Bichat-Claude Bernard, Assistance Publique Hôpitaux de Paris, F-75877 Paris, France; gerard.zalcman@aphp.fr
6 U830 INSERM "Cancer, Heterogeneity Instability and Plasticity (CHIP), STRESS Group", Curie Institute Research Center, F-75005 Paris, France
7 Department of Pathology, GHU AP-HP University Hospital Henri-Mondor, F-94010 Creteil, France
8 INSERM U955 Institut Mondor de Recherche Biomédicale (IMRB), Paris Est Créteil University, F-94010 Creteil, France
* Correspondence: guenaelle.levallet@unicaen.fr

Citation: Levallet, G.; Dubois, F.; Leclerc, A.; Petit, E.; Bekaert, L.; Faisant, M.; Creveuil, C.; Emery, E.; Zalcman, G.; Lechapt-Zalcman, E. The Use of Pro-Angiogenic and/or Pro-Hypoxic miRNAs as Tools to Monitor Patients with Diffuse Gliomas. *Int. J. Mol. Sci.* **2022**, 23, 6042. https://doi.org/10.3390/ijms23116042

Academic Editor: Nives Pećina-Šlaus

Received: 30 April 2022
Accepted: 25 May 2022
Published: 27 May 2022

Abstract: IDH (isocitrate dehydrogenase) mutation, hypoxia, and neo-angiogenesis, three hallmarks of diffuse gliomas, modulate the expression of small non-coding RNAs (miRNA). In this paper, we tested whether pro-angiogenic and/or pro-hypoxic miRNAs could be used to monitor patients with glioma. The miRNAs were extracted from tumoral surgical specimens embedded in the paraffin of 97 patients with diffuse gliomas and, for 7 patients, from a blood sample too. The expression of 10 pro-angiogenic and/or pro-hypoxic miRNAs was assayed by qRT-PCR and normalized to the miRNA expression of non-tumoral brain tissues. We confirmed in vitro that IDH in hypoxia (1% O_2, 24 h) alters pro-angiogenic and/or pro-hypoxic miRNA expression in HBT-14 (U-87 MG) cells. Then, we reported that the expression of these miRNAs is (i) strongly affected in patients with glioma compared to that in a non-tumoral brain; (ii) correlated with the histology/grade of glioma according to the 2016 WHO classification; and (iii) predicts the overall and/or progression-free survival of patients with glioma in univariate but not in a multivariate analysis after adjusting for sex, age at diagnosis, and WHO classification. Finally, the expression of miRNAs was found to be the same between the plasma and glial tumor of the same patient. This study highlights a panel of seven pro-angiogenic and/or pro-hypoxic miRNAs as a potential tool for monitoring patients with glioma.

Keywords: miRNA; hypoxia; angiogenesis; glioma

1. Introduction

Adult diffuse gliomas are the most common primary malignant brain tumors, accounting for approximately 60% of all central nervous system tumors. These tumors are characterized by a number of criteria, either morphological/histological (tumoral cells mitoses, microvascular proliferation, hypoxia/necrosis as infiltration (for review: [1])) or molecular (isocitrate dehydrogenase (IDH) mutations, 1p/19q codeletion, ATRX (alpha thalassemia/mental retardation syndrome X-linked) mutations, mutations in the promoter of *TERT* (telomerase reverse transcriptase), etc., [2–4]). Since the revision of the classification of brain tumors according to the World Health Organization (WHO) in 2016 [2], which was reviewed in 2021 [5], all these morphological/histological and molecular criteria have been integrated by pathologists to establish a histoprognostic grade of these tumors for each histological diagnosis. Some molecular abnormalities (for example, trisomy 7 associated with monosomy 10, mutation of the *TERT* promoter, biallelic deletion of the *CDK2NA* gene) alone allow us to classify the tumor into grade 4, independently of the histomorphological criteria [5].

Diffuse gliomas in adults are therefore now well characterized from a molecular point of view; however, there is still no molecular tool for monitoring these patients longitudinally, even though a low-grade mutated *IDH* tumor will invariably evolve into a higher-grade tumor [6]. It is therefore necessary to improve the care and follow-up of patients with diffuse gliomas by identifying such tools. It is acknowledged that the occurrence of the *IDH1/2* mutation remains the upstream genetic event in two diffuse glioma lineages: diffuse astrocytomas and oligodendrogliomas [3,4]. It is also known that the *IDH1/2* mutation, hypoxia leading to necrosis, and microvascular proliferation are interrelated [1,2,7,8]. *IDH1/2* mutation can control these phenomena by modulating the expression of miRNAs. Indeed, *IDH1/2* mutation is related to epigenetic modification by both DNA hypermethylation [9,10] and a change in miRNA (miRNA) expression (for review [8]). Interestingly, in lower-grade gliomas, the *IDH1/2* mutation has more of an impact on miRNA expression than histological and other genomic features [11].

By repressing transcription or inducing the degradation of their target mRNA molecules, miRNAs can control cell growth, proliferation, metabolism, and apoptosis [12]. Many miRNAs are dysregulated in gliomas and are linked to their development and progression [13,14] (for review: [15]). Among these, some appear oncogenic, such as tumor suppressor miRNAs [16] as well as miRNAs correlated with the grade and/or histology of the glial tumor and/or with the outcome of glioma patients, as reviewed in a recent meta-analysis based on the data of 4708 glioma patients [17]. For example, a high expression of miR-15b, 21, 148a, 196, 210, and 221 or a low expression of miR-106a and 124 predicts a poor prognosis in glioma patients, while the expression of miR-10b, 17, 20a, 155, 182, 200b, and 222 fails to predict such survival [17]. Among these miRNAs, some are induced by hypoxia and/or neoangiogenesis, the two hallmarks of glioma history [1], such as mir210 [18,19]. Interestingly, the detection of stable miRNA expression in cerebrospinal fluid, blood serum, and other bodily fluids has led to the possibility of using miRNAs as non-invasive biomarkers for clinical applications [20,21]. However, specific miRNAs still need to be elucidated in the diagnosis of a glioma, especially in the early screening stage [22].

As IDH genes play important roles in the mechanism of glioma, here, we tested the diagnostic and prognostic values of miRNAs, reflecting on the features of gliomas in the WHO 2016 classification and patients' survival in a series of 97 grade II to IV gliomas. We focused on ten miRNAs: has-mir-200b-3p, -200c-3p, -210-3p, -100-5p, -126-5p, -132-3p, -221-3p, -424-5p, -128-3p, and -451-5p. These are miRNAs which could be involved in the regulation of hypoxia/cell proliferation/differentiation of glioma cells, as described for other cell types [23–33] (for review: [15,34]).

2. Results

2.1. Expression of the Pro-Angiogenic and/or Pro-Hypoxic miRNAs Studied Is Affected by IDH Mutation and Hypoxia In Vitro

We first validated the influence of IDH1 mutation on proangiogenic miRNA expression by comparing the expression of these miRNAs between two isogenic HBT-14 (U-87 MG) cell lines only differing in their expression of either wild-type IDH1 (HBT-14 (U-87 MG) $IDH1^{WT}$) or mutated IDH1 (HBT-14 (U-87 MG) $IDH1^{R132H}$) (Figures 1 and 2). We report that four of the miRNAs studied had significantly altered expressions (has-mir-100-5p, 128-3p, -221-3p and -451-5p) when the cell line expressed the IDH1 R132H-mutant, while the others showed an expression variation that did not reach significance, although some looked to be substantial (for example, has-mir-210-3p and -424-5p). To note, the mir-126-5p was undetectable in these lines.

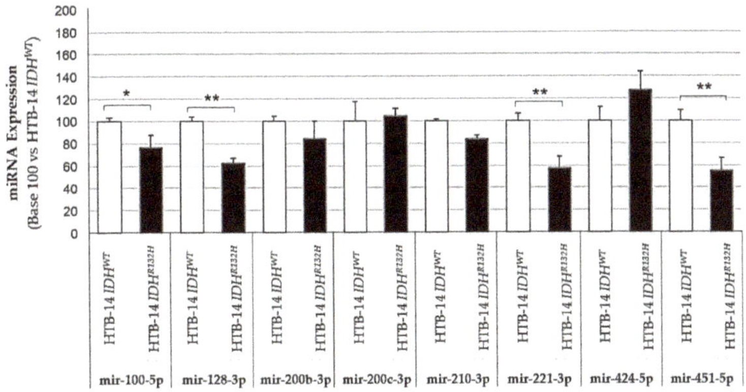

Figure 1. Expression of pro-angiogenic or pro-hypoxic miRNAs according to the presence (HBT-14 (U-87 MG) $IDH1^{R132H}$ cells) or absence (HBT-14 (U-87 MG) $IDH1^{WT}$ cells) of the IDH1 R132H mutation. The miRNAs were extracted from cell lines using miRNAeasy (Qiagen™), then retro-transcribed (RT) and amplified (PCR) using the TaqMan MiRNA Reverse transcription kit (Applied Biosystem). The RT-PCR data were normalized to the small nucleolar house-keeping RNA, RNA RNU48 (SNORD48) (assay ID 001006). Each miRNA was expressed in base 100 (100 being attributed to the delta-CT of the miRNA measured in HBT-14 (U-87 MG) $IDH1^{WT}$ cells) ($n = 3$, ANOVA followed by a post hoc Dunnett's test, *: $p < 0.05$, **: $p < 0.01$). The mir-126-5p was undetectable in these lines.

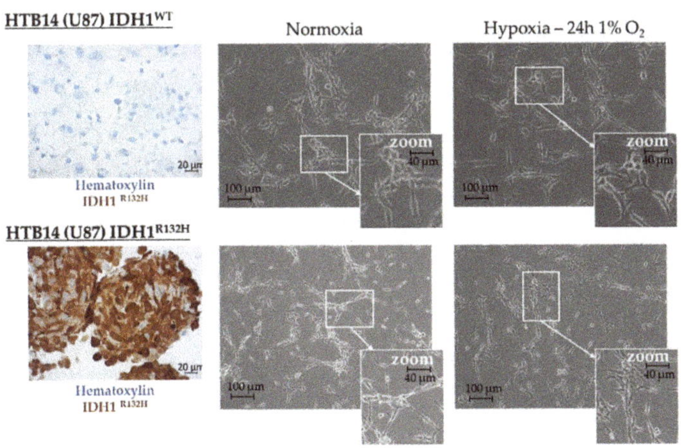

Figure 2. Morphological appearance of HBT-14 (U-87 MG) $IDH1^{WT}$ and HBT-14 (U-87 MG) $IDH1^{R132H}$ grown in normoxia or hypoxia (1% O_2, 24 h). HBT-14 (U-87 MG) $IDH1^{WT}$ and HBT-14 (U-87 MG) $IDH1^{R132H}$ cells, as validated by immunohistochemical staining carried out according to standard procedures against $IDH1^{R132H}$ (left panel), reaching 60% confluence, were cultivated for an additional 24 h in physoxia or hypoxia (0.1% O_2). The appearance of these cells was imaged under a phase contrast microscope: the right panel presents representative photos of these cells according to the culture condition (normoxia/hypoxia).

Microvascular proliferation is linked to hypoxia, a hallmark of high-grade gliomas, which is itself counterbalanced by this microvascular proliferation, although the neo-vessels formed are defective [1]. Thus, we next tested the influence of hypoxia by incubating HBT-14 (U-87 MG) $IDH1^{WT}$ and HBT-14 (U-87 MG) $IDH1^{R132H}$ cells for 24 h under 1% O_2. Morphologically, we observed that the cells survived in these growing conditions, although they appeared less numerous than their homologues cultivated in physoxia, showed a more star-shaped form, and featured more interconnections (Figure 2).

The quantification of miRNAs from HBT-14 (U-87 MG) *IDH1*^{WT} and HBT-14 (U-87 MG) *IDH1*^{R132H} grown in normoxia or hypoxia (1% O_2, 24 h) revealed that hypoxia could interfere with the IDH1 mutation and influence miRNA expression when compared to cells grown in normoxia (Figure 3).

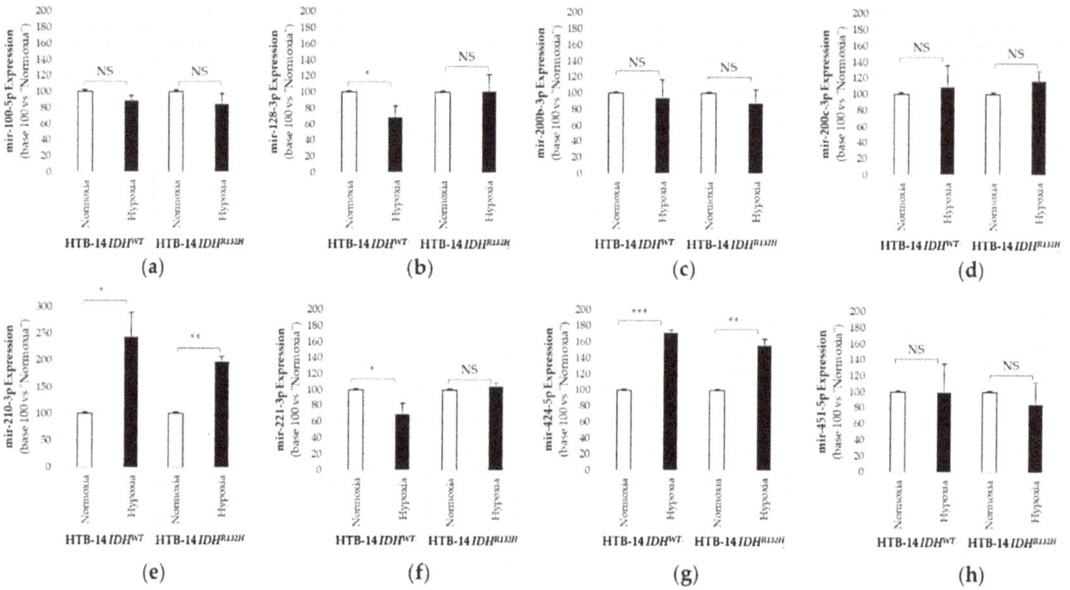

Figure 3. Expression of pro-angiogenic or pro-hypoxic miRNAs in HBT-14 (U-87 MG) *IDH1*^{R132H} or in HBT-14 (U-87 MG) *IDH1*^{WT} cells grown in hypoxia (1% O_2, 24 h). HBT-14 (U-87 MG) *IDH1*^{WT} and HBT-14 (U-87 MG) *IDH1*^{R132H} cells at 60% confluence were cultivated for an additional 24 h in normoxia or hypoxia (0.1% O_2). The miRNAs from HBT-14 (U-87 MG) *IDH1*^{WT} and HBT-14 (U-87 MG) *IDH1*^{R132H} cells were extracted using miRNAeasy (Qiagen™), then retrotranscribed (RT) and amplified (PCR) using the TaqMan MiRNA Reverse transcription kit (Applied Biosystem). The RT-PCR data were normalized to the small nucleolar house-keeping RNA, RNAS RNU48 (SNORD48) (assay ID 001006). Each miRNA ((**a**): mir-100-5p; (**b**): mir-128-3p; (**c**): mir-200b-3p; (**d**): mir-200c-3p; (**e**): mir-210-3p; (**f**): mir-221-3p; (**g**): mir-424-5p; (**h**): mir-451-5p) was thus finally expressed in base 100 (100 being attributed to the delta-CT of the miRNA measured in HBT-14 (U-87 MG) *IDH1*^{WT} or HBT-14 (U-87 MG) *IDH1*^{R132H} cells grown in normoxia) ($n = 3$, ANOVA followed by a post hoc Dunnett's test, *: $p < 0.05$; **: $p < 0.01$; ***: $p < 0.001$, NS: non-significant).

We measured significant increases in mir-210-3p and mir-424-5p in both HBT-14 (U-87 MG) *IDH1*^{WT} and HBT-14 (U-87 MG) *IDH1*^{R132H} when grown for 24h under 1% O_2 (Figure 3). We also reported significant decreases in mir-128-3p and mir-221-5p in HBT-14 (U-87 MG) *IDH1*^{WT} grown for 24h under 1% O_2 (Figure 3), but not in IDH1-mutant cells. Therefore, hypoxia modified the expression of mir-210-3p and mir-424-5p studied here in HTB-14 (U-87 MG) cells regardless of *IDH1/2* status, unlike the mir-128-3p and mir-221-5p whose variation in expression under hypoxia seems to be linked by the presence of a wild-type IDH.

2.2. The Expression of Pro-Angiogenic and/or Pro-Hypoxic miRNAs Is Strongly Affected in Patients with Glioma

The characteristics, treatment history, and pathologic data pertaining to 97 glioma samples from the 97 patients studied are summarized in Table S1. The median age was 53.9 years [range: 22.3–79.3]. There were 37 females and 60 males. The median follow-up period was 28.85 months [range: 0.26–304.23 months]. According to the 2016 WHO classification [2], the 97 glioma samples were classified as follows: 11 Grade II, isocitrate dehydrogenase 1/2 (IDH)-mutant and 1p19q-codeleted oligodendrogliomas (O); 16 Grade III, IDH-mutant and 1p19q-codeleted anaplastic oligodendrogliomas (AO); 18 Grade II, diffuse and IDH-mutant astrocytomas (A-IDH^{MUT}); 7 Grade III, IDH-mutant anaplastic astrocytomas (AA-IDH^{MUT}); 8 Grade IV, IDH-mutant glioblastomas (GB-IDH^{MUT}); 37 Grade IV, IDH-wild-type glioblastomas (GB-IDH^{WT}) (Table S1).

We first assessed the level of expression of has-miR-200b-3p, -200c-3p, -210-3p, -100-5p, -126-5p, -132-3p, -221-3p, -424-5p, -128-3p, and -451-5p between glial tumors regardless of their WHO classification. We observed that the expression of each miRNA was modified by at least >2 fold (increase or decrease) as compared with non-tumoral brain tissue in 62.8 to 84.5% of patients with glioma according to the concerned miRNA (Figure 4).

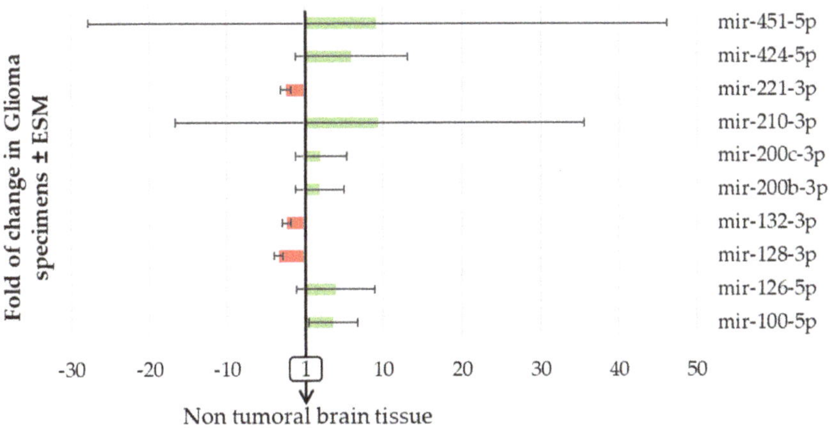

Figure 4. The expression of pro-angiogenic or pro-hypoxic miRNAs varied markedly in patients with glioma. miRNA from the 97 FFPE surgically resected tumor specimens or the 8 healthy brain tissues were extracted using miRNAeasy-FFPE kit (Qiagen™), then retrotranscribed (RT) and amplified (PCR) using the TaqMan MiRNA reverse transcription kit (Applied Biosystem). The RT-PCR data were normalized to the small nucleolar house-keeping RNA, RNAS RNU48 (SNORD48) (assay ID 001006). Results are expressed as the fold change in glioma samples ± ESM compared to healthy brain tissue.

We report that three miRNAs (mir-221-3p, mir-132-3p, and mir-128-3p, histograms in red in Figure 4) were decreased in glioma specimens compared to normal brain tissue, while the seven other miRNAs studied were all increased (histograms in green in Figure 4). We also observed that two miRNAs (mir-210-3p and mir-451-5p) varied markedly between different subtypes of glial tumors (Figure 4). In the same patient's tumor specimen, the changes in miRNAs expression seen were not exclusively increases or decreases; some miRNAs increased when the others decreased. There was also no change in miRNA exclusive to the other miRNAs which was expected since these miRNAs reflect interrelated features in gliomas. We even report that, as detailed in Table 1, these miRNA expressions appeared to be strongly correlated with each other, except for mir-128-3p and mir-210-3p.

Table 1. Inter-miRNA correlation (Spearman rank correlation coefficients).

	mir-200c-3p	mir-210-3p	mir-100-5p	mir-126-5p	mir-132-3p	mir-221-3p	mir-424-5p	mir-128-3p	mir-451-5p
mir-200b-3p	0.707	0.596	0.713	0.740	0.700	0.784	0.793	0.385	0.619
mir-200c-3p		0.494	0.591	0.778	0.726	0.791	0.764	0.355	0.641
mir-210-3p			0.435	0.550	0.344	0.543	0.549	0.078	0.464
mir-100-5p				0.642	0.627	0.607	0.741	0.535	0.514
mir-126-5p					0.729	0.797	0.781	0.434	0.749
mir-132-3p						0.822	0.725	0.592	0.546
mir-221-3p							0.728	0.381	0.621
mir-424-5p								0.549	0.702
mir-128-3p									0.539

Correlations were highly significant for all miRNAs ($p < 0.001$), except mir-128-3p and mir-210-3p ($p = 0.45$).

Indeed, as seen in Table 1, the Spearman rank correlation coefficient testing the interaction between each pair of miRNAs varied from 0.344 to 0.822 and was strongly significant ($p < 0.001$) except for mir-128-3p and mir-210-3p (Spearman rank correlation coefficient: 0.078, $p = 0.45$).

2.3. *The Expression of Pro-Angiogenic and/or Pro-Hypoxic miRNAs Is Correlated with the 2016 WHO Classification*

Having observed that the expression of microRNAs was affected in patients with glial tumors (Figure 4), we next investigated which feature of glioma could have influenced the expression of the miRNAs studied here. We thus looked at the miRNA expression according to (1) the mutational status of IDH (Table 2), (2) the microvascular proliferation (Table 3), and (3) the histological subgroup (Table 4).

Table 2. MiRNA expression according to the *IDH1/2* mutation.

	IDH1/2-^{WT} ($n = 37$)			*IDH1/2*-^{MUT} ($n = 60$)			
	Median (%)	**First Quartile (%)**	**Third Quartile (%)**	**Median (%)**	**First Quartile (%)**	**Third Quartile (%)**	*p*
mir-200b-3p	250.5	82.9	450.0	59.3	21.8	97.8	<0.001
mir-200c-3p	262.4	122.0	449.0	64.8	20.9	138.9	<0.001
mir-210-3p	1109.0	380.1	1891.2	126.7	67.2	201.3	<0.001
mir-100-5p	362.9	189.1	597.7	238.6	141.2	410.4	0.086
mir-126-5p	451.0	175.7	1194.3	119.6	38.0	294.6	<0.001
mir-132-3p	32.1	18.4	56.0	20.0	9.2	36.7	0.024
mir-221-3p	35.7	17.4	90.4	8.5	3.0	18.5	<0.001
mir-424-5p	699.8	257.9	1324.0	215.4	104.8	410.5	<0.001
mir-128-3p	4.3	1.2	21.9	9.9	5.4	37.8	0.027
mir-451-5p	322.7	67.4	1104.3	73.5	25.9	310.6	0.0010

The 100 value was attributed to the miRNA expression in non-tumoral tissue. IDH: isocitrate dehydrogenase.

Table 3. MiRNA expression according to microvascular proliferation.

	No Microvascular Proliferation (n = 40)			Microvascular Proliferation (n = 57)			p
	Median (%)	First Quartile (%)	Third Quartile (%)	Median (%)	First Quartile (%)	Third Quartile (%)	
mir-200b-3p	62.0	28.6	115.0	99.0	26.0	386.4	0.021
mir-200c-3p	57.7	21.0	162.4	138.7	71.3	359.7	0.010
mir-210-3p	117.4	53.3	154.0	498.0	192.7	1568.4	<0.001
mir-100-5p	251.9	176.8	420.4	267.5	124.8	579.3	0.96
mir-126-5p	120.2	43.8	334.4	275.7	117.5	688.3	0.0073
mir-132-3p	26.4	9.5	42.3	25.5	13.0	49.3	0.51
mir-221-3p	8.6	3.8	20.6	19.5	7.8	63.2	0.0063
mir-424-5p	192.2	114.6	366.2	447.5	179.9	1075.5	0.0063
mir-128-3p	10.0	5.4	42.5	5.8	2.2	26.4	0.087
mir-451-5p	59.9	25.9	261.2	256.1	51.8	879.9	0.0068

The 100 value was attributed to the miR expression in non-tumoral tissue.

Table 4. MiRNA expression according to the WHO 2016 classification.

	O (n = 11)	AO (n = 16)	A (n = 18)	AA (n = 7)	GB-IDH^{MUT} (n = 8)	GB-IDH^{WT} (n = 37)	p
	Median (%)						
mir-200b-3p	37.9	35.0	58.9	96.6	63.5	250.5	<0.001
mir-200c-3p	45.0	88.9	66.7	166.1	53.0	262.4	<0.001
mir-210-3p	98.1	139.1	110.5	118.2	228.1	1109.0	<0.001
mir-100-5p	208.3	241.2	350.8	242.7	231.6	362.9	0.28
mir-126-5p	105.9	274.3	113.3	131.8	103.5	451.0	<0.001
mir-132-3p	15.6	18.6	23.1	47.8	15.2	32.1	0.10
mir-221-3p	4.6	8.1	8.5	13.6	8.8	35.7	<0.001
mir-424-5p	115.0	271.1	203.9	574.3	274.9	699.8	0.0012
mir-128-3p	7.1	12.9	15.2	9.7	10.6	4.3	0.28
mir-451-5p	71.9	313.9	53.9	58.2	116.3	322.7	0.0077

A value of 100 value was attributed to the miR expression in non-tumoral tissue. A: diffuse astrocytoma, IDH-mutant; AA: anaplastic astrocytoma, IDH-mutant; AO: anaplastic oligodendroglioma, IDH-mutant, and 1p19q-codeleted; GB-IDHMUT: glioblastoma, IDH-mutant; GB-IDHWT: glioblastoma, IDH-wild-type; IDH: isocitrate dehydrogenase; O: oligodendroglioma, IDH-mutant, and 1p19q-codeleted; WHO: World Health Organization.

As shown in Table 2, 7/10 miRNAs studied here were more strongly expressed in glioma *IDH1/2*-^{WT} (expression > 100) than in non-tumor brain tissue, while only 2/10 miRNAs were more strongly expressed in glioma *IDH1/2*-^{MUT} than in non-tumor brain tissue. Thus, the *IDH1/2* mutation significantly decreased the expression of the miRNAs studied here by 4.2 to 8.8-fold according to the miRNA and except for mir-128-3p (increased by *IDH1/2* mutation) and mir-100-5p (decrease not significant).

As shown in Table 3, we further report that the microvascular proliferation also significantly increased the expression of pro-angiogenic or pro-hypoxic miRNAs up to 4.3 folds (except for mir-128-3p, mir-132-3p, and mir-100-5p (not significant). Thus, the expression of 7/10 miRNAs studied here was stronger when the glial tumor exhibited a microvascular proliferation than when it did not.

Finally, as shown in Table 4, the histological subgroup influenced the expression of the pro-angiogenic or pro-hypoxic miRNAs studied here. Indeed, the level of expression of the microRNAs mir-100-5p, -126-3p, -128-3p, -132-3p, -210-5p, and -221-3p was comparable between astrocytic tumors (A, AA) and oligodendroglial tumors (O, AO).

Conversely, the mir-200b-3p was found to be more expressed in astrocytic (A, AA) than in oligodendroglial tumors (O, AO). Similarly, the expression of mir-451-5p increased between O and AO, but not between A and AA. Other miRNAs could be used for comparison: the expression of mir-200c-3p, -424-5p, and -451-5p increased with the tumor evolution (A > AA, O > AO).

2.4. The Expression of Pro-Angiogenic and/or Pro-Hypoxic miRNAs Predicts Overall (OS) and Progression-Free (PFS) Survival in Patients with Glioma in Univariate Analysis

We next tested the influence of the expression of pro-angiogenic and/or pro-hypoxic miRNAs on the OS and PFS using univariate and multivariate Cox-proportional hazard models. As detailed in Table 5, except for mir-128-3p, mir-132-3p, and mir-100-5p, the expression of the miRNAs studied here predicted the OS and PFS of patients with glioma in the univariate analysis ($p < 0.001$) but not in the multivariate analysis, following adjustments for sex, age at diagnosis, and WHO 2016 classification.

Table 5. MiRNA expression and overall and progression-free survival of patients with glioma.

	HR †	IC95%		*p*	Adjusted HR ‡	IC95%		*p*
OS								
mir-200b-3p	1.08	1.05	1.12	<0.001	1.03	0.98	1.08	0.30
mir-200c-3p	1.06	1.03	1.09	<0.001	1.02	0.98	1.05	0.42
mir-210-3p	1.015	1.008	1.02	<0.001	1.01	0.997	1.01	0.23
mir-100-5p	1.02	0.90	1.15	0.80	0.92	0.81	1.06	0.25
mir-126-5p	1.08	1.03	1.12	<0.001	0.99	0.94	1.05	0.83
mir-132-3p	1.04	1.00	1.08	0.055	1.02	0.98	1.07	0.38
mir-221-3p	1.07	1.04	1.11	<0.001	1.01	0.97	1.05	0.50
mir-424-5p	1.07	1.02	1.13	0.0038	1.00	0.95	1.06	0.94
mir-128-3p	1.02	0.97	1.07	0.40	1.00	0.96	1.04	0.99
mir-451-5p	1.008	1.003	1.01	0.0022	1.003	0.997	1.01	0.31
PFS								
mir-200b-3p	1.07	1.03	1.11	<0.001	1.02	0.96	1.07	0.56
mir-200c-3p	1.06	1.03	1.09	<0.001	1.02	0.98	1.06	0.44
mir-210-3p	1.013	1.007	1.02	<0.001	1.01	0.997	1.01	0.23
mir-100-5p	1.01	0.90	1.12	0.92	0.97	0.86	1.09	0.62
mir-126-5p	1.06	1.02	1.11	0.007	0.99	0.93	1.04	0.63
mir-132-3p	1.04	1.00	1.08	0.053	1.02	0.98	1.06	0.41
mir-221-3p	1.06	1.03	1.09	<0.001	1.00	0.97	1.04	0.90
mir-424-5p	1.07	1.02	1.12	0.0034	1.00	0.95	1.06	0.93
mir-128-3p	1.01	0.98	1.05	0.45	0.99	0.96	1.03	0.76
mir-451-5p	1.007	1.00	1.01	0.0077	1.003	0.996	1.01	0.40

† mir-132-3p, mir-221-3p, and mir-128-3p for a 10-point increase in the level of expression; mir-200b-3p and mir-200c-3p for a 50-point increase in the level of expression; mir-210-3p, mir-126-5p, and mir-451-5p for a 100-point increase in the level of expression; mir-100-5p and mir-424-5p for a 150-point increase in the level of expression. ‡ Adjustments for sex, age at diagnosis, and WHO 2016 classification.

2.5. Mir-128-3p Predicts a Poorer PFS in Patients with AA-IDH^{MUT} or AO-IDH^{MUT} and mir-100-5p Predicts a Poorer PFS in Patients with AA-IDH^{MUT}

For each miRNA, we used an interaction test to compare the hazard ratios between the WHO 2016 classes for OS and PFS. Interactions were not statistically significant for OS, but for PFS, significant differences were found for mir-128-3p and mir-100-5p ($p = 0.033$ and $p = 0.013$, respectively; Table S4).

We thus further calculated the prognostic value of these two miRNAs for each WHO subgroup and reported that mir-128-3p predicts a poorer PFS in patients with AA-IDH^{MUT} or AO-IDH^{MUT} and that mir-100-5p predicts a poorer PFS in patients with AA-IDH^{MUT} (Table 6). However, the low number of patients by class (sometimes less than 10 subjects) means that it is necessary to interpret such results with caution.

Table 6. Value of mir-100-5p or mir-128-3p within WHO 2016 classes for PFS.

	mir-100-5p				mir-128-3p			
	HR	IC95%		*p*	HR	IC95%		*p*
O	0.75	0.28	2.00	0.56	0.95	0.75	1.21	0.66
A	0.79	0.57	1.11	0.18	0.97	0.72	1.30	0.82
AO	0.99	0.776	1.26	0.94	1.33	1.046	1.68	0.020
AA	5.12	1.84	14.24	0.0018	1.31	1.06	1.60	0.010
GB-IDH^{MUT}	0.68	0.42	1.08	0.10	0.93	0.65	1.31	0.66
GB-IDH^{WT}	1.05	0.90	1.23	0.52	0.99	0.95	1.03	0.55

A: diffuse astrocytoma, IDH-mutant; AA: anaplastic astrocytoma, IDH-mutant; AO: anaplastic oligodendroglioma, IDH-mutant, and 1p19q-codeleted; GB-IDHMUT: glioblastoma, IDH-mutant; GB-IDHWT: glioblastoma, IDH-wild-type; IDH: isocitrate dehydrogenase; O: oligodendroglioma, IDH-mutant, and 1p19q-codeleted; WHO: World Health Organization.

2.6. For the Same Patient, Expression of Plasma miRNAs Coincides with the Expression of Tumoral miRNAs

We know that stable miRNA expression is measurable from bodily fluids such as cerebrospinal fluid or blood serum [20,21] but that specific miRNAs in the diagnosis of a glioma, in particular at the early stage, are still missing [22]. We thus next tested the possibility of directly assaying the microRNAs studied in this study from the patient's blood sample.

A blood sample was collected from seven patients, from which we also extracted and quantified the pro-angiogenic miRNAs. All the miRNAs could be assayed from these blood samples (no amplification failure and the number of CTs after real-time PCR amplification was <35), even if, for some miRNAs in the circulatory system, their expression patterns were at a slightly lower concentration compared to the tumoral tissues. As illustrated in Figure 5 for four miRNAs (has-miR-100-5p, -132-3p, -200b-3p, and -221-3p), the amount of each miRNA detected in the glial tumor was correlated with that detected in the blood of the same patient, except for the miR-100-5p in patient 3.

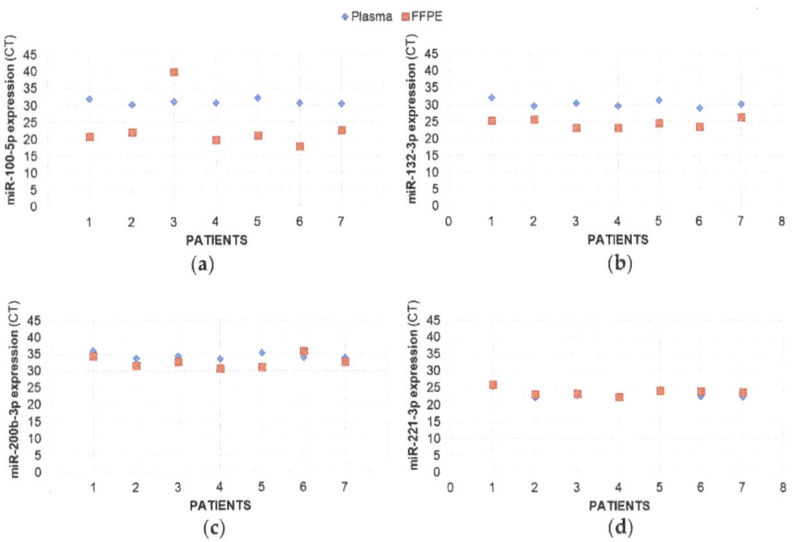

Figure 5. The expression levels of pro-angiogenic or pro-hypoxic miRNAs were correlated between the tumor sample and plasma from patients with glioma. For 7 patients with glioma, the miRNAs from the FFPE surgically resected tumor specimens or the plasma samples were extracted using, miRNAeasy-FFPE kit (Qiagen™) and NucleoSpin miRNA Plasma (Macherey-Nagel™), respectively,

then retrotranscribed (RT) and amplified (PCR) using the TaqMan MiRNA Reverse transcription kit (Applied Biosystem). Results illustrated here for (**a**) mir-100-5p, (**b**) mir-132-3p, (**c**) mir-200b-3p, and (**d**) mir-221-3p are expressed as cycle thresholds (CTs) assayed for each miRNA normalized to the small nucleolar house-keeping RNA, RNAS RNU48 (SNORD48) (assay ID 001006).

3. Discussion

It is now well known that miRNAs are involved in tumor initiation and development [35]. Such miRNAs thus appear to be interesting tools for use in the diagnosis and/or monitoring of cancers, whether in gliomas or other tumors, particularly due to their non-invasive nature since they can be measured from a patient's blood sample. Indeed, the expression of miRNAs in blood and tissues has tumor-related and tissue-specific features, and their expression is remarkably stable [36]. Nevertheless, miRNAs have not yet come to be used in daily clinical practice. To relaunch the debate on their possible clinical interest, particularly for the diagnosis of patients with gliomas, a meta-analysis recently tested the diagnostic performance of circulating miRNAs for gliomas [22]. After analyzing 18 articles covering 24 studies containing 2170 glioma patients and 1456 healthy participants, the authors concluded that circulating miRNAs have the potential to serve as diagnostic biomarkers for gliomas. Our results are in agreement with their conclusion; indeed, we prove here that we can use miRNAs as tools for monitoring patients with gliomas by selecting several miRNAs involved in intertwined phenomena and accounting for the natural history of the disease. Indeed, we tested the diagnostic and prognostic values of miRNAs reflecting the features of gliomas (IDH mutation, microvascular proliferation, hypoxia) and patients' survival in a series of 97 gliomas of grades II to IV to determine whether miRNAs could be used as tools for monitoring patients with gliomas. As detailed in the Section 1, we thus chose the miRNAs involved in the regulation of hypoxia/cell proliferation/differentiation of glioma cells, as described for other cell types [23–33] (for review: [15,34]). As expected, since the miRNAs studied here reflected the mechanisms related to gliomagenesis, these miRNAs appeared to be strongly correlated with each other, except for mir-128-3p and mir-210-3p.

First, we successively reported that the expression of some of the pro-angiogenic and/or pro-hypoxic miRNAs studied could actually be affected by IDH mutation and hypoxia in vitro, and then in tumor specimens from patients with glioma that the expression of some pro-angiogenic and/or pro-hypoxic miRNAs could be strongly affected by and correlated with the 2016 WHO classification (*IDH1/2* mutation, microvascular proliferation, histoprognostic group according to WHO 2016 classification). Interestingly, the induced hypoxia of IDH HTB-14 (U-87 MG) cells modified the expression of four miRNAs, for two (mir-210-3p and mir-424-5p, which decreased under hypoxic conditions) independently of the IDH mutation, while for the two others (128-3p and mir-221-5p, which decreased under hypoxic conditions), only in wild-type IDH HTB-14 (U-87 MG) cells. Such results could explain why IDHWT glioma are more aggressive tumors with a higher microvascular proliferation than IDHMUT glioma; indeed, the mir-128-3p is a tumor suppressor [37] and both the mir-128-3p and the mir-221-5p sometimes display anti-angiogenic behavior by targeting, in particular, VEGFC [38] and the Hypoxia-inducible factor 1 alpha, respectively [39]. Then, in a small sample of seven patients for whom both tissue and blood were available, we then reported that the expression of pro-angiogenic and/or pro-hypoxic miRNAs in plasma coincided with the expression of tumor miRNAs and that the tumor expression of some miRNAs could predict OS and PFS in patients with glioma, at least in a univariate analysis. We thus suggest the use of pro-angiogenic and/or pro-hypoxic miRNAs as tools for monitoring patients specifically with *IDH1/2*-mutated gliomas. Indeed, we also observed in this study that the variations in the expression of miRNAs within the group of patients with GB IDH WT was specific to this group, which is consistent with the report that miRNA profiles play a more significant prognostic role in *IDH*-mutant tumors than in *IDH* wild-type tumors [11,40]. Moreover, the *IDH1/2* mutation status had a greater impact than the histological and other genomic features on miRNA expression patterns;

361/487 (74%) of the miRNAs were differentially expressed according to their *IDH1/2* mutation status [11].

Our results, which indicate that we could use miRNAs as tools for monitoring patients with *IDH1/2*-mutated gliomas, are consistent with the study of Chen and co-workers, based on another miRNA, mir-720, assayed on 122 patients with glioma (Stage I: 20; Stage II: 17; Stage III: 35; and Stage IV: 50, according to the WHO 2016 classification). Chen an al. reported that the plasma miR-720 was associated with the tumor grade and associated with recurrence or development in patients with glioma but that the sensitivity and specificity results indicated that the diagnostic ability of miR-720 for glioma was only moderate [41]. This result raises the question of what methodological choices should be made when one wants to use miRNAs as diagnostic and/or monitoring tools for patients with tumors and more particularly gliomas. Although circulating miRNAs are promising diagnostic biomarkers for patients with glioma, the serum miRNAs and miRNA panels presented a superior diagnostic performance compared to the use of only one miRNA [22]. Moreover, variations in the plasma concentrations of only one miRNA could alternatively reflect another cancer pathology in the same patient and not be specific to glioma, thus compromising the potential clinical utility in diagnosis and follow-up by leading to false detection. To overcome these possible biases, it is therefore more relevant to assay a signature of miRNAs—moreover, a signature of miRNAs reflecting the specific characteristics of the tumor that one seeks to characterize. As an example, the use of a four-miRNA risk classifier (miR-10b, miR-130b, miR-1304, and miR-302b), involved in the proliferation, invasion, and survival of glioma or other tumors cells [42–45], would allow one to independently distinguish cases as either at a high or low risk of poor prognosis in *IDH1/2*-mut lower-grade glioma [11].

Finally, the fact that the plasma dosage of miRNAs could reflect the tumor dosage is also consistent with what other authors have been able to report in the literature, such as during Spinal Cord Glioma Progression [46]. Indeed, the brain is among the tissues with the strongest correlation for microRNA for both plasma and serum [47].

In our work, mir-128-3p, mir-132-3p, and mir-100-5p often behaved differently from the other miRNAs that we chose to study; these miRNAs were not linked to the WHO 2016 classification and did not predict the OS and PFS of patients with glioma in univariate analysis. Conversely, mir-128-3p predicted a poorer PFS in patients with AA-IDH^{MUT} or AO-IDH^{MUT}, while mir-100-5p predicted a worse PFS in patients with AA-IDH^{MUT}. This, again, is consistent with the tumor suppressor role attributed to mir-128-3p [37], which we confirmed by identifying a drop in its expression compared to the level quantified in non-tumor brain tissue, and with the work of Zhang et al., 2019, who reported that a low miR-100 expression correlated with worse clinicopathological characteristics such as Karnofsky Performance Scale and *IDH1/2* mutation status [48]. It remains difficult to explain why these three miRNAs behave differently: it is possible that the variations in such miRNAs are more difficult to highlight since they are constitutively highly expressed in the brain [49,50] and that, regardless of the processes in which they participate, their variations are masked and under-evaluated in this tissue. In any case, it would be preferable to exclude these miRNAs from the panel of miRNAs used for characterizing angiogenesis and for hypoxia to be analyzed to evaluate the progression of the disease in patients with *IDH*-mutated gliomas.

4. Materials and Methods

4.1. Tissue Samples and Patient Characteristics

Ninety-seven patients, aged 18 years or older with a tissue diagnosis of WHO grade II, III, and IV diffuse gliomas made between September 2001 and March 2012, were identified from the brain tumor registry of Caen University Hospital, France. Characteristics, treatment history, and pathologic data from these patients are summarized in Table S1. For 7 patients, a blood sample was also available. The 8 non-tumoral brain tissues were from patients without glioma who underwent brain surgery.

All patients provided informed consent regarding the collection of tumor specimens and their molecular evaluation, as required by French law. The study was approved by the institutional ethics committee of Caen University Hospital, France (DC-2008-588). All tumor specimens were reviewed by a neuropathologist (ELZ) to confirm the diagnosis and grade according to the new classification system adopted by the World Health Organization (WHO) in 2016, as described in [10]. Indeed, we used the classification in place at the time of the constitution of this cohort, i.e., the WHO 2016 classification and could not update on the classification of 2021 due to the exhaustion of numerous tumor specimens.

4.2. *Cell Culture and Hypoxia Treatment*

The human glioblastoma cell lines HBT-14 (U-87 MG) $IDH1^{WT}$ and HBT-14 (U-87 MG) $IDH1^{R132H}$ cells obtained from the generous gift of Pr. Marc Sanson (Hospital Group Pitié-Salpêtrière, Paris, France) [51], were maintained in Dulbecco's modified Eagle's medium (DMEM) supplemented with 10% fetal bovine serum (FBS) and 1% penicillin/streptomycin. Normoxic cells (21% O_2) were grown in a humidified air atmosphere incubator containing 95% air/5% CO_2 at 37 °C. Hypoxia experiments were performed in a controlled atmosphere chamber (INVIVO2 1000, Ruskinn, Awel, France) set at 1% O_2, 94% N_2, and 5% CO_2 at 37 °C for 24 h.

4.3. *DNA and miRNA Extraction*

DNA from HBT-14 (U-87 MG) cells was extracted using the QIAmp DNA kit (Qiagen™) according to the manufacturer's recommendation.

The miRNAs from HBT-14 (U-87 MG) cells were extracted using miRNAeasy (Qiagen™). The miRNAs from the 97 FFPE (formalin-fixed paraffin-embedded) surgically resected tumor specimens or the 8 non-tumoral brain tissues were extracted on 3 adjacent 15 µm cuts using the miRNAeasy-FFPE kit (Qiagen™, Hilden, Germany). For the tumoral specimens, morphological control was systematically carried out beforehand by a neuropathologist (ELZ) in order to guarantee that the percentage of tumor cells was greater than 70% and the absence of areas of necrosis or hemorrhage. When this was not the case, a macro-dissection of the samples was performed to determine these quality criteria. miRNAs were extracted from the seven plasma samples using NucleoSpin miRNA Plasma (Macherey-Nagel™, Düren, Luxembourg). For each sample, miRNA extraction was carried out according to the respective manufacturer's instructions.

The integrity and quality of the purified DNA were assessed by 1% agarose gel electrophoresis, and the DNA/miRNA concentration was measured with the NanoDrop 2000 spectrophotometer (Thermo Fisher Scientific, Asnières-sur-Seine, France).

4.4. *Quantitative Real-Time Reverse Transcription: PCR*

The miRNAs were retrotranscribed and amplified (PCR) using the TaqMan MiRNA Reverse transcription kit (Applied Biosystem, Birchwood, United Kingdom). An amount of 5 ng of miRNA was used for every miRNA that we tested. MiRNA expression was analyzed using the following TaqMan MiRNA assays (Applied Biosystem Birchwood, United Kingdom): has-miR-200b-3p (Assay ID: 002251), has-miR-200c-3p (002300), has-miR-210-3p (000512), has-miR-100-5p (000437), has-miR-126-5p (000451), has-mir-132-3p (000457), hsa-miR-221-3p (000524), has-miR-424-5p (000604), has-miR-128-3p (002216), and has-miR-451-5p (001141) (targets for each miRNA are listed in Table S2).

The RT-PCR data were normalized to the small nucleolar house-keeping RNA, RNAS RNU48 (SNORD48) (assay ID 001006). Positive standards and reaction mixtures lacking the reverse transcriptase were used routinely as controls for each miRNA sample. Relative quantification was conducted using the deltaCt method, where deltaCt is CtmiRX-CtRNU48. To facilitate comparison between conditions (histology, grade), the miRNAs were further normalized to the miRNA expression in healthy brain tissue (quantification averaged from the normal brain tissue of 8 subjects, operated on to cure their epilepsy). As presented in Table S3, the delta-CT averages of each miRNA of these 8 patients were calculated along with the SD (standard deviation) and did not vary more than one CT between the brain tissue specimens of these patients.

Each miRNA was thus finally expressed in base 100, with the value 100 being attributed to the delta-CT of the miRNA measured in normal brain tissue.

4.5. IDH1 and IDH2 Mutations Assay

HBT-14 (U-87 MG) cells with wild-type IDH1 (HBT-14 (U-87 MG) $IDH1^{WT}$) or mutated IDH1 (HBT-14 (U-87 MG) $IDH1^{R132H}$) were certificated by $IDH1^{R132H}$ staining (Figure S1). Cells were collected, washed with PBS, and fixed in neutral buffered formalin for 20 min. After centrifugation, the pellet was embedded in 6–10 drops of melted Bio-Agar (Bio-Optica, Milano, Italy, cod. 05-98035) and chilled in the freezer (−20 °C) until complete solidification was achieved. Next, the sample pellet was putted between two pads of a bio-cassette and then processed and paraffin-embedded according to the histologic routine. For IDH1 (R132H) detection, 3 µm sections were cut and placed on poly-L-lysine-coated slides. Immunohistochemical staining was performed using the Ventana Discovery XT automated immunostainer (Ventana Medical Systems, Tucson, AZ, USA). Slides were subjected to deparaffinization in xylene and hydration through a series of decreasing alcohol concentrations, following standard procedures. Antigen retrieval was performed using a high-pH Tris-based solution (CC1; VMS) for 64 min at 100 °C. The slides were incubated with the anti-IDH1 R132H primary antibody (Diagomics, Blagnac, France, IHC132-100) at a 1:100 dilution and then with the Ventana UltraView detection kit and Ventana DAB. Counterstaining with hematoxylin was performed on the Leica ST 5020. The slides were finally washed in running water for 10 min, dehydrated, cleared, and mounted with resinous mounting medium.

4.6. Statistical Analyses

In vitro data are presented as means ± SEM ($n \geq 3$). Statistical differences were determined by one-way analysis of variance (ANOVA) followed by Dunnett's Multiple Comparison Test (GraphPad Software, Inc., San Diego, CA, USA). Statistical significance was set at $p \leq 0.05$.

Spearman's rank correlation coefficient was used to test the correlation between the miRNAs. Comparisons of the expression of miRNAs according to *IDH1/2* mutation, microvascular proliferation, and WHO2016 classification were made using the Mann–Whitney and Kruskal–Wallis tests. Univariate and multivariate Cox-proportional hazard models were used to assess the prognostic value of the miRNA expression. Hazard ratios (HR) were estimated with 95% confidence intervals (95%CI). For each miRNA, the hazard ratios were compared between the WHO 2016 classes by including an interaction term in the Cox models. Statistical significance was set at $p < 0.05$. The data were analyzed with the IBM SPSS software (New York, NY, USA), Version 22.

5. Conclusions

With this work, we reported that the expression of a panel of seven pro-angiogenic and/or pro-hypoxic miRNAs (has-miR-200b-3p, -200c-3p, -210-3p, -126-5p, -221-3p, -424-5p, and -451-5p) was affected in patients with glioma and related to the histology/grade of glioma according to the 2016 WHO classification. We also suggested that, by predicting glioma patients' overall and/or progression-free survival in a univariate analysis, pro-angiogenic and/or pro-hypoxic miRNAs can be used as tools for monitoring patients, specifically with IDH-mutated low-grade tumors, since they are also measurable in plasma. However, our results remain preliminary because of the small sample size, the lack of a longitudinal follow-up for patients with plasma sampling, and the need to verify these results in prospective dedicated studies before being used for diagnosis, to monitor response to treatment, or to assess the risk of residual disease and relapse after surgical resection.

Supplementary Materials: The following supporting information can be downloaded at https://www.mdpi.com/article/10.3390/ijms23116042/s1.

Author Contributions: Conceptualization, G.L., G.Z. and E.L.-Z.; resources, G.L., F.D., A.L., M.F., E.P., L.B., C.C., E.E., G.Z. and E.L.-Z.; data curation, G.L., F.D., A.L., M.F., C.C., E.E., G.Z. and E.L.-Z.; writing—original draft preparation, G.L. and E.L.-Z.; writing—review and editing, G.L., G.Z. and E.L.-Z. All authors have read and agreed to the published version of the manuscript.

Funding: This research was funded by APRI from the at the Caen University Hospital center, grant number 2012 and the "Ligue contre le cancer—Normandie", grant number 2012.

Institutional Review Board Statement: The study was conducted according to the guidelines of the Declaration of Helsinki and approved by the Institutional Ethics Committee of the Caen University Hospital center (protocol code DC-2008-588; date of approval: 12 April 2010) authorizing the collection, conservation, and preparation activities for scientific purposes included in this collection of human biological samples at the Caen University Hospital center.

Informed Consent Statement: Informed consent was obtained from all subjects involved in the study.

Data Availability Statement: All data are stored at the CHU of Caen and CFB center and can be made available upon request.

Acknowledgments: The authors thank Marc Sanson for the generous gift of the HBT-14-*IDH1*^{WT} and HBT-14-*IDH1*^{R132H} cells.

Conflicts of Interest: The authors declare no conflict of interest.

References

1. Torrisi, F.; Alberghina, C.; D'Aprile, S.; Pavone, A.M.; Longhitano, L.; Giallongo, S.; Tibullo, D.; Di Rosa, M.; Zappalà, A.; Cammarata, F.P.; et al. The Hallmarks of Glioblastoma: Heterogeneity, Intercellular Crosstalk and Molecular Signature of Invasiveness and Progression. *Biomedicines* **2022**, *10*, 806. [CrossRef]
2. Louis, D.N.; Perry, A.; Reifenberger, G.; von Deimling, A.; Figarella-Branger, D.; Cavenee, W.K.; Ohgaki, H.; Wiestler, O.D.; Kleihues, P.; Ellison, D.W. The 2016 World Health Organization Classification of Tumors of the Central Nervous System: A summary. *Acta Neuropathol.* **2016**, *131*, 803–820. [CrossRef]
3. Komori, T. The 2016 WHO Classification of Tumours of the Central Nervous System: The major points of revision. *Neurol Med. Chir.* **2017**, *57*, 301–311. [CrossRef]
4. Iuchi, T.; Sugiyama, T.; Ohira, M.; Kageyama, H.; Yokoi, S.; Sakaida, T.; Hasegawa, Y.; Setoguchi, T.; Itami, M. Clinical significance of the 2016 WHO classification in Japanese patients with gliomas. *Brain Tumor. Pathol.* **2018**, *35*, 71–80. [CrossRef] [PubMed]
5. Louis, D.N.; Perry, A.; Wesseling, P.; Brat, D.J.; Cree, I.A.; Figarella-Branger, D.; Hawkins, C.; Ng, H.K.; Pfister, S.M.; Reifenberger, G.; et al. The 2021 WHO Classification of Tumors of the Central Nervous System: A summary. *Neuro-Oncology* **2021**, *23*, 1231–1251. [CrossRef] [PubMed]
6. Han, S.; Liu, Y.; Cai, S.J.; Qian, M.; Ding, J.; Larion, M.; Gilbert, M.R.; Yang, C. IDH mutation in glioma: Molecular mechanisms and potential therapeutic targets. *Br. J. Cancer* **2020**, *122*, 1580–1589. [CrossRef] [PubMed]
7. van den Bent, M.J.; Dubbink, H.J.; Marie, Y.; Brandes, A.A.; Taphoorn, M.J.; Wesseling, P.; Frenay, M.; Tijssen, C.C.; Lacombe, D.; Idbaih, A.; et al. IDH1 and IDH2 mutations are prognostic but not predictive for outcome in anaplastic oligodendroglial tumors: A report of the European Organization for Research and Treatment of Cancer Brain Tumor Group. *Clin. Cancer Res.* **2010**, *16*, 1597–1604. [CrossRef] [PubMed]
8. Beyer, S.; Fleming, J.; Meng, W.; Singh, R.; Haque, S.J.; Chakravarti, A. The Role of miRNAs in Angiogenesis, Invasion and Metabolism and Their Therapeutic Implications in Gliomas. *Cancers* **2017**, *9*, 85. [CrossRef] [PubMed]
9. Turcan, S.; Rohle, D.; Goenka, A.; Walsh, L.A.; Fang, F.; Yilmaz, E.; Campos, C.; Fabius, A.W.; Lu, C.; Ward, P.S.; et al. IDH1 mutation is sufficient to establish the glioma hypermethylator phenotype. *Nature* **2012**, *483*, 479–483. [CrossRef]
10. Levallet, G.; Creveuil, C.; Bekaert, L.; Péres, E.; Planchard, G.; Lecot-Cotigny, S.; Guillamo, J.S.; Emery, E.; Zalcman, G.; Lechapt-Zalcman, E. Promoter Hypermethylation of Genes Encoding for RASSF/Hippo Pathway Members Reveals Specific Alteration Pattern in Diffuse Gliomas. *J. Mol. Diagn.* **2019**, *21*, 695–704. [CrossRef]
11. Cheng, W.; Ren, X.; Zhang, C.; Han, S.; Wu, A. Expression and prognostic value of microRNAs in lower-grade glioma depends on IDH1/2 status. *J. Neurooncol.* **2017**, *132*, 207–218. [CrossRef] [PubMed]
12. Miska, E.A. How microRNAs control cell division, differentiation and death. *Curr. Opin. Genet. Dev.* **2005**, *15*, 563–568. [CrossRef] [PubMed]
13. Pang, J.C.; Kwok, W.K.; Chen, Z.; Ng, H.K. Oncogenic role of microRNAs in brain tumors. *Acta Neuropathol.* **2009**, *117*, 599–611. [CrossRef] [PubMed]
14. Yan, W.; Li, R.; Liu, Y.; Yang, P.; Wang, Z.; Zhang, C.; Bao, Z.; Zhang, W.; You, Y.; Jiang, T. MicroRNA expression patterns in the malignant progression of gliomas and a 5-micro-RNA signature for prognosis. *Oncotarget* **2014**, *5*, 12908–12915. [CrossRef]

15. Rezaei, O.; Honarmand, K.; Nateghinia, S.; Taheri, M.; Ghafouri-Fard, S. miRNA signature in glioblastoma: Potential biomarkers and therapeutic targets. *Exp. Mol. Pathol.* **2020**, *117*, 104550. [CrossRef]
16. Ames, H.; Halushka, M.K.; Rodriguez, F.J. miRNA Regulation in Gliomas: Usual Suspects in Glial Tumorigenesis and Evolving Clinical Applications. *J. Neuropathol. Exp. Neurol.* **2017**, *76*, 246–254. [CrossRef]
17. Zhang, Y.; Chen, J.; Xue, Q.; Wang, J.; Zhao, L.; Han, K.; Zhang, D.; Hou, L. Prognostic Significance of MicroRNAs in Glioma: A Systematic Review and Meta-Analysis. *Biomed. Res. Int.* **2019**, *2019*, 4015969. [CrossRef]
18. Fasanaro, P.; Greco, S.; Lorenzi, M.; Pescatori, M.; Brioschi, M.; Kulshreshtha, R.; Banfi, C.; Stubbs, A.; Calin, G.A.; Ivan, M.; et al. An integrated approach for experimental target identification of hypoxia-induced miR-210. *J. Biol Chem.* **2009**, *284*, 35134–35143. [CrossRef]
19. Alaiti, M.A.; Ishikawa, M.; Masuda, H.; Simon, D.I.; Jain, M.K.; Asahara, T.; Costa, M.A. Up-regulation of miR-210 by vascular endothelial growth factor in ex vivo expanded CD34+ cells enhances cell-mediated angiogenesis. *J. Cell Mol. Med.* **2012**, *16*, 2413–2421. [CrossRef]
20. Chen, X.; Ba, Y.; Ma, L.; Cai, X.; Yin, Y.; Wang, K.; Guo, J.; Zhang, Y.; Chen, J.; Guo, X.; et al. Characterization of microRNAs in serum: A novel class of biomarkers for diagnosis of cancer and other diseases. *Cell Res.* **2008**, *18*, 997–1006. [CrossRef]
21. Teplyuk, N.M.; Mollenhauer, B.; Gabriely, G.; Giese, A.; Kim, E.; Smolsky, M.; Kim, R.Y.; Saria, M.G.; Pastorino, S.; Kesari, S.; et al. MicroRNAs in cerebrospinal fluid identify glioblastoma and metastatic brain cancers and reflect disease activity. *Neuro-oncology* **2012**, *14*, 689–700. [CrossRef] [PubMed]
22. He, J.; Jiang, Y.; Liu, L.; Zuo, Z.; Zeng, C. Circulating MicroRNAs as Promising Diagnostic Biomarkers for Patients with Glioma: A Meta-Analysis. *Front. Neurol.* **2021**, *11*, 610163. [CrossRef] [PubMed]
23. Shimono, Y.; Zabala, M.; Cho, R.W.; Lobo, N.; Dalerba, P.; Qian, D.; Diehn, M.; Liu, H.; Panula, S.P.; Chiao, E.; et al. Downregulation of miRNA-200c links breast cancer stem cells with normal stem cells. *Cell* **2009**, *138*, 592–603. [CrossRef] [PubMed]
24. Ghosh, G.; Subramanian, I.V.; Adhikari, N.; Zhang, X.; Joshi, H.P.; Basi, D.; Chandrashekhar, Y.S.; Hall, J.L.; Roy, S.; Zeng, Y.; et al. Hypoxia-induced microRNA-424 expression in human endothelial cells regulates HIF-α isoforms and promotes angiogenesis. *J. Clin. Investig.* **2010**, *120*, 4141–4154. [CrossRef] [PubMed]
25. Choi, Y.C.; Yoon, S.; Jeong, Y.; Yoon, J.; Baek, K. Regulation of vascular endothelial growth factor signaling by miR-200b. *Mol. Cells* **2011**, *32*, 77–82. [CrossRef]
26. Katare, R.; Riu, F.; Mitchell, K.; Gubernator, M.; Campagnolo, P.; Cui, Y.; Fortunato, O.; Avolio, E.; Cesselli, D.; Beltrami, A.P.; et al. Transplantation of human pericyte progenitor cells improves the repair of infarcted heart through activation of an angiogenic program involving micro-RNA-132. *Circ. Res.* **2011**, *109*, 894–906. [CrossRef]
27. González-Gómez, P.; Sánchez, P.; Mira, H. MicroRNAs as Regulators of Neural Stem Cell-Related Pathways in Glioblastoma Multiforme. *Mol. Neurobiol.* **2011**, *44*, 235–249. [CrossRef]
28. Quintavalle, C.; Garofalo, M.; Zanca, C.; Romano, G.; Iaboni, M.; del Basso De Caro, M.; Martinez-Montero, J.C.; Incoronato, M.; Nuovo, G.; Croce, C.M.; et al. miR-221/222 overexpession in human glioblastoma increases invasiveness by targeting the protein phosphate PTPμ. *Oncogene* **2012**, *31*, 858–868. [CrossRef]
29. Ma, X.; Yoshimoto, K.; Guan, Y.; Hata, N.; Mizoguchi, M.; Sagata, N.; Murata, H.; Kuga, D.; Amano, T.; Nakamizo, A.; et al. Associations between microRNA expression and mesenchymal marker gene expression in glioblastoma. *Neuro-Oncology* **2012**, *14*, 1153–1162. [CrossRef]
30. Shang, C.; Hong, Y.; Guo, Y.; Liu, Y.H.; Xue, Y.X. MiR-210 up-regulation inhibits proliferation and induces apoptosis in glioma cells by targeting SIN3A. *Med. Sci. Monit.* **2014**, *20*, 2571–2577.
31. Luan, Y.; Zhang, S.; Zuo, L.; Zhou, L. Overexpression of miR-100 inhibits cell proliferation, migration, and chemosensitivity in human glioblastoma through FGFR3. *Onco. Targets Ther.* **2015**, *8*, 3391–3400. [PubMed]
32. Han, I.B.; Kim, M.; Lee, S.H.; Kim, J.K.; Kim, S.H.; Chang, J.H.; Teng, Y.D. Down-regulation of MicroRNA-126 in Glioblastoma and its Correlation with Patient Prognosis: A Pilot Study. *Anticancer Res.* **2016**, *36*, 6691–6697. [CrossRef] [PubMed]
33. Zhao, K.; Wang, L.; Li, T.; Zhu, M.; Zhang, C.; Chen, L.; Zhao, P.; Zhou, H.; Yu, S.; Yang, X. The role of miR-451 in the switching between proliferation and migration in malignant glioma cells: AMPK signaling, mTOR modulation and Rac1 activation required. *Int. J. Oncol.* **2017**, *50*, 1989–1999. [CrossRef] [PubMed]
34. Salinas-Vera, Y.M.; Marchat, L.A.; Gallardo-Rincón, D.; Ruiz-García, E.; Astudillo-De La Vega, H.; Echavarría-Zepeda, R.; López-Camarillo, C. AngiomiRs: MicroRNAs driving angiogenesis in cancer (Review). *Int. J. Mol. Med.* **2019**, *43*, 657–670. [CrossRef]
35. Behl, T.; Kumar, C.; Makkar, R.; Gupta, A.; Sachdeva, M. Intercalating the Role of MicroRNAs in Cancer: As Enemy or Protector. *Asian Pac. J. Cancer Prev.* **2020**, *21*, 593–598. [CrossRef]
36. Molina-Pinelo, S.; Suárez, R.; Pastor, M.D.; Nogal, A.; Márquez-Martín, E.; Martín-Juan, J.; Carnero, A.; Paz-Ares, L. Association between the miRNA signatures in plasma and bronchoalveolar fluid in respiratory pathologies. *Dis. Markers* **2012**, *32*, 221–230. [CrossRef]
37. Papagiannakopoulos, T.; Friedmann-Morvinski, D.; Neveu, P.; Dugas, J.C.; Gill, R.M.; Huillard, E.; Liu, C.; Zong, H.; Rowitch, D.H.; Barres, B.A.; et al. Pro-neural miR-128 is a glioma tumor suppressor that targets mitogenic kinases. *Oncogene* **2012**, *15*, 1884–1895. [CrossRef]
38. Li, Y.; Yan, C.; Fan, J.; Hou, Z.; Han, Y. MiR-221-3p targets Hif-1α to inhibit angiogenesis in heart failure. *Lab. Investig.* **2021**, *101*, 104–115. [CrossRef]

39. Zhou, X.U.; Qi, L.; Tong, S.; Cui, Y.U.; Chen, J.; Huang, T.; Chen, Z.; Zu, X.B. miR-128 downregulation promotes growth and metastasis of bladder cancer cells and involves VEGF-C upregulation. *Oncol. Lett.* **2015**, *10*, 3183–3190. [CrossRef]
40. Foote, M.B.; Papadopoulos, N.; Diaz, L.A., Jr. Genetic classification of gliomas: Refining histopathology. *Cancer Cell* **2015**, *28*, 9–11. [CrossRef]
41. Chen, P.; Zhang, G.; Zhou, Q.; Li, Z. Plasma microRNA-720 may predict prognosis and diagnosis in glioma patients. *Biosci. Rep.* **2020**, *40*, BSR20201449. [CrossRef] [PubMed]
42. Agostini, M.; Knight, R.A. Mir-34: From bench to bedside. *Oncotarget* **2014**, *5*, 872–881. [CrossRef] [PubMed]
43. Lee, C.C.; Chen, P.H.; Ho, K.H.; Shih, C.M.; Cheng, C.H.; Lin, C.W.; Cheng, K.T.; Liu, A.J.; Chen, K.C. The microRNA-302b-inhibited insulin-like growth factor-binding protein 2 signaling pathway induces glioma cell apoptosis by targeting nuclear factor IA. *PLoS ONE* **2017**, *12*, e0173890. [CrossRef] [PubMed]
44. Li, C.G.; Pu, M.F.; Li, C.Z.; Gao, M.; Liu, M.X.; Yu, C.Z.; Yan, H.; Peng, C.; Zhao, Y.; Li, Y.; et al. MicroRNA-1304 suppresses human non-small cell lung cancer cell growth in vitro by targeting heme oxygenase-1. *Acta Pharmacol. Sin.* **2017**, *38*, 110–119. [CrossRef] [PubMed]
45. Xiao, Z.Q.; Yin, T.K.; Li, Y.X.; Zhang, J.H.; Gu, J.J. miR-130b regulates the proliferation, invasion and apoptosis of glioma cells via targeting of CYLD. *Oncol. Rep.* **2017**, *38*, 167–174. [CrossRef]
46. An, T.; Fan, T.; Zhang, X.Q.; Liu, Y.F.; Huang, J.; Liang, C.; Lv, B.H.; Wang, Y.Q.; Zhao, X.G.; Liu, J.X.; et al. Comparison of Alterations in miRNA Expression in Matched Tissue and Blood Samples during Spinal Cord Glioma Progression. *Sci. Rep.* **2019**, *9*, 9169–9184. [CrossRef]
47. Cui, C.; Cui, Q. The relationship of human tissue microRNAs with those from body fluids. *Sci. Rep.* **2020**, *10*, 5644–5657. [CrossRef]
48. Zhang, H.; Wang, J.; Wang, Z.; Ruan, C.; Wang, L.; Guo, H. Serum miR-100 is a potential biomarker for detection and outcome prediction of glioblastoma patients. *Cancer Biomark.* **2019**, *24*, 43–49. [CrossRef]
49. Adlakha, Y.K.; Saini, N. Brain microRNAs and insights into biological functions and therapeutic potential of brain enriched miRNA128. *Mol. Cancer* **2014**, *21*, 13–33.
50. TissueAtlas. Available online: https://ccb-web.cs.uni-saarland.de/tissueatlas/patterns (accessed on 27 April 2022).
51. Wang, X.W.; Labussière, M.; Valable, S.; Pérès, E.A.; Guillamo, J.S.; Bernaudin, M.; Sanson, M. IDH1(R132H) mutation increases U87 glioma cell sensitivity to radiation therapy in hypoxia. *Biomed. Res. Int.* **2014**, *2014*, 198697.

International Journal of
Molecular Sciences

MDPI

Article

Sedoheptulose Kinase SHPK Expression in Glioblastoma: Emerging Role of the Nonoxidative Pentose Phosphate Pathway in Tumor Proliferation

Sara Franceschi [1,*], Francesca Lessi [1], Mariangela Morelli [1], Michele Menicagli [1], Francesco Pasqualetti [2,3], Paolo Aretini [1] and Chiara Maria Mazzanti [1]

1 Fondazione Pisana per la Scienza, 56017 Pisa, Italy; f.lessi@fpscience.it (F.L.); m.morelli@fpscience.it (M.M.); m.menicagli@fpscience.it (M.M.); p.aretini@fpscience.it (P.A.); c.mazzanti@fpscience.it (C.M.M.)
2 Department of Radiation Oncology, Azienda Ospedaliera Universitaria Pisana, University of Pisa, 56126 Pisa, Italy; f.pasqualetti@ao-pisa.toscana.it
3 Department of Oncology, University of Oxford, Oxford OX3 7DQ, UK
* Correspondence: s.franceschi@fpscience.it

Abstract: Glioblastoma (GBM) is the most common form of malignant brain cancer and is considered the deadliest human cancer. Because of poor outcomes in this disease, there is an urgent need for progress in understanding the molecular mechanisms of GBM therapeutic resistance, as well as novel and innovative therapies for cancer prevention and treatment. The pentose phosphate pathway (PPP) is a metabolic pathway complementary to glycolysis, and several PPP enzymes have already been demonstrated as potential targets in cancer therapy. In this work, we aimed to evaluate the role of sedoheptulose kinase (SHPK), a key regulator of carbon flux that catalyzes the phosphorylation of sedoheptulose in the nonoxidative arm of the PPP. SHPK expression was investigated in patients with GBM using microarray data. SHPK was also overexpressed in GBM cells, and functional studies were conducted. SHPK expression in GBM shows a significant correlation with histology, prognosis, and survival. In particular, its increased expression is associated with a worse prognosis. Furthermore, its overexpression in GBM cells confirms an increase in cell proliferation. This work highlights for the first time the importance of SHPK in GBM for tumor progression and proposes this enzyme and the nonoxidative PPP as possible therapeutic targets.

Keywords: glioblastoma; cancer metabolism; pentose phosphate pathway; cell proliferation

Citation: Franceschi, S.; Lessi, F.; Morelli, M.; Menicagli, M.; Pasqualetti, F.; Aretini, P.; Mazzanti, C.M. Sedoheptulose Kinase SHPK Expression in Glioblastoma: Emerging Role of the Nonoxidative Pentose Phosphate Pathway in Tumor Proliferation. *Int. J. Mol. Sci.* **2022**, *23*, 5978. https://doi.org/10.3390/ijms23115978

Academic Editors: Nives Pećina-Šlaus and Ivana Jovčevska

Received: 29 April 2022
Accepted: 25 May 2022
Published: 26 May 2022

1. Introduction

Glioblastoma (GBM) is one of the most aggressive and deadly types of central nervous system tumors [1,2]. GBMs make their way into surrounding brain tissue in widespread and unpredictable ways, making surgical resection a nearly impossible mission [3–5]. The rapid evolution and progression of the tumor and the extreme heterogeneity of GBM, even within the same tumor, means that most currently available cancer treatments fail to be effective [6,7]. Survival rates from GBM enjoyed a modest bump in the 1980s when radiation became a standard part of the treatment protocol [8,9]. Patients could expect to live for almost another year after diagnosis, from just four to six months. The introduction of the chemotherapy drug temozolomide in the 2000s increased survival by another few months [10]. However, since then, patient survival rates have stalled. Because of the poor outcomes of this disease, primarily due to its daunting resistance to almost all forms of treatment, new and innovative therapies are urgently needed. Impairment of physiological metabolism is one of the most striking hallmarks of GBM [11,12], and the cancer cells' propensity to use glycolysis instead of oxidative phosphorylation (Warburg effect) [13,14] led to a focus on glucose metabolism to stop tumor progression. In addition to glycolysis, which is considered the main energy source in tumors, other metabolic pathways are specifically exploited in GBM [13] to meet the demand of a rapidly proliferating tumor. The pentose phosphate pathway (PPP) is a metabolic pathway parallel to glycolysis and represents the first committed step of glucose metabolism [15]. The PPP plays a critical role in sustaining cancer cell survival and growth by producing ribose-5-phosphate (R5P) for nucleic acid synthesis and providing nicotinamide adenine dinucleotide phosphate (NADPH), which is necessary for fatty acid synthesis and cell survival under high-stress conditions [16,17]. Indeed, NADPH and R5P play critical roles in the regulation of metabolism, proliferation, and DNA damage response in cancer cells, and several PPP enzymes have already been studied, highlighting their potential role as molecular targets for the development of cancer therapies [18]. In GBM in particular, most studies have investigated the enzyme glucose-6-phosphate dehydrogenase (G6PD), which is part of the oxidative arm of the PPP, demonstrating a prognostic relevance [19]. Although the role of the nonoxidative arm of the PPP has been little studied in GBM, its enzymes and the Sedo-heptulose-7-phosphate (S7P) intermediate appear to play an important role in other malignancies [20,21]. In this work, we aimed to evaluate the role of sedoheptulose kinase (SHPK), a key regulator of carbon flux that catalyzes the phosphorylation of sedoheptulose into S7P, making it available to cells. As S7P is the substrate of two enzymes in the nonoxidative arm, Transketolase (TKT) and Transaldolase (TALDO1), the action of SHPK is important in regulating flux through the PPP. SHPK expression in GBM shows a significant correlation with histology, patient prognosis, and survival. Moreover, its overexpression in GBM cells confirms an increase in tumor proliferation.

2. Results

2.1. Correlation between SHPK Expression and Clinical Characteristics of Glioma Patients

Clinical and gene expression data from 219 GBMs, 225 low-grade tumors, primary tumors, and 28 normal specimens were used to correlate SHPK mRNA expression levels with histopathological features through the Rembrandt dataset [22] via GlioVis, a web-based data visualization and analysis application for exploring brain tumor expression datasets [23].

Results show a significantly positive correlation between histology and WHO grade and SHPK expression (Figure 1A,B). Using 219 GBMs, we further investigated how SHPK mRNA expression correlates with a specific molecular subtype (defined by Wang [24]) and patient survival. As shown in Figure 1C, SHPK was significantly upregulated in the classical and mesenchymal subtypes compared with the proneural subtype. To evaluate the prognostic value of SHPK in GBM samples, Kaplan–Meier (KM) survival curves were plotted. We observed that higher SHPK mRNA expression predicted a significantly shorter survival, as shown in Figure 1E.

Moreover, we verified the protein levels of SHPK in glioma tissues using the Human Protein Atlas (HPA) database (available from http://www.proteinatlas.org, accessed on 10 January 2022). The immunohistochemistry data in the HPA database reveal that the immunoreactive score (IRS) of SHPK was significantly higher in glioma tissues (both low- and high-grade gliomas) than in the normal cerebral cortex tissues (Figure 1D).

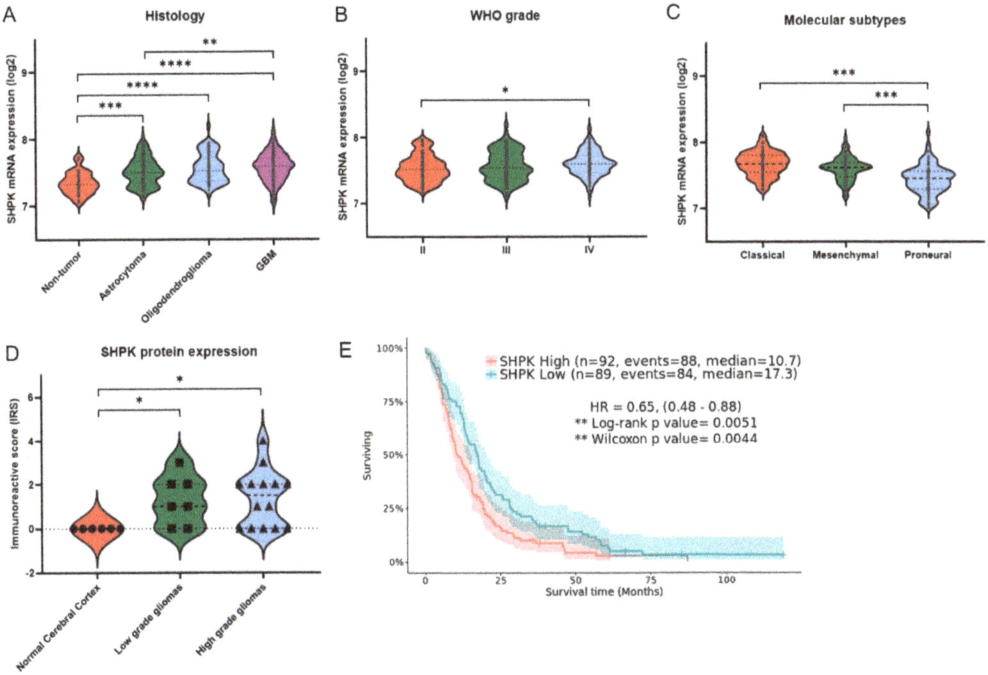

Figure 1. SHPK expression correlates with clinical characteristics and prognosis of GBM patients. (**A**) SHPK mRNA expression within healthy cerebral tissues and different brain tumor histologies. (**B**) SHPK mRNA expression of different WHO-grade brain tumors. (**C**) SHPK mRNA expression in the three different molecular subtypes of brain tumors. (**D**) SHPK protein expression of histological sections from normal and cancer tissues obtained by immunohistochemistry. (**E**) Survival analysis with Kaplan–Meier estimator and visualization of confidence intervals of GBM samples using the median of SHPK mRNA expression values as the cutoff. Hazard Ratio (HR) and p-values (Log-Rang and Wilcox) are also shown. Violin plot p-values were calculated using an unpaired nonparametric test, the two-tailed Mann–Whitney, with GraphPad Prism 9.3.1. **** $p < 0.0001$; *** $p < 0.001$; ** $p < 0.01$; * $p < 0.05$.

2.2. SHPK-Related Biological Process

Expression data of the 219 GBMs were employed to conduct a differential expression analysis. In total, 263 differentially expressed genes (DEGs) were identified by dividing GBM samples according to their SHPK expression (high vs. low: split into two groups, 25% lower-expressing vs. 25% higher-expressing). To better understand the function of the identified DEGs, GO analysis was performed in g:Profiler [25] as g:GOSt functional profiling. Table 1 shows the results of the functional enrichment analysis of g:Profiler exploiting the Reactome [26] database.

Table 1. Functional enrichment analysis of DEGs.

abs(LogFC) < 1

Reactome	ID	*p*-Value
Transmission across Chemical Synapses	112315	8.79×10^{-13}
Neuronal System	112316	1.15×10^{-12}
Serotonin Neurotransmitter Release Cycle	181429	3.97×10^{-8}
Dopamine Neurotransmitter Release Cycle	212676	3.35×10^{-7}
Neurotransmitter release cycle	112310	7.56×10^{-6}
Neurotransmitter receptors and postsynaptic signal transmission	112314	1.47×10^{-5}
Glutamate Neurotransmitter Release Cycle	210500	2.18×10^{-5}
Acetylcholine Neurotransmitter Release Cycle	264642	1.10×10^{-4}
Norepinephrine Neurotransmitter Release Cycle	181430	1.54×10^{-4}
GABA receptor activation	977443	5.65×10^{-4}
Protein–protein interactions at synapses	6794362	6.42×10^{-3}
GABA synthesis, release, reuptake and degradation	888590	8.24×10^{-3}
Long-term potentiation	9620244	1.58×10^{-2}
Neurotoxicity of clostridium toxins	168799	3.70×10^{-2}

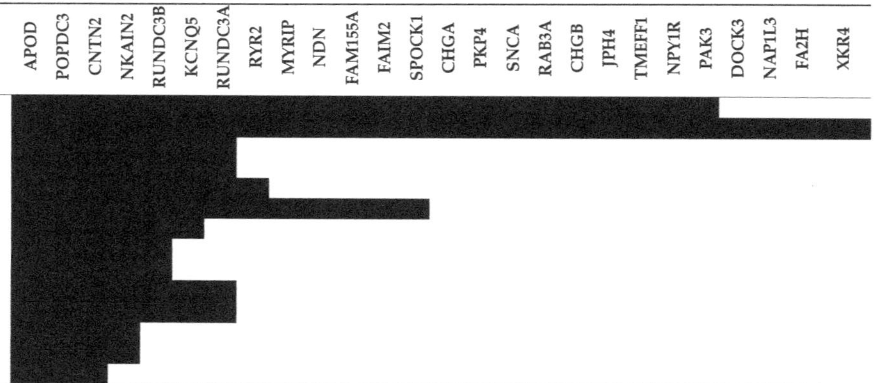

abs(LogFC) > 1

Reactome	ID	*p*-Value
Integrin cell surface interactions	216083	8.21×10^{-6}
Nonintegrin membrane-ECM interactions	3000171	2.45×10^{-5}
ECM proteoglycans	3000178	1.16×10^{-4}
Crosslinking of collagen fibrils	2243919	1.84×10^{-4}
Scavenging by Class A Receptors	3000480	2.32×10^{-4}
Collagen formation	1474290	3.23×10^{-4}
Extracellular matrix organization	1474244	4.40×10^{-4}
Syndecan interactions	3000170	8.80×10^{-4}
Assembly of collagen fibrils and other multimeric structures	2022090	9.94×10^{-4}
Collagen biosynthesis and modifying enzymes	1650814	1.72×10^{-3}
Collagen chain trimerization	8948216	7.61×10^{-3}
Anchoring fibril formation	2214320	7.90×10^{-3}
Collagen degradation	1442490	3.37×10^{-2}

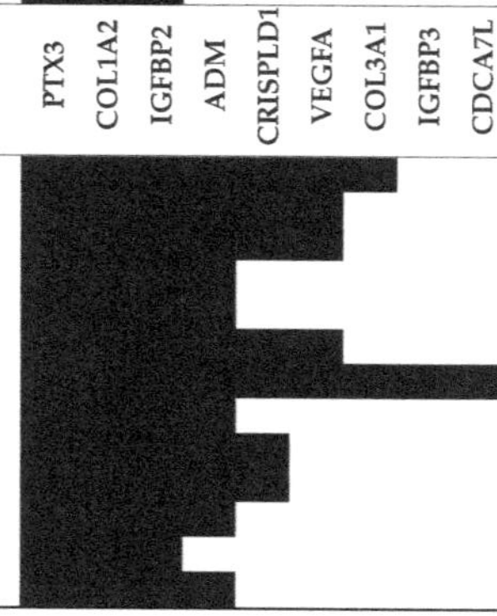

GBM samples were divided into two groups according to their SHPK mRNA expression (high and low). The top table shows the biological pathways in which genes overexpressed in samples with the lowest SHPK mRNA values (25% lower-expressing) are involved, while the bottom shows the biological pathways of genes overexpressed in samples with the highest SHPK mRNA values (25% higher-expressing). Reactome, biological pathway description; ID, biological pathway identifier (REAC:R-HSA); *p*-value, adjusted enrichment *p*-values in negative log10 scale; Black square indicates which gene belongs to that pathway.

The biological processes most represented by genes overexpressed in GBMs with low SHPK mRNA values are nervous system development, synaptic signaling, cell–cell signaling, and neurogenesis. The most significant biological processes to which overexpressed genes in GBMs with high SHPK mRNA expression belong are extracellular matrix and structure organization, response to endogenous stimuli, and regulation of cell population proliferation.

2.3. SHPK Correlation with Other PPP Enzymes

We then conducted a correlation analysis with the expression data from the 219 GBMs. In particular, we conducted Pearson correlation analysis of the mRNA expression of SHPK with the mRNA expression of the other enzymes constituting the PPP (Figure 2A,B). Specifically, Figure 2A shows the correlations performed among the enzymes in the nonoxidative arm of the PPP while Figure 2B shows the correlations performed among the enzymes comprising the oxidative arm of the PPP. SHPK correlated significantly and positively with Ribose 5-Phosphate Isomerase A (RPIA, nonoxidative branch of PPP) and with 6-Phosphogluconolactonase, Phosphogluconate Dehydrogenase, and Hexose-6-Phosphate Dehydrogenase/Glucose 1-Dehydrogenase (PGLS, PGD, and H6PD, oxidative branch of PPP).

2.4. SHPK Expression and Mutational Status Association

The LinkFinder module of LinkedOmics [27] was used to conduct multi-omics analyses within another cohort of 595 GBM patients. We conducted an association analysis between mutational status (whole-exome data) and SHPK mRNA expression levels (RNAseq data). Association analysis results are shown in Figure 2C (volcano plot). SHPK mRNA expression levels were assessed for each gene for its mutational status (mutated or WT) with the Wilcoxon test. The mutational status of nine genes, Isocitrate Dehydrogenase 1 (IDH1), Tumor Protein P53 (TP53), ATRX Chromatin Remodeler (ATRX), Sodium Voltage-Gated Channel Alpha Subunit 9 (SCN9A), Phosphoinositide-3-Kinase Regulatory Subunit 1 (PIK3R1), Frizzled Class Receptor 10 (FZD10), Polycystic Kidney Additionally, Hepatic Disease 1 (Autosomal Recessive)-Like 1 (PKHD1L1), Armadillo Repeat-Containing 3 (ARMC3), and Acyl-CoA Synthetase Medium Chain Family Member 2B (ACSM2B), was significantly associated with a lower SHPK mRNA expression (Figure 2C and Table 2). In contrast, 13 other genes' mutated state was significantly associated with increased SHPK mRNA expression (Figure 2C and Table 2). These were Epidermal Growth Factor Receptor (EGFR), Vacuolar protein sorting-associated protein 8 homolog (VPS8), Rho Guanine Nucleotide Exchange Factor 16 (ARHGEF16), Striated-Muscle-Enriched Protein Kinase (SPEG), Cadherin 9 (CDH9), Transformation/Transcription-Domain-Associated Protein (TRRAP), Ryanodine Receptor 2 (RYR2), Solute Carrier Family 4 Member 1 (SLC4A1), Molybdenum Cofactor Synthesis 3 (MOCS3), Dynein Axonemal Heavy Chain 2 (DNAH2), Lysine Methyltransferase 2D (MLL2), Xin-Actin-Binding Repeat-Containing 2 (XIRP2), and APC Membrane Recruitment Protein 3 (FAM123C).

Table 2. Gene with mutational status significantly associated with SHPK mRNA expression level.

Gene	Log2FC (Median)	*p*-Value	FDR (BH)	Event_SD	Event_TD
PKHD1L1	−0.8033	0.0304	0.8846	141	3
FZD10	−0.7171	0.0227	0.8846	141	3
IDH1	−0.6885	0.0005	0.2030	141	8
ARMC3	−0.6270	0.0327	0.8846	141	3
SCN9A	−0.6201	0.0052	0.4584	141	5
ATRX	−0.5071	0.0042	0.4584	141	8
ACSM2B	−0.4953	0.0418	0.8846	141	3
PIK3R1	−0.2773	0.0201	0.8846	141	12
TP53	−0.1720	0.0026	0.4584	141	45
EGFR	0.1987	0.0053	0.4584	141	45
RYR2	0.2172	0.0296	0.8846	141	12
MOCS3	0.3278	0.0403	0.8846	141	3
FAM123C	0.3311	0.0490	0.8846	141	4
SLC4A1	0.3325	0.0332	0.8846	141	4
CDH9	0.3434	0.0277	0.8846	141	5
DNAH2	0.3726	0.0419	0.8846	141	5
SPEG	0.3916	0.0234	0.8846	141	4
TRRAP	0.4470	0.0284	0.8846	141	4
MLL2	0.4534	0.0449	0.8846	141	4
XIRP2	0.4578	0.0463	0.8846	141	3
ARHGEF16	0.4721	0.0218	0.8846	141	3
VPS8	0.6664	0.0108	0.7768	141	3

Gene, gene in given target dataset whose association with SHPK expression has been performed. Log2FC (median), change in gene expression level expressed in log2 of mutated/WT ratio. *p*-value, *p*-value obtained from the Wilcoxon statistical test. FDR (BH), false discovery rate calculated by BH (Benjamini–Hochberg method). Event_SD, Number of observations in search dataset attribute without NA's and Zero's. Event_TD, Number of observations in target dataset attribute without NA's and Zero's.

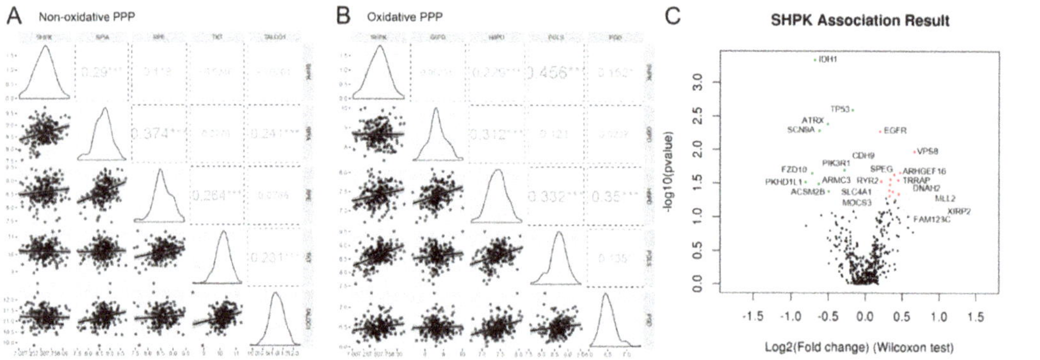

Figure 2. SHPK correlation and association analysis. Pearson correlation analysis between mRNA expression of enzymes constituting the nonoxidative (**A**) and oxidative (**B**) arms of PPP. The figure shows the scatter plot with regression line for each correlation (bottom diagonal), the density plot (middle diagonal), and the Pearson correlation coefficient with significance: *** $p < 0.001$, * $p < 0.05$ (upper diagonal). (**C**) Volcano plot showing the log2 (fold change) vs. log10 (*p*-value) obtained from the analysis. In green, with a negative Log2FC, statistically significant associations between the absence of mutation and SHPK mRNA overexpression and between the presence of mutation and downregulation of SHPK mRNA expression are highlighted. In red, with a positive Log2FC, statistically significant associations between the presence of the mutation and SHPK mRNA overexpression and between the absence of the mutation and downregulation of SHPK mRNA expression are highlighted.

2.5. SHPK Overexpression and Cell Functional Studies

SHPK has been successfully cloned and overexpressed in three commercial human GBM lines (T98G, U118, and U87) with an average transfection efficiency of 68% for the vector and 59% for the vector containing SHPK (data not shown). Both SHPK mRNA and protein expression were significantly higher in SHPK-overexpressed cells than in vector cells (Figure 3A,B).

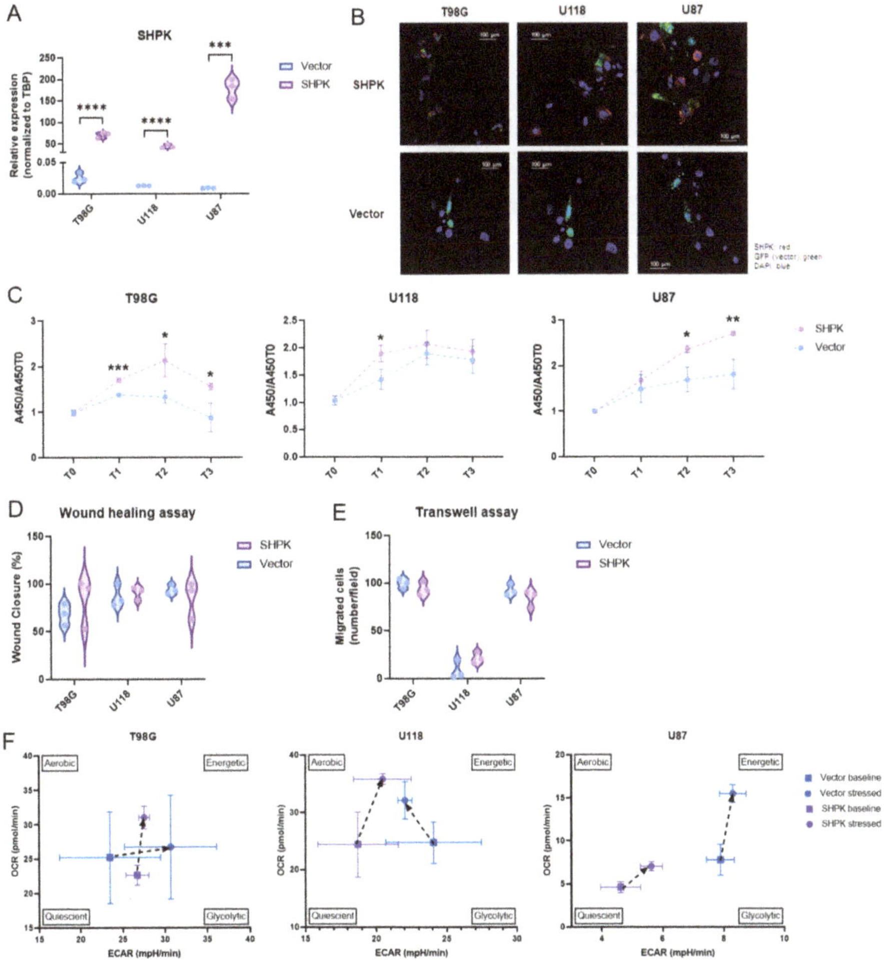

Figure 3. SHPK overexpression and cell functional studies. (**A**) SHPK mRNA expression in T98G, U87, and U118 cells after SHPK overexpression and relative controls. (**B**) Immunofluorescence staining of SHPK protein (red) in T98G, U87, and U118 cells overexpressing SHPK and relative controls (vector alone, green). Nuclei were stained with DAPI (blue). (**C**) Viability of T98G, U87, and U118 cells after SHPK overexpression and their controls at the time of seeding (T0) and after 24 h (T1), 48 h (T2), and 72 h (T3). A450 absorbance values relative to T0 are shown in the vertical axis (y). (**D**) Wound healing assay of T98G, U87, and U118 cells after SHPK overexpression and relative controls. (**E**) Transwell migration assay of T98G, U87, and U118 cells after SHPK overexpression and relative controls. Cells that crossed the membrane were counted in five visual fields as migrated cells.

(**F**) differences in the metabolic phenotype of T98G, U118, and U87 cells under both basal and stress conditions in the presence or absence of SHPK overexpression. Each measure of OCR and ECAR was calculated by averaging the measurements made in triplicate (SD shown) for three different measurements (9 total measurements) in both the baseline and stressed states. p-values were calculated using a two-tailed, unpaired t-test with GraphPad Prism 9.3.1. **** $p < 0.0001$; *** $p < 0.001$; ** $p < 0.01$; * $p < 0.05$.

The WST-1 cell proliferation test was used to determine the effect of SHPK overexpression in T98G, U118, and U87 cells at four different time points, T0, T1 (24 h), T2 (48 h), and T3 (72 h). It was found that SHPK overexpression significantly enhanced cell proliferation (Figure 3C). The effect of SHPK overexpression on cell migration was studied with the wound healing assay. In all three cell lines (U118, U87, and T98G), four different time points were taken, T0, T1 (24 h), T2 (48 h), and T3 (72 h). Wound healing test results showed that SHPK overexpression did not change the wound healing ability compared to transfected empty vector cells (Figure 3D). To determine the effect of SHPK overexpression on T98G, U118, and U87 cell invasion, a transwell invasion assay was performed. The number of migrating cells was unchanged in SHPK-overexpressed cells compared to control cells (Figure 3E). Colony formation assays were performed to evaluate the SHPK overexpression effect on clonogenic survival. SHPK overexpression did not change the T98G, U118, or U87 colony formation ability (data not shown). To gain insight into metabolic differences in cells after SHPK overexpression, we analyzed metabolic phenotypes and the metabolic potential of live cells with the Seahorse XFp extracellular flow analyzer. We examined the effects of SHPK overexpression on the oxygen consumption rate (OCR), which is a measure of the rate of mitochondrial respiration of cells, and the extracellular acidification rate (ECAR), a measure of the rate of glycolysis of cells using the cell energy phenotype assay (Figure 3F). Although there were differences in basal metabolic status due to SHPK overexpression, this did not occur equally in the three cell lines (Figure 3F). In particular, the basal metabolic status appeared to shift toward glycolysis when SHPK was overexpressed in T98G (Figure 3F). In U118 and U87, however, it shifted to lower values of glycolytic activity; even in U87, it went toward a more quiescent metabolic state (Figure 3F). Regarding the metabolic differences of the stressed phenotype (cells under an induced energy demand), we can observe in Figure 3F that stress increased mitochondrial respiration in all three cell lines (OCR) independently of SHPK overexpression, whereas cell glycolytic activity (ECAR) appeared to increase under stress in all lines and under both SHPK expression conditions except for U118 cells without overexpression (Figure 3F).

3. Discussion

GBM is the most common and malignant primary brain tumor, characterized by high morbidity and poor survival [28]. Despite advances in diagnosis and treatment, life expectancy remains at approximately 12–18 months [1,29]. Brain invasion, motility, and rapid proliferation are characteristic of GBM cells, and this ability to invade surrounding tissue is a major determinant for malignant tumor progression [30,31]. The dispersion of tumor cells from the primary tumor site into adjacent brain tissue results in rapid and almost inevitable recurrence [29,32].

GBM cells are characterized by a preference for aerobic glycolysis rather than oxidative phosphorylation, which is more active in normal cells [33,34]. However, in addition to elevated glycolysis, proliferating and cancer cells must also divert carbon from glycolysis to the pentose phosphate pathway (PPP) to satisfy their anabolic demands and maintain the redox homeostasis of cells [21]. The PPP plays a key role in the regulation of cancer cell growth by producing ribose-5-phosphate and NADPH for detoxification of intracellular ROS, reductive biosynthesis, and ribose biogenesis [35]. Thus, the PPP is directly related to cell proliferation, survival, and senescence. It has been reported that in GBM cells, migrating cells are characterized by up-regulation of many key glycolysis enzymes at the expense of PPP enzyme expression, whereas in rapidly dividing GBM cells, the opposite occurs: PPP enzyme expression increases, and glycolysis enzymes decrease their expression [16,35,36]. The PPP is mainly used during proliferation, and glycolysis is used as the energy source during migration [37]. The metabolism acts as a mutual switch between the two pathways, glycolysis, and the PPP, and the mechanisms of cancer cell invasion and proliferation are thought to be mutually exclusive behaviors, called the "migration–proliferation dichotomy" or "go or grow" [16,35,38].

The PPP is composed of two functionally interrelated branches: the oxidative and the nonoxidative. The oxidative branch consists of three irreversible reactions leading to the generation of NADPH and ribonucleotides [15,21]. Additional glycolytic intermediates such as fructose-6-phosphate (F6P) and glyceraldehyde-3-phosphate (G3P) are recruited into the nonoxidative branch through a series of reversible reactions [18,21,39]. This reversible nature of the nonoxidative branch of the PPP makes it capable of adapting to the metabolic demands of cells, acting in a multitude of ways [21,40].

Previous studies have examined the relative gene expression of enzymes involved in the oxidative phase of the PPP, finding an overall increase in glycolytic and PPP genes driving ATP production and excess nucleotides resulting in uncontrolled proliferation in GBM cells [36,41,42]. In addition, inhibitors of the oxidative branch of the PPP have been studied, the findings of which indicate increased radiosensitivity in human gliomas [43]. Although the oxidative branch of the PPP has already been studied in GBM by correlating enzyme expression with tumor aggressiveness and patient survival and proposing their targeting as a promising therapeutic target, the nonoxidative branch of the PPP has never been sufficiently evaluated in GBM. However, there are studies on other cancer types that confer an important role in tumor progression on the nonoxidative branch of the PPP and overexpression of its enzymes mainly due to increased proliferation of cancer cells [15,21,34,44–46].

In this work, we focused on the sedoheptulokinase (SHPK) protein belonging to the nonoxidative branch enzymes of the PPP. SHPK is the carbohydrate kinase that catalyzes the phosphorylation of sedoheptulose into sedoheptulose 7-phosphate (S7P), which then enters the PPP stream [18]. S7P represents a glycolysis-independent entry and exit point into/out of nonoxidative PPP [40] and has been shown to play an important role in other malignancies [20,21,47] and found in greater amounts in high-grade than low-grade gliomas [48]. Since studies suggest that a role of SHPK might be to provide increased PPP flow during an increased need for energy [40], we investigated the expression of this enzyme in GBM. With this work, we want to draw attention to the importance of the increased production of S7P by SHPK by correlating the expression of this enzyme with progression and tumor aggressiveness in GBM. We initially evaluated how SHPK expression correlated with the clinical characteristics of patients. SHPK mRNA expression was significantly higher in tumor tissues than in nontumor tissues and among different histologies. In particular, SHPK was highest in GBM, then in oligodendrogliomas, and significantly lower in astrocytomas. Within the adult-type diffuse gliomas (WHO-grade II-IV [49]) there was a significant difference between GBM (grade IV), in which SHPK was overexpressed, and grade II. In addition, a significant overexpression of SHPK protein in gliomas compared with normal tissues was found. The expression of SHPK within GBM samples was evaluated among the three different molecular subtypes defined by Wang [24]. Proneural GBMs show a significant decrease in SHPK mRNA expression. Compared to the other three subtypes, proneural subtypes have better survival rates [50]. This finding was also confirmed by Kaplan–Meier survival analysis, showing a significant difference between the survival of GBM patients characterized by high SHPK mRNA levels compared to those with low SHPK mRNA expression.

To understand which cellular pathways were related to the different expressions of SHPK in GBMs and what this increased flux in the PPP of the S7P intermediate led to, we conducted a functional enrichment analysis. In agreement with that described in the literature [16,21,35–37,41,42], increased SHPK mRNA expression and a subsequent S7P flux in the PPP triggered a number of pathways involved in cell proliferation. In contrast, GBMs characterized by a low level of SHPK mRNA showed active physiological cell signaling pathways and had no differentially expressed genes involved in extracellular matrix remodeling and proliferation pathways. Next, we also assessed how SHPK mRNA expression correlated with the expression of the other PPP enzymes in GBM, both the nonoxidative and oxidative branches. Significant correlations of SHPK with other enzymes in the PPP were all positive. In the nonoxidative branch, only one enzyme correlated with SHPK (RPIA), while in the oxidative branch, three enzymes positively correlated with SHPK (PGLS, PGD, and H6PD). This may suggest that the two pathways do not operate separately but rather that they may work in synergy, or one may compensate for the reduced work of the other. Metabolic control analyses performed on PPP regulatory enzymes have been carried out previously, which revealed that the nonoxidative branch of the PPP is more important for tumor growth than the oxidative one [51]. On the other hand, the results obtained in other studies [52,53] demonstrate the importance of a forced balance of the activity of the two branches in the direction of the oxidative one to sustain high tumor cell proliferation. When dividing the GBM population into high and low SHPK expression, we could observe differences in the mutational status of some important genes in GBM. For example, the most significant genes in each group were IDH1, which associates its mutated state with lower SHPK expression, and EGFR, which instead associates its mutated state with higher SHPK expression. This agrees perfectly with what we found in the literature, where strong evidence shows that IDH1 mutation is associated with a better prognosis for GBM patients [54–56], and on the other hand, EGFR mutation with a worse one [57].

Finally, we evaluated the effects of SHPK overexpression in three different GBM cell lines (T98G, U87, and U118) to try to associate a particular cell behavior with increased SHPK and the respective intermediate of the S7P PPP. In all three cell lines, the increase in SHPK alone was able to enhance their viability/proliferation. On the other hand, no other cellular capacity (invasion, migration, and clonogenicity) was altered by increasing SHPK. This finding further confirms the key role of the nonoxidative PPP in tumor proliferation but especially highlights and proposes a possible leading role of SHPK in the activation of the nonoxidative branch. To conclude the functional studies, we evaluated metabolic phenotypes and the metabolic potential of GBM lines after SHPK overexpression. With this type of assay, we were able to observe whether SHPK expression can affect the metabolic state of GBM cells both before (basal) and after an induced energy demand. Although there were differences, we could not find one that was reflected in all three cell lines. This is most likely because SHPK alone is unable to trigger specific metabolic switches. However, it should be kept in mind that these assays were conducted within 48 days of transfection with one day of settling. Surely real-time studies would be helpful to better understand if there are indeed metabolic differences following SHPK overexpression, and permanent transfection studies could be more informative.

In conclusion, we can state that expression and therefore the activity of SHPK to produce S7P that enters the PPP stream could trigger the activity of the nonoxidative branch. This results in an increase in cell proliferation that, through functional studies, has been attributed to the activity of SHPK alone. Moreover, SHPK expression is also significantly correlated with multiple clinical data proposing the correlation of its expression with a worse prognosis. This work highlights for the first time in GBM the importance of the SHPK enzyme, although the mechanisms and flow direction by which the PPP is activated and which enzymes are primarily involved are still unclear. Further studies are needed to better understand the flux of metabolites through the nonoxidative PPP and in particular to understand how we can exploit the SHPK enzyme as a therapeutic target.

4. Materials and Methods

4.1. Glioma Samples and Normal Controls

Clinical and gene expression data from 219 GBMs, 225 low-grade tumors, primary tumors, and 28 normal samples were analyzed using microarray data from the Rembrandt cohort through the data visualization and analysis tool GLIOVIS [23] (http://GLIOVIS.bioinfo.cnio.es/, accessed on 11 January 2021) to correlate SHPK expression with histopathological characteristics. Correlation of SHPK with molecular subtype (classic, mesenchymal, and proneural), and survival was conducted considering only the 219 GBMs. Multiomics analysis to associate gene mutation states with SHPK mRNA expression in GBM was conducted on the TCGA (The Cancer Genome Atlas) cohort of 595 samples, using the LinkFinder module of LinkedOmics [27]. We selected RNAseq data as the select search (query) dataset (Illumina HiSeq 2000 RNA Sequencing) and SHPK as the gene of interest. Mutation datatype (Illumina GA IIx) was chosen as the target dataset. T-test was chosen as the statistical analysis method.

4.2. *Immunohistochemistry of Histological Sections*

Information used for immunohistochemistry and annotation data was provided by Human Protein Atlas (proteinatlas.org, accessed on 10 January 2022). Basic annotation parameters included an assessment of staining intensity (negative, weak, moderate, or strong) and a fraction of stained cells (<25%, 25–75%, or >75%). As shown in Table 3, each tumor was assigned a score based on staining intensity (no staining = 0; weak staining = 1; moderate staining = 2; strong staining = 3) and extent of stained cells (0% = 0; <25% = 1; 25–75% = 2; >75% = 3). The final immunoreactivity score (IRS) was determined by multiplying the intensity and extent of stained-cell positivity scores, with a minimum score of 0 and a maximum score of 9. In gliomas, tumor cell staining was considered, whereas in healthy tissues, glial cell staining was considered.

Table 3. The immunoreactive score (IRS).

A (Percentage of Positive Cells)	B (Intensity of Staining)	IRS Score (A × B)
0 = no positive cells	0 = no color reaction	0 = negative
1 = <25% of positive cells	1 = weak reaction	1–2 = mild
2 = 25–75% of positve cells	2 = moderate reaction	3–6 = moderate
3 = >75% of positive cells	3 = intense reaction	7–9 = strong

IRS is calculated as the product of multiplication between the score of the proportion of positive cells (0–4) and the score of the staining intensity (0–3). IRS value ranges between 0 and 9.

4.3. *Cell Lines and Transfection*

T98G, U87, and U118 GBM cell lines were obtained from American Type Culture Collection (ATCC, Rockville, MD, USA). To ensure the quality and integrity of the human cell lines, STR analysis was conducted using the GenePrint 10 system (Promega, Madison, WI, USA). Cells were grown as monolayers in high-glucose Dulbecco's Modified Eagle Medium (DMEM) supplemented with 10% FBS and 1% penicillin–streptomycin. Cells were tested for the presence of mycoplasma (EZ-PCR Mycoplasma Test Kit; Biological Industries, Kibbutz Beit-Haemek, Israel) with negative results. SHPK was overexpressed using the pCMV6-AC-GFP vector with the molecular sequence of its clone (NM_013276) (OriGene Technologies, Rockville, MD, USA) cloned within, as a control. The SHPK sequence was confirmed by Sanger sequencing. Transfection of the plasmid was performed with Lipofectamine 3000 reagent, following the manufacturer's instructions. Cells were incubated for 48 h after SHPK overexpression before characterization and functional experiments.

4.4. *SHPK mRNA Expression of Cell Lines*

Total cellular RNA was extracted from GBM cells using the Maxwell 16 LEV simplyRNA kit (Madison, WI, USA) according to the manufacturer's instructions, and quantified using a Qubit 2.0 Fluorometer (Thermo Fisher Scientific, Waltham, MA, USA). Total RNA was inversely transcribed into cDNA using the RT-NanoScript kit (PrimerDesign, Southampton, UK). Real-time PCR was performed following the manufacturer's instructions for the SsoAdvanced SYBR Green Supermix kit (Bio-Rad, Hercules, CA, USA) on the CFX96 instrument (Bio-Rad). TBP expression values were used for normalization. Real-time PCR primer assays (Bio-Rad) for SHPK (Assay ID: qHsaCID0016666) and TBP

(Assay ID: qHsaCID0007122) were used. Gene expression analysis was performed using CFX Manager software (Bio-Rad). All expression experiments were performed in triplicate.

4.5. Immunofluorescence

Cells were grown on cell culture chamber slides and fixed in 1.5% paraformaldehyde for 15 min. Cells were permeabilized with 0.1% Triton X-100 for 15 min and blocked with 2% BSA for 45 min. Primary SHPK antibody (HPA024361, Sigma Aldrich, St. Louis, MO, USA) was diluted at 1:50 and incubated for 60 min at RT. Phycoerythrin-conjugated secondary antibody (P9287, Sigma Aldrich) was diluted at 1:20 and incubated for 30 min. Cells were counterstained with DAPI (Thermo Fisher Scientific) and visualized using the CARL ZEISS Axio Observer 3 Z1FLMot inverted microscope (Zeiss, Gina, Germany).

4.6. Cell Viability Assay

Cell viability was determined using the WST1 assay (Clontech Laboratories, Mountain View, CA, USA). A total of 5000 cells per well were seeded in a 96-well plate. At the time of seeding (T0) and after 24 h (T1), 48 h (T2), and 72 h (T3), the WST1 reagent was added and incubated for an additional 60 min before reading the plate. Each assay was conducted in triplicate. The amount of formazan dye was directly related to the number of metabolically active cells and was quantified by measuring absorbance at 450 nm in a multiwell plate reader (Tecan, Mannedorf, Switzerland). OD values at 24 h (T1), 48 h (T2), and 72 h (T3) were normalized to T0.

4.7. Wound Healing Assay

Cells were plated in Culture-Insert 2 Well in 35 mm µ-Dish (IBIDI, Martinsried, Germany) until cells were confluent or nearly confluent (>90%). After removal of the insert, cell migration in the wound area was observed and digitally photographed. Wound healing was measured on the images using the free and open-source ImageJ software [58], and % closure was calculated at each time (T0–T3, 0–72 h) as the area to be healed divided by the area of the original wound $\times$ 100. Experiments were performed in triplicate.

4.8. Transwell Assay

Cell invasion was assessed using 24-well inserts (Sarstedt, Nuembrecht, Germany) with 5 µm pores according to the manufacturer's instructions. Briefly, 1×10^5 cells were seeded in the upper chamber with 1% FBS medium and were allowed to invade the lower reservoir, containing 10% FBS, at 37 °C for 24 h. Noninvasive cells in the upper surface of the filters were removed with a cotton swab. The remaining cells were fixed in 70% ethanol and stained with 0.01% crystal violet for 30 min. Cells that crossed the membrane were counted in five visual fields as migrated cells. The experiment was performed in triplicate.

4.9. Clonogenic Survival Assay

Cells were seeded at 500 cells/well in 6-well plates and incubated for 2 weeks. Cells were fixed with 70% ethanol and stained with 0.01% crystal violet for 30 min. The mean $\pm$ SD number of colonies >50 µm in diameter was counted microscopically in five nonoverlapping fields in three independent experiments.

4.10. Cell Energy Phenotype Test

Cell mitochondrial function was evaluated by using the Seahorse XFp Cell Energy Phenotype Test Kit on the Seahorse XFp Analyzer (Agilent Technologies, Santa Clara, CA, USA). Cells were seeded at 20,000 cells per well into XFp well cell culture plates and incubated overnight at 37 °C in a 5% CO_2-humidified atmosphere in Seahorse XF Base Medium (Agilent Technologies) with 1 mM pyruvate, 2 mM glutamine, and 10 mM glucose. Cartridge compounds were loaded to obtain a final concentration of 1 µM Oligomycin and 1 µM FCCP. Data were analyzed and visualized using Wave 2.3.0 software (Agilent Technologies), and values of OCR and ECAR were normalized to the total protein levels

(Bradford Reagent assay, Sigma-Aldrich) in each well. The experiment was performed with three replicates.

Author Contributions: Conceptualization, S.F. and C.M.M.; methodology, F.L. and M.M. (Michele Menicagli); software, P.A.; investigation, S.F.; resources, F.P.; data curation, M.M. (Mariangela Morelli); writing—original draft preparation, S.F.; writing—review and editing, S.F. and C.M.M.; supervision, C.M.M. All authors have read and agreed to the published version of the manuscript.

Funding: This research received no external funding.

Institutional Review Board Statement: Not applicable.

Informed Consent Statement: Not applicable.

Data Availability Statement: All data generated and analyzed are included in the published article.

Conflicts of Interest: The authors declare no conflict of interest.

References

1. Hanif, F.; Muzaffar, K.; Perveen, K.; Malhi, S.M.; Simjee, S.U. Glioblastoma Multiforme: A Review of its Epidemiology and Pathogenesis through Clinical Presentation and Treatment. *Asian Pac. J. Cancer Prev. APJCP* **2017**, *18*, 3–9. [CrossRef] [PubMed]
2. Davis, M.E. Glioblastoma: Overview of Disease and Treatment. *Clin. J. Oncol. Nurs.* **2016**, *20* (Suppl. S5), S2–S8. [CrossRef] [PubMed]
3. Lara-Velazquez, M.; Al-Kharboosh, R.; Jeanneret, S.; Vazquez-Ramos, C.; Mahato, D.; Tavanaiepour, D.; Rahmathulla, G.; Quinones-Hinojosa, A. Advances in Brain Tumor Surgery for Glioblastoma in Adults. *Brain Sci.* **2017**, *7*, 166. [CrossRef] [PubMed]
4. McCutcheon, I.E.; Preul, M.C. Historical Perspective on Surgery and Survival with Glioblastoma: How Far Have We Come? In *World Neurosurgery*; Elsevier: Amsterdam, The Netherlands, 2021; Volume 149, pp. 148–168. [CrossRef]
5. Verburg, N.; de Witt Hamer, P.C. State-of-the-art imaging for glioma surgery. *Neurosurg. Rev.* **2020**, *44*, 1331–1343. [CrossRef]
6. Perrin, S.L.; Samuel, M.S.; Koszyca, B.; Brown, M.P.; Ebert, L.M.; Oksdath, M.; Gomez, G.A. Glioblastoma heterogeneity and the tumour microenvironment: Implications for preclinical research and development of new treatments. *Biochem. Soc. Trans.* **2019**, *47*, 625–638. [CrossRef]
7. Oliver, L.; Lalier, L.; Salaud, C.; Heymann, D.; Cartron, P.F.; Vallette, F.M. Drug resistance in glioblastoma: Are persisters the key to therapy? *Cancer Drug Resist.* **2020**, *3*, 287–301. [CrossRef]
8. Gianfaldoni, S.; Gianfaldoni, R.; Wollina, U.; Lotti, J.; Tchernev, G.; Lotti, T. An Overview on Radiotherapy: From Its History to Its Current Applications in Dermatology. *Open Access Maced. J. Med Sci.* **2017**, *5*, 521–525. [CrossRef]
9. Corso, C.D.; Bindra, R.S.; Mehta, M.P. The role of radiation in treating glioblastoma: Here to stay. *J. Neuro-Oncol.* **2017**, *44*, 479–485. [CrossRef]
10. Stupp, R.; Mason, W.P.; van den Bent, M.J.; Weller, M.; Fisher, B.; Taphoorn, M.J.B.; Belanger, K.; Brandes, A.A.; Marosi, C.; Bogdahn, U.; et al. Radiotherapy plus Concomitant and Adjuvant Temozolomide for Glioblastoma. *N. Engl. J. Med.* **2005**, *352*, 987–996. [CrossRef]
11. Di Ianni, N.; Musio, S.; Pellegatta, S. Altered Metabolism in Glioblastoma: Myeloid-Derived Suppressor Cell (MDSC) Fitness and Tumor-Infiltrating Lymphocyte (TIL) Dysfunction. *Int. J. Mol. Sci.* **2021**, *22*, 4460. [CrossRef]
12. Agnihotri, S.; Zadeh, G. Metabolic reprogramming in glioblastoma: The influence of cancer metabolism on epigenetics and unanswered questions. *Neuro-Oncol.* **2015**, *18*, 160–172. [CrossRef] [PubMed]
13. Caniglia, J.L.; Jalasutram, A.; Asuthkar, S.; Sahagun, J.; Park, S.; Ravindra, A.; Tsung, A.J.; Guda, M.R.; Velpula, K.K. Beyond glucose: Alternative sources of energy in glioblastoma. *Theranostics* **2021**, *11*, 2048–2057. [CrossRef] [PubMed]
14. Vander Heiden, M.G.; Cantley, L.C.; Thompson, C.B. Understanding the Warburg Effect: The Metabolic Requirements of Cell Proliferation. *Science* **2009**, *324*, 1029–1033. [CrossRef] [PubMed]
15. Jin, L.; Zhou, Y. Crucial role of the pentose phosphate pathway in malignant tumors (Review). *Oncol. Lett.* **2019**, *17*, 4213–4221. [CrossRef] [PubMed]
16. Kathagen-Buhmann, A.; Schulte, A.; Weller, J.; Holz, M.; Herold-Mende, C.; Glass, R.; Lamszus, K. Glycolysis and the pentose phosphate pathway are differentially associated with the dichotomous regulation of glioblastoma cell migration versus proliferation. *Neuro-Oncol.* **2016**, *18*, 1219–1229. [CrossRef]
17. Zhou, W.; Wahl, D.R. Metabolic Abnormalities in Glioblastoma and Metabolic Strategies to Overcome Treatment Resistance. *Cancers* **2019**, *11*, 1231. [CrossRef]
18. Ge, T.; Yang, J.; Zhou, S.; Wang, Y.; Li, Y.; Tong, X. The Role of the Pentose Phosphate Pathway in Diabetes and Cancer. *Front. Endocrinol.* **2020**, *11*, 365. [CrossRef]
19. Yang, C.-A.; Huang, H.-Y.; Lin, C.-L.; Chang, J.-G. G6PD as a predictive marker for glioma risk, prognosis and chemosensitivity. *J. Neuro-Oncol.* **2018**, *139*, 661–670. [CrossRef]

20. Ying, H.; Kimmelman, A.C.; Lyssiotis, C.A.; Hua, S.; Chu, G.C.; Fletcher-Sananikone, E.; Locasale, J.W.; Son, J.; Zhang, H.; Coloff, J.L.; et al. Oncogenic Kras Maintains Pancreatic Tumors through Regulation of Anabolic Glucose Metabolism. *Cell* **2012**, *149*, 656–670. [CrossRef]
21. Patra, K.C.; Hay, N. The pentose phosphate pathway and cancer. *Trends Biochem. Sci.* **2014**, *39*, 347–354. [CrossRef]
22. Madhavan, S.; Zenklusen, J.-C.; Kotliarov, Y.; Sahni, H.; Fine, H.A.; Buetow, K. Rembrandt: Helping Personalized Medicine Become a Reality through Integrative Translational Research. *Mol. Cancer Res.* **2009**, *7*, 157–167. [CrossRef] [PubMed]
23. Bowman, R.L.; Wang, Q.; Carro, A.; Verhaak, R.G.W.; Squatrito, M. GlioVis data portal for visualization and analysis of brain tumor expression datasets. *Neuro-Oncol.* **2017**, *19*, 139–141. [CrossRef] [PubMed]
24. Wang, Q.; Hu, B.; Hu, X.; Kim, H.; Squatrito, M.; Scarpace, L.; Decarvalho, A.C.; Lyu, S.; Li, P.; Li, Y.; et al. Tumor Evolution of Glioma-Intrinsic Gene Expression Subtypes Associates with Immunological Changes in the Microenvironment. *Cancer Cell* **2017**, *32*, 42–56.e6. [CrossRef] [PubMed]
25. Raudvere, U.; Kolberg, L.; Kuzmin, I.; Arak, T.; Adler, P.; Peterson, H.; Vilo, J. g:Profiler: A web server for functional enrichment analysis and conversions of gene lists (2019 update). *Nucleic Acids Res.* **2019**, *47*, W191–W198. [CrossRef] [PubMed]
26. Assal, B.; Matthews, L.; Viteri, G.; Gong, C.; Lorente, P.; Fabregat, A.; Sidiropoulos, K.; Cook, J.; Gillespie, M.; Haw, R.; et al. The reactome pathway knowledgebase. *Nucleic Acids Res.* **2020**, *48*, D498–D503. [CrossRef]
27. Vasaikar, S.V.; Straub, P.; Wang, J.; Zhang, B. LinkedOmics: Analyzing multi-omics data within and across 32 cancer types. *Nucleic Acids Res.* **2018**, *46*, D956–D963. [CrossRef]
28. Tan, A.C.; Ashley, D.M.; López, G.Y.; Malinzak, M.; Friedman, H.S.; Khasraw, M. Management of glioblastoma: State of the art and future directions. *CA Cancer J. Clin.* **2020**, *70*, 299–312. [CrossRef]
29. Seker-Polat, F.; Degirmenci, N.P.; Solaroglu, I.; Bagci-Onder, T. Tumor Cell Infiltration into the Brain in Glioblastoma: From Mechanisms to Clinical Perspectives. *Cancers* **2022**, *14*, 443. [CrossRef]
30. Schiera, G.; Di Liegro, C.M.; Di Liegro, I. Molecular Determinants of Malignant Brain Cancers: From Intracellular Alterations to Invasion Mediated by Extracellular Vesicles. *Int. J. Mol. Sci.* **2017**, *18*, 2774. [CrossRef]
31. Koh, I.; Cha, J.; Park, J.; Choi, J.; Kang, S.-G.; Kim, P. The mode and dynamics of glioblastoma cell invasion into a decellularized tissue-derived extracellular matrix-based three-dimensional tumor model. *Sci. Rep.* **2018**, *8*, 4608. [CrossRef]
32. Osuka, S.; Van Meir, E.G. Overcoming therapeutic resistance in glioblastoma: The way forward. *J. Clin. Investig.* **2017**, *127*, 415–426. [CrossRef] [PubMed]
33. Biserova, K.; Jakovlevs, A.; Uljanovs, R.; Strumfa, I. Cancer Stem Cells: Significance in Origin, Pathogenesis and Treatment of Glioblastoma. *Cells* **2021**, *10*, 621. [CrossRef] [PubMed]
34. Ghanem, N.; El-Baba, C.; Araji, K.; El-Khoury, R.; Usta, J.; Darwiche, N. The Pentose Phosphate Pathway in Cancer: Regulation and Therapeutic Opportunities. *Chemotherapy* **2021**, *66*, 179–191. [CrossRef]
35. Kathagen-Buhmann, A.; Maire, C.L.; Weller, J.; Schulte, A.; Matschke, J.; Holz, M.; Ligon, K.L.; Glatzel, M.; Westphal, M.; Lamszus, K. The secreted glycolytic enzyme GPI/AMF stimulates glioblastoma cell migration and invasion in an autocrine fashion but can have anti-proliferative effects. *Neuro-Oncol.* **2018**, *20*, 1594–1605. [CrossRef]
36. Stanke, K.M.; Wilson, C.; Kidambi, S. High Expression of Glycolytic Genes in Clinical Glioblastoma Patients Correlates with Lower Survival. *Front. Mol. Biosci.* **2021**, *8*, 1281. [CrossRef] [PubMed]
37. Armento, A.; Ehlers, J.; Schötterl, S.; Naumann, U. Molecular Mechanisms of Glioma Cell Motility. In *Glioblastoma*; De Vleeschouwer, S., Ed.; Codon Publications: Brisbane, Australia, 2017; pp. 73–93. [CrossRef]
38. Mosier, J.A.; Schwager, S.C.; Boyajian, D.A.; Reinhart-King, C.A. Cancer cell metabolic plasticity in migration and metastasis. *Clin. Exp. Metastasis* **2021**, *38*, 343–359. [CrossRef] [PubMed]
39. Hay, N. Reprogramming glucose metabolism in cancer: Can it be exploited for cancer therapy? *Nat. Rev. Cancer* **2016**, *16*, 635–649. [CrossRef]
40. Stincone, A.; Prigione, A.; Cramer, T.; Wamelink, M.M.C.; Campbell, K.; Cheung, E.; Olin-Sandoval, V.; Grüning, N.-M.; Krüger, A.; Tauqeer Alam, M.; et al. The return of metabolism: Biochemistry and physiology of the pentose phosphate pathway. *Biol. Rev.* **2015**, *90*, 927–963. [CrossRef]
41. Batsios, G.; Taglang, C.; Cao, P.; Gillespie, A.M.; Najac, C.; Subramani, E.; Wilson, D.M.; Flavell, R.R.; Larson, P.E.Z.; Ronen, S.M.; et al. Imaging 6-Phosphogluconolactonase Activity in Brain Tumors In Vivo Using Hyperpolarized δ-[1-13C] gluconolactone. *Front. Oncol.* **2021**, *11*, 1194. [CrossRef]
42. Xu, X.; Wang, L.; Zang, Q.; Li, S.; Li, L.; Wang, Z.; He, J.; Qiang, B.; Han, W.; Zhang, R.; et al. Rewiring of purine metabolism in response to acidosis stress in glioma stem cells. *Cell Death Dis.* **2021**, *12*, 227. [CrossRef]
43. Jiang, P.; Du, W.; Wu, M. Regulation of the pentose phosphate pathway in cancer. *Protein Cell* **2014**, *5*, 592–602. [CrossRef] [PubMed]
44. Kowalik, M.A.; Columbano, A.; Perra, A. Emerging Role of the Pentose Phosphate Pathway in Hepatocellular Carcinoma. *Front. Oncol.* **2017**, *7*, 87. [CrossRef] [PubMed]
45. Cossu, V.; Bonanomi, M.; Bauckneht, M.; Ravera, S.; Righi, N.; Miceli, A.; Morbelli, S.; Orengo, A.M.; Piccioli, P.; Bruno, S.; et al. Two high-rate pentose-phosphate pathways in cancer cells. *Sci. Rep.* **2020**, *10*, 22111. [CrossRef] [PubMed]
46. Qin, Z.; Xiang, C.; Zhong, F.; Liu, Y.; Dong, Q.; Li, K.; Shi, W.; Ding, C.; Qin, L.; He, F. Transketolase (TKT) activity and nuclear localization promote hepatocellular carcinoma in a metabolic and a non-metabolic manner. *J. Exp. Clin. Cancer Res.* **2019**, *38*, 154. [CrossRef]

47. Dos Santos, G.C.; Saldanha-Gama, R.; De Brito, N.M.; Renovato-Martins, M.; Barja-Fidalgo, C. Metabolomics in cancer and cancer-associated inflammatory cells. *J. Cancer Metastasis Treat.* **2021**, *7*, 1. [CrossRef]
48. Fack, F.; Tardito, S.; Hochart, G.; Oudin, A.; Zheng, L.; Fritah, S.; Golebiewska, A.; Nazarov, P.; Bernard, A.; Hau, A.; et al. Altered metabolic landscape in IDH -mutant gliomas affects phospholipid, energy, and oxidative stress pathways. *EMBO Mol. Med.* **2017**, *9*, 1681–1695. [CrossRef]
49. Louis, D.N.; Perry, A.; Wesseling, P.; Brat, D.J.; Cree, I.A.; Figarella-Branger, D.; Hawkins, C.; Ng, H.K.; Pfister, S.M.; Reifenberger, G.; et al. The 2021 WHO Classification of Tumors of the Central Nervous System: A summary. *Neuro-Oncol.* **2021**, *23*, 1231–1251. [CrossRef]
50. Zhang, P.; Xia, Q.; Liu, L.; Li, S.; Dong, L. Current Opinion on Molecular Characterization for GBM Classification in Guiding Clinical Diagnosis, Prognosis, and Therapy. *Front. Mol. Biosci.* **2020**, *7*, 562798. [CrossRef]
51. Boren, J.; Montoya, A.R.; De Atauri, P.; Comin-Anduix, B.; Cortes, A.; Centelles, J.J.; Frederiks, W.M.; Van Noorden, C.J.; Cascante, M. Metabolic control analysis aimed at the ribose synthesis pathways of tumor cells: A new strategy for antitumor drug development. *Mol. Biol. Rep.* **2002**, *29*, 7–12. [CrossRef]
52. Ramos-Montoya, A.; Lee, W.-N.P.; Bassilian, S.; Lim, S.; Trebukhina, R.V.; Kazhyna, M.V.; Ciudad, C.J.; Noé, V.; Centelles, J.J.; Cascante, M. Pentose phosphate cycle oxidative and nonoxidative balance: A new vulnerable target for overcoming drug resistance in cancer. *Int. J. Cancer* **2006**, *119*, 2733–2741. [CrossRef]
53. Frederiks, W.M.; Vizan, P.; Bosch, K.S.; Vreeling-Sindelárová, H.; Boren, J.; Cascante, M. Elevated activity of the oxidative and non-oxidative pentose phosphate pathway in (pre)neoplastic lesions in rat liver. *Int. J. Exp. Pathol.* **2008**, *89*, 232–240. [CrossRef] [PubMed]
54. Pandith, A.A.; Qasim, I.; Baba, S.M.; Koul, A.; Zahoor, W.; Afroze, D.; Lateef, A.; Manzoor, U.; A Bhat, I.; Sanadhya, D.; et al. Favorable role of IDH1/2 mutations aided with MGMT promoter gene methylation in the outcome of patients with malignant glioma. *Futur. Sci. OA* **2020**, *7*, FSO663. [CrossRef] [PubMed]
55. Chen, J.-R.; Yao, Y.; Xu, H.-Z.; Qin, Z.-Y. Isocitrate Dehydrogenase (IDH)1/2 Mutations as Prognostic Markers in Patients with Glioblastomas. *Medicine* **2016**, *95*, e2583. [CrossRef] [PubMed]
56. Murugan, A.K.; Alzahrani, A.S. Isocitrate Dehydrogenase IDH1 and IDH2 Mutations in Human Cancer: Prognostic Implications for Gliomas. *Br. J. Biomed. Sci.* **2022**, *79*, 10208. [CrossRef]
57. Wang, H.; Wang, X.; Xu, L.; Zhang, J.; Cao, H. Analysis of the EGFR Amplification and CDKN2A Deletion Regulated Transcriptomic Signatures Reveals the Prognostic Significance of SPATS2L in Patients with Glioma. *Front. Oncol.* **2021**, *11*, 713. [CrossRef]
58. Schneider, C.A.; Rasband, W.S.; Eliceiri, K.W. NIH Image to ImageJ: 25 Years of image analysis. *Nat. Methods* **2012**, *9*, 671–675. [CrossRef]

International Journal of *Molecular Sciences*

MDPI

Article

Different Effects of RNAi-Mediated Downregulation or Chemical Inhibition of NAMPT in an Isogenic IDH Mutant and Wild-Type Glioma Cell Model

Maximilian Clausing [1,2,3], Doreen William [1,3,4], Matthias Preussler [1,2,3], Julia Biedermann [1], Konrad Grützmann [3,4], Susan Richter [5], Frank Buchholz [2,3,6], Achim Temme [2,7], Evelin Schröck [1,3,4] and Barbara Klink [1,3,4,8,*]

1 Institute for Clinical Genetics, University Hospital Carl Gustav Carus, Technische Universität Dresden, ERN-GENTURIS, Hereditary Cancer Syndrome Center Dresden, 01307 Dresden, Germany; maximilian.clausing@uniklinikum-dresden.de (M.C.); doreen.william@uniklinikum-dresden.de (D.W.); matthias.preussler@uniklinikum-dresden.de (M.P.); julia.biedermann@uniklinikum-dresden.de (J.B.); evelin.schroeck@uniklinikum-dresden.de (E.S.)
2 National Center for Tumor Diseases Partner Site Dresden (NCT/UCC), German Cancer Consortium (DKTK), 01307 Dresden, Germany; frank.buchholz@tu-dresden.de (F.B.); achim.temme@uniklinikum-dresden.de (A.T.)
3 German Cancer Research Center (DKFZ), 69120 Heidelberg, Germany; konrad.gruetzmann@tu-dresden.de
4 Core Unit for Molecular Tumor Diagnostics (CMTD), National Center for Tumor Diseases (NCT/UCC), German Cancer Consortium (DKTK), 01307 Dresden, Germany
5 Institute for Clinical Chemistry and Laboratory Medicine, University Hospital Carl Gustav Carus, Technische Universität Dresden, 01307 Dresden, Germany; susan.richter@uniklinikum-dresden.de
6 Medical Systems Biology, Faculty of Medicine, University Hospital Carl Gustav Carus, Technische Universität Dresden, 01307 Dresden, Germany
7 Section Experimental Neurosurgery/Tumor Immunology, Department of Neurosurgery, University Hospital Carl Gustav Carus, Technische Universität Dresden, 01307 Dresden, Germany
8 National Center of Genetics (NCG), Laboratoire National de Santé (LNS), L-3555 Luxembourg, Luxembourg
* Correspondence: barbara.klink@lns.etat.lu

Citation: Clausing, M.; William, D.; Preussler, M.; Biedermann, J.; Grützmann, K.; Richter, S.; Buchholz, F.; Temme, A.; Schröck, E.; Klink, B. Different Effects of RNAi-Mediated Downregulation or Chemical Inhibition of NAMPT in an Isogenic IDH Mutant and Wild-Type Glioma Cell Model. *Int. J. Mol. Sci.* **2022**, *23*, 5787. https://doi.org/10.3390/ijms23105787

Academic Editor: Nives Pećina-Šlaus

Received: 13 April 2022
Accepted: 13 May 2022
Published: 21 May 2022

Abstract: The IDH1^{R132H} mutation in glioma results in the neoenzymatic function of IDH1, leading to the production of the oncometabolite 2-hydroxyglutarate (2-HG), alterations in energy metabolism and changes in the cellular redox household. Although shifts in the redox ratio NADPH/NADP$^+$ were described, the consequences for the NAD$^+$ synthesis pathways and potential therapeutic interventions were largely unexplored. Here, we describe the effects of heterozygous IDH1^{R132H} on the redox system in a CRISPR/Cas edited glioblastoma model and compare them with IDH1 wild-type (IDH1^{wt}) cells. Besides an increase in 2-HG and decrease in NADPH, we observed an increase in NAD$^+$ in IDH1^{R132H} glioblastoma cells. RT-qPCR analysis revealed the upregulation of the expression of the NAD$^+$ synthesis enzyme nicotinamide phosphoribosyltransferase (NAMPT). Knockdown of NAMPT resulted in significantly reduced viability in IDH1^{R132H} glioblastoma cells. Given this dependence of IDH1^{R132H} cells on NAMPT expression, we explored the effects of the NAMPT inhibitors FK866, GMX1778 and GNE-617. Surprisingly, these agents were equally cytotoxic to IDH1^{R132H} and IDH1^{wt} cells. Altogether, our results indicate that targeting the NAD$^+$ synthesis pathway is a promising therapeutic strategy in IDH mutant gliomas; however, the agent should be carefully considered since three small-molecule inhibitors of NAMPT tested in this study were not suitable for this purpose.

Keywords: IDH1 mutation; glioma; redox household; nicotinamide phosphoribosyltransferase; NAD$^+$ synthesis

1. Introduction

Gliomas are the most prevalent histological type of primary malignant central nervous system tumors and one of the most malignant types of cancer according to their aggressive invasive potential [1,2]. Current therapies using combinations of surgery, radiotherapy and chemotherapy have limited success with 10-year overall survival rates of less than 1% for glioblastoma [3,4]. The non-specific nature of current treatments might be one factor. Therefore, the identification of new therapeutic strategies more specifically targeting tumor cells is of great interest.

Mutations of the key Krebs cycle enzyme isocitrate dehydrogenase (IDH) are considered to be vital for the genesis of low-grade gliomas and secondary glioblastomas [5], prompting the World Health Organization (WHO) to separate gliomas according to IDH mutation status into IDH mutant and IDH wild-type ($IDH1^{wt}$) entities in the "2016 WHO Classification of Tumors of the Central Nervous System" [6]. The most frequent of these mutations is $IDH1^{R132H}$—the point mutation of arginine to histidine at residue 132, accounting for more than 80% of IDH mutations [7]. It results in a neoenzymatic function of IDH1, leading to the near-complete elimination of the oxidation of isocitrate to α-ketoglutarate (α-KG) catalyzed by $IDH1^{wt}$. Instead, $IDH1^{R132H}$ leads to the NADPH-consuming production of 2-hydroxyglutarate (2-HG) via the reduction of α-KG [8,9]. 2-HG is accepted as an oncometabolite, with many studies having been focused on oncogenic effects through the inhibition of α-KG dependent dioxygenases [10,11], such as hypoxia-inducible factor (HIF) hydroxylases and methylcytosine dioxygenase TET2, resulting in altered HIF activity and CpG island-methylator phenotype (G-CIMP) [12,13].

Besides these 2-HG-mediated oncogenic effects, other metabolic changes were described in IDH1 mutated gliomas. These alterations include decreases in glutathione metabolites [14], activation of glutaminolysis [15], aberrations in lipid metabolism and reduced glucose turnover [16]. Moreover, IDH1 mutations were shown to alter redox metabolism in glioma cells. Reported drops in NADPH levels may be a direct result of the NADPH-consuming reaction catalyzed by mutated IDH1, whereas wild-type IDH1 was identified as the main source of cytosolic NADPH in glia cells and glioblastoma [17,18]. We and others recently also described a significant decrease in NAD^+ levels in glioma cells with $IDH1^{R132H}$, leading to the hypothesis of NAPDH restoration via the phosphorylation of NAD^+ and the identification of NAD^+ synthesis inhibition as a possible treatment [17,19]. Intriguingly, we found that NAD^+ levels are not altered in IDH1 mutated astrocytes [17], indicating successful compensatory mechanisms in those cells and, therefore, potentially limiting the vulnerability of cancer cells to NAD^+ synthesis inhibition.

Nicotinamide phosphoribosyltransferase (NAMPT) catalyzes the rate-limiting conversion of nicotinamide to nicotinamide mononucleotide (NMN) as part of the NAD^+ salvage pathway [20]. Of the various NAD^+ synthesis enzymes, we found NAMPT to be the only one ubiquitously expressed among different patient-derived glioma cell lines, making it a natural target in those cells [17]. NAMPT was identified as a promising anticancer target due to the absence of other NAD+ synthesis enzymes in several tumors (e.g., prostate carcinoma, sarcomas, neuroblastomas and glioblastomas) [21–23], its association with worse prognosis in glioblastomas [24] and its overexpression in several types of tumor cells including gliomas [25]. NAMPT small-molecule inhibitors were shown to induce cytotoxicity through NAD^+ depletion in a wide range of tumor models in vitro and in vivo (e.g., colorectal carcinoma, acute myeloid leukemia and glioblastoma) [26].

This study focused on redox metabolism as a treatment option in IDH1 mutated gliomas. We created a novel cell model by inducing the heterozygous $IDH1^{R132H}$ point mutation at the target genomic sequence into a primary patient-derived glioblastoma cell line using CRISPR/Cas9. In light of the difficulties of culturing IDH1 mutated patient-derived glioma cells in vitro [27,28] and the lack of patient-derived IDH mutant in vitro and in vivo models [29,30], our approach provides a suitable alternative to investigate $IDH1^{R132H}$-dependent alterations in tumor cell metabolism. Our endogenous $IDH1^{R132H}$ cell model showed 100-fold elevated 2-HG levels, confirming the functionality of the mutated IDH1 enzyme and making it comparable to patient samples [8]. In this work, we explored the effects of $IDH1^{R132H}$ on NAD^+ metabolism and the therapeutic potential of NAMPT knockdown and NAMPT small-molecule inhibitors in $IDH1^{wt}$ and $IDH1^{R132H}$ glioma cells.

2. Results

2.1. Alterations in NAD⁺ and NADP⁺ Metabolism in IDH1^{R132H} Cells Could Be Directly Attributed to the Neoenzymatic Function of IDH1^{R132H}

We successfully introduced the IDH1 c.395G > A point mutation heterozygously in a primary patient-derived glioblastoma cell line (HT7606 [31]) using CRISPR/Cas9. Three IDH1^{R132H} clones, as well as three IDH1^{wt} clones as controls, were created. The mutation was stable in all three IDH1^{R132H} clones in a long-time culture for at least 50 passages (Supplementary Figure S1a). We confirmed the IDH1 and IDH1^{R132H} protein expression (Supplementary Figure S1b) and the functionality of the IDH1^{R132H} enzyme. The 2-HG level was approximately 100-fold elevated in IDH1^{R132H} mutated cells compared with the IDH1^{wt} cells. Treatment with the mutant IDH1 inhibitor AGI-5198 normalized 2-HG levels in IDH1^{R132H} cells, confirming the functionality of the inhibitor and suitability of our cell line model (Supplementary Figure S1c). The results for the CRISPR/Cas-edited cell lines are comparable with patient samples [8] and verified the functionality of the mutated enzyme.

Intracellular NAD^+ and $NADPH^+$ levels were previously reported to differ in gliomas in a manner that was dependent on the IDH1 status [16,17,19,32]. To investigate alterations in the $NADP^+$/H and NAD^+/H metabolism in our isogenic IDH1^{R132H} cell models, we measured the NADPH/$NADP^+$ and NADH/NAD^+ ratios, as well as the $NADP^+$, NADPH, NAD^+ and NADH levels. The NADPH/$NADP^+$ ratios were significantly lower in IDH1^{R132H} cells compared with IDH1^{wt} cells due to a significant decrease in NADPH levels (Figure 1A). Furthermore, the NADH/NAD^+ ratios were also significantly lower in IDH1^{R132H} cells (Figure 1B). This effect was caused by increases in the NAD^+ levels in the mutated cells, indicating an upregulation of NAD^+ synthesis in IDH1^{R132H} cells. To confirm that the observed changes in redox metabolism were directly linked to the neo-enzymatic activity of IDH1^{R132H}, the cells were treated with the specific mutant IDH1 inhibitor AGI-5198. Rescue of the NADPH level and NADPH/$NADP^+$ ratio upon treatment with AGI-5198 in IDH1 mutant cells confirmed this to be a direct effect due to the increased consumption of NADPH by the IDH1^{R132H} enzyme (Figure 1C). Notably, treatment with AGI-5198 also normalized the NAD^+ levels in IDH1^{R132H} mutated cells (Figure 1D). Therefore, the observed changes in redox metabolism could be directly attributed to the functionality of IDH1^{R132H}. We hypothesize that IDH1^{R132H} leads to upregulation of NAD^+ synthesis via NAMPT to restore consumed NADPH via the phosphorylation of NAD^+.

2.2. IDH1^{R132H} Altered NAMPT Expression

We previously identified NAMPT as the only NAD^+ synthesis enzyme that is ubiquitously expressed in IDH mutant and IDH wild-type gliomas and that NAMPT protein expression is lower in IDH mutant gliomas [17]. We, therefore, analyzed the expression of NAMPT on mRNA and protein levels in our IDH1^{wt} and IDH1^{R132H} cells. NAMPT mRNA levels of IDH1^{wt} cells were comparable to those in normal brain tissue (Figure 2A). IDH1^{R132H} cells exhibited approximately three-fold higher NAMPT mRNA expression, supporting our hypothesis of upregulated NAD^+ synthesis. However, analysis of the NAMPT protein levels revealed similar expressions in IDH1^{R132H} and IDH1^{wt} cells (Figure 2B). The discrepancy between changes in NAMPT mRNA and protein expression suggests that posttranslational regulation of NAMPT expression differs between IDH1^{R132H} and IDH1^{wt} cells.

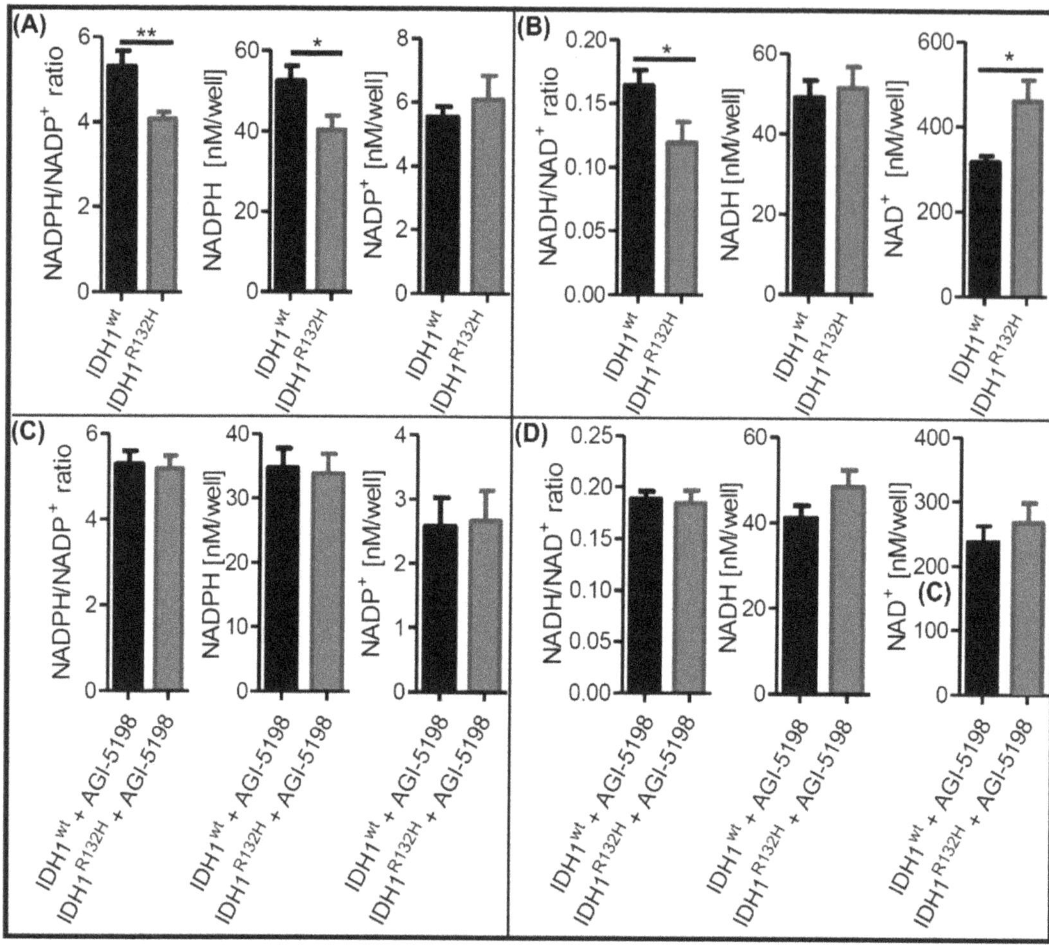

Figure 1. Alterations in NAD+ and NADP+ metabolism can be directly attributed to the neoenzymatic function of IDH1^{R132H}. (**A**,**C**) Intracellular NADPH/NADP+ ratios, as well as NADPH and NADP+ levels, in untreated IDH1^{wt} and IDH1^{R132H} cells before (**A**) and after treatment with the selective mutant IDH1 inhibitor AGI-5198 for 48 h (**C**). (**B**,**D**) Intracellular NADH/NAD+ ratios, as well as NADH and NAD+ levels, in untreated IDH1^{wt} and IDH1^{R132H} cells before (**B**) and after treatment with the selective mutant IDH1 inhibitor AGI-5198 for 48 h (**D**) (nb = 3 per group; nt = 3; * $p \leq 0.05$, ** $p \leq 0.01$).

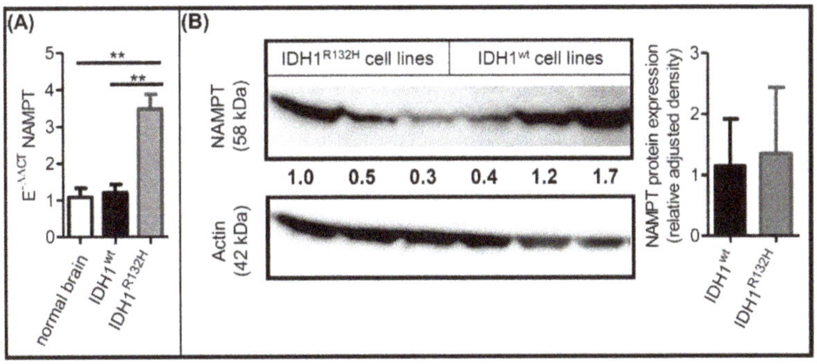

Figure 2. IDH1^{R132H} altered NAMPT expression. (**A**) Relative NAMPT mRNA levels in comparison to reference genes GAPDH and ARF1 ($E^{-\Delta CT}$). Relative NAMPT mRNA levels in IDH1^{wt} (nb = 2) and IDH1^{R132H} cells (nb = 3) were compared with the relative NAMPT mRNA levels in a normal brain ($E^{-\Delta\Delta CT}$) (nt = 3). (**B**) Western blot analyses of NAMPT in IDH1^{wt} cell lines (HT7606-IDH1$^{wt/wt-40}$, HT7606-IDH1$^{wt/wt-141}$, HT7606) and IDH1^{R132H} cell lines (HT7606-IDH1$^{R132H/wt-1}$, HT7606-IDH1$^{R132H/wt-16}$, HT7606-IDH1$^{R132H/wt-88}$). Data are presented as the ratio of NAMPT to the reference protein GAPDH (nb = 3; nt = 2) (** $p \leq 0.01$).

2.3. Knockdown of NAMPT and other NAD$^+$ Synthesis Enzymes Selectively Reduced the Viability of IDH1^{R132H} Cells

To investigate the effects of NAMPT gene knockdown as a potential target for the treatment of IDH mutant glioma, IDH1^{wt} and IDH1^{R132H} cells were transfected using two different esiRNAs targeting different sequences of NAMPT mRNA (esiNAMPT-A and esiNAMPT-B) to confirm the esiRNA specificity. The RT-qPCR results revealed knockdown efficiencies of approximately 95% for esiNAMPT-A and 75% for esiNAMPT-B 72 h post-transfection (Figure 3A,C). NAMPT protein levels were considerably lower 48 h post-transfection of esiNAMPT-A in IDH1^{wt} and IDH1^{R132H} compared with the mock control treatment cells and no longer detectable 72 h after transfection (Figure 3B). Likewise, NAMPT protein levels were not detectable 72 h post-transfection of esiNAMPT-B (Figure 3D).

Forty-eight hours after the NAMPT knockdown, the NAD(P)H-dependent WST-1-reducing capability was significantly diminished in both the IDH1^{wt} and IDH1^{R132H} cells (Figure 4A), indicating a decrease in the reducing agents due to a lack of functional NAMPT. NAD$^+$ levels were higher in untreated IDH1^{R132H} cells compared with IDH1^{wt} but reached similar levels after NAMPT knockdown (Figure 4B). NADP$^+$ levels were unaltered by NAMPT knockdown. NADPH levels, on the other hand, dropped in both cell lines, reducing the NADPH levels in IDH1^{R132H} cells even further to about one-third of that of IDH1^{wt}. NADH levels dropped in IDH1^{R132H} cells after NAMPT knockdown by nearly 80%, but not in IDH1^{wt}. NAMPT knockdown resulted in a decrease in cell viability of 45% for esiNAMPT-A and 59% for esiNAMPT-B in IDH1^{R132H} cells, as determined by fluorescent-based quantification of DNA content (Figure 4C). Despite the observed change in WST-1-reducing capacity, cell viability was not affected in IDH1^{wt} cells. These results indicated an enhanced dependence of IDH1^{R132H} cells on NAD$^+$ synthesis compared with IDH1^{wt} cells.

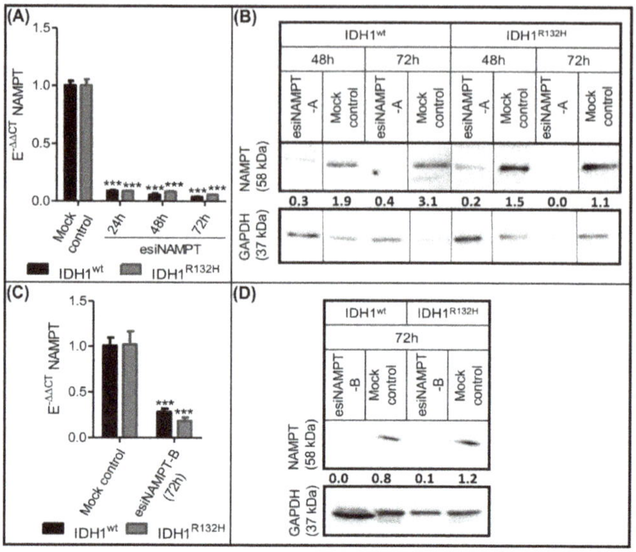

Figure 3. Successful knockdown of NAMPT. (**A**,**C**) Relative NAMPT mRNA levels compared with the reference genes GAPDH and ARF1 ($E^{-\Delta CT}$) in IDH1^{wt} and IDH1^{R132H} cells after treatment with esiNAMPT-A (**A**) or esiNAMPT-B (**C**) were compared with the relative NAMPT mRNA levels in the respective cells after treatment with the mock control ($E^{-\Delta\Delta CT}$). Data are presented as the ratio to treatment with the mock control. (nt = 3; *** $p \leq 0.001$ compared with the mock control). (**B**,**D**) Western blot analyses of NAMPT in IDH1^{wt} (HT7606) and IDH1^{R132H} (HT7606-IDH1$^{R132H/wt-1}$) cells after treatment with the mock control, esiNAMPT-A (**B**) or esiNAMPT-B (**D**) for 48 or 72 h; black lines indicate where bands from the same gel were ordered differently for improved clarity. Data are presented as the ratio of NAMPT to GAPDH.

To confirm the selective dependence of IDH1^{R132H} cells on NAD$^+$ synthesis, we analyzed the effects of NAMPT knockdown using live cell counting and extended our study to other enzymes involved in NAD$^+$ synthesis pathways (Figure 4D). We previously analyzed the expression levels of the rate-limiting enzymes of the four NAD$^+$ synthesis pathways in glioma cells and astrocytes and found the two salvage pathways via NMRK1 and NAMPT to be the sole contributors to NAD$^+$ synthesis [17]. Thus, we performed a knockdown of NMRK1, the rate-limiting enzyme of the second salvage pathway, as well as a knockdown of NMNAT1, which is the last step in NAD synthesis in all four NAD$^+$ synthesis pathways, in our cell models. We confirmed the successful knockdown after esiRNA treatment using Western blotting (Supplementary Materials Figure S2). Live cell counting with the Operetta High-Content Imaging System confirmed the selective reduction in cell viability in IDH1^{R132H} cells after the knockdown of NAMPT with esiNAMPT-A and esiNAMPT-B (Figure 4E). The knockdown of NMRK1 or NMNAT1 resulted in a similar selective effect on IDH1^{R132H} cells.

To investigate whether NAD$^+$ synthesis might be upregulated in IDH1^{R132H} cells to restore consumed NADPH via phosphorylation of NAD$^+$, we performed an esiRNA-mediated knockdown of NADK, which is the enzyme catalyzing the ATP-dependent phosphorylation of NAD$^+$ to NADP$^+$. In line with our hypothesis, NADK knockdown reproduced the effects of NAMPT and NMRK1 knockdowns (Figure 4E). Taken together, these data showed an increased dependence of IDH1^{R132H} glioma cells on NAD$^+$ synthesis and phosphorylation, suggesting that the investigated pathways might be promising therapeutic targets for gliomas with an IDH1^{R132H} mutation.

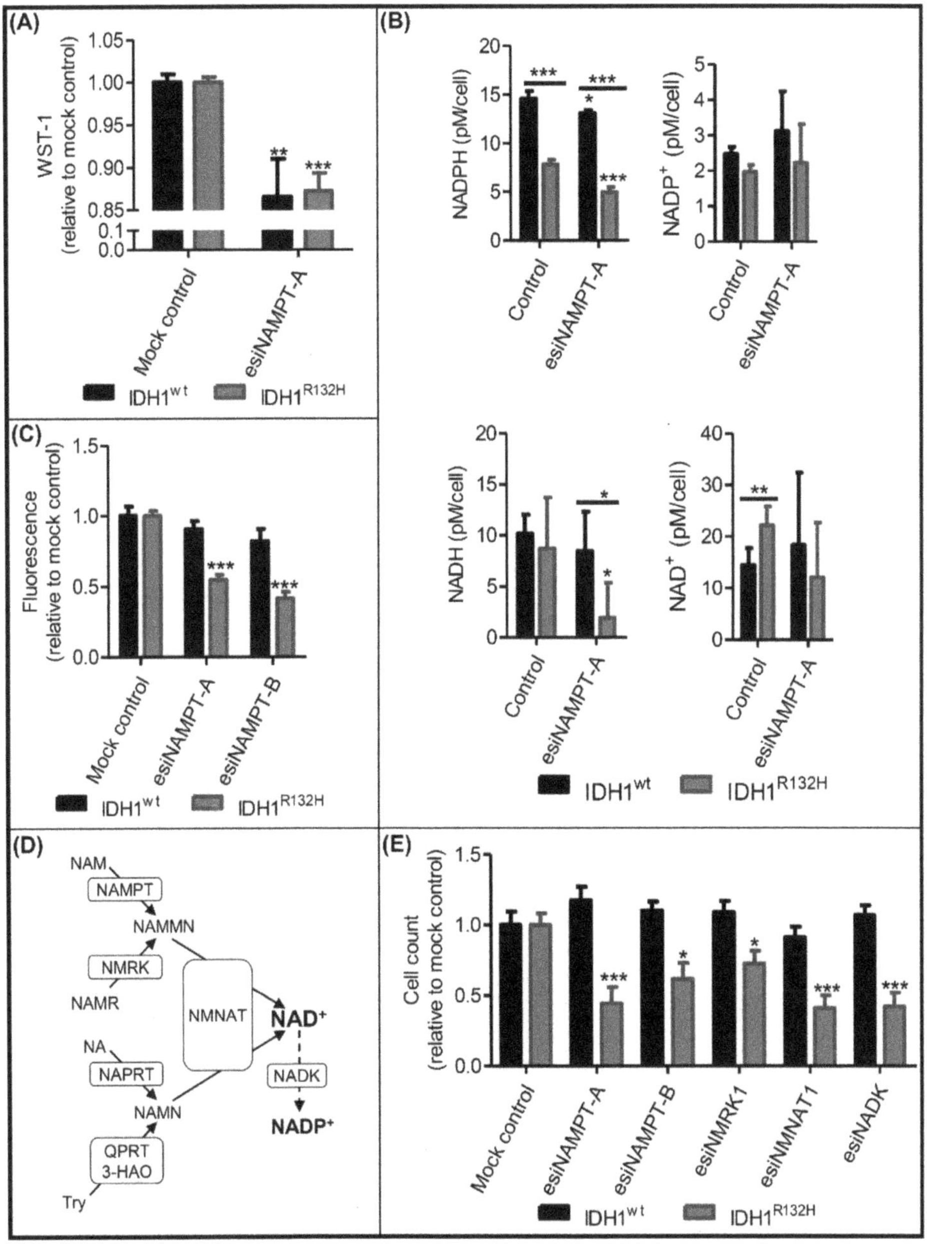

Figure 4. Knockdown of NAMPT and other NAD$^+$ synthesis enzymes selectively reduced the viability of IDH1^{R132H} cells. (**A**) Metabolic activity in IDH1^{wt} and IDH1^{R132H} cells after treatment with esiNAMPT-A for 48 h (nb = 3). (**B**) NADPH, NADP$^+$, NADH and NAD$^+$ levels after 48 h of treatment with DMSO (control) or esiNAMPT-A. Each condition was normalized to the cell count of each sample (nb = 2 per group). (**C**) Cell viability after treatment with esiNAMPT-A or esiNAMPT-B for 72 h (nb = 2 for IDH1^{wt}; nb = 3 for IDH1^{R132H}). (**D**) NAD$^+$ synthesis and salvage pathways. NAD$^+$ is synthesized de novo from tryptophan (Try) or salvaged from nicotinamide (NAM), nicotinamide riboside (NAMR) or nicotinic acid (NA). NAD$^+$ kinase (NADK) generates NADP$^+$ from NAD$^+$ and

ATP. NAMPT: nicotinaminde phosphoribosyltransferase, NMRK: nicotinamide riboside kinase, NAPRT: nicotinic acid phosphoribosyltransferase, 3-HAO: quinolinic acid-synthesis-enzyme 3-hydroxyanthranilate 3,4-dioxygenase, QPRT: quinolinic acid phosphoribosyltransferase, NMNAT: nicotinamide mononucleotide adenylyltransferase, NAMMN: nicotinamide mononucleotide, NAMN: nicotinic acid mononucleotide. (**E**) Cell count after treatment of $IDH1^{wt}$ and $IDH1^{R132H}$ cells with esiNAMPT-A, esiNAMPT B, esiNMRK1, esiNMNAT or esiNADK for 72 h (nb = 3 per group). All data are presented as a ratio to treatment with the mock control/control (nt = 3; * $p \leq 0.05$, ** $p \leq 0.01$, *** $p \leq 0.001$ compared with the mock control/control unless indicated otherwise).

2.4. Effects of NAMPT Small-Molecule Inhibitors on Cell Viability Were Independent of IDH1 Status

Several NAMPT small-molecule inhibitors were recently described as potential anti-cancer agents [26]. To test whether pharmacological NAMPT inhibition resulted in similar cytotoxicity in $IDH1^{R132H}$ cells to the esiRNA-mediated NAMPT knockdown, we exposed our cells to the chemically distinct, specific NAMPT inhibitors FK866, GMX1778 and GNE-617. Each inhibitor reduced the metabolic activity—quantified via the NAD(P)H-dependent WST-1-reduction rate—after 48 h of treatment in a concentration-dependent manner (Figure 5A). The IC50 values were determined to be 36.8 nM, 19.9 nM and 27.9 nM for FK866, GMX1778 and GNE-617, respectively. The viability of our cell models was unaffected after 48 h treatment with NAMPT inhibitors at concentrations of 100 nM, which corresponded to the maximal inhibitory concentrations measured using a WST-1 assay (Figure 5B). After 72 h, cytotoxic effects were observed in both $IDH1^{wt}$ and $IDH1^{R132H}$ cells, with an additional decrease in viability after 96 h of incubation. These effects were confirmed in another cell viability assay based on fluorescence quantification (Figure 5C). After 96 h of NAMPT inhibitor treatment, all investigated cells showed significantly reduced cell viability. In fact, the cytotoxicity of NAMPT inhibitors was even more severe in $IDH1^{wt}$ cells, contrasting with the higher sensitivity of $IDH1^{R132H}$ cells to esiRNA-mediated NAMPT knockdown. Lower inhibitor concentrations of 25 nM, approximately corresponding to the IC50, resulted in similar effects; the cell viability was more reduced in $IDH1^{wt}$ compared to $IDH1^{R132H}$ cells. NAMPT inhibitor concentrations of 1 nM were not sufficient to impair the cell viability, whereas cytotoxicity at very high concentrations (10 µM) did not substantially differ from the described results at 100 nM (Supplementary Figure S3). In summary, the $IDH1^{R132H}$ selective effect of NAMPT-knockdown was not reproducible using the NAMPT small-molecule inhibitors FK866, GMX1778 and GNE-617.

2.5. Combinatorial esiNAMPT and GMX1778 Treatment Indicated Unspecific Effects of Small Molecule NAMPT Inhibitors

To further investigate the effects of NAMPT small-molecule inhibitors in our cell models, we determined the NAD(H) and NADP(H) levels after 48 h of treatment with 25 nM (near IC50) GMX1778 (Figure 6A). Independently of the IDH1 status, the GMX1778 treatment reduced NAD^+ and NADH to unmeasurable levels, explaining the unselective cytotoxicity in our cell models. Accordingly, the NADPH levels were significantly and strongly reduced in both the IDH mutant and wild-type cells. Since different NAD^+ and $NADP^+$ synthesis pathways exist, complete loss of NAD(H) after treatment with GMX1778 indicated that inhibition of NAD^+ synthesis might be non-selective and not NAMPT specific.

We, therefore, explored the effects of treatment with GMX1778 on cell viability after the knockdown of NAMPT using esiNAMPT-A. Silencing with esiNAMPT-A led to a 95% effective NAMPT knockdown at 48 h and 72 h after the treatment (Figure 3B). For combinatory esiNAMPT-A and GMX1778 treatment, we applied GMX1778 48 h after the esiRNA transfection (Figure 6B). If GMX1778 specifically inhibited NAMPT, we would not expect an additional effect of GMX1778 in NAMPT knockdown cells. The GMX1778 treatment reduced viability in $IDH1^{wt}$ and $IDH1^{R132H}$ cells after 96 h (Figure 6B). In this experimental setup, $IDH1^{R132H}$ cells showed a higher reduction in cell viability than

IDH1^{wt}. Intriguingly, the GMX1778 treatment caused the same non-selective effect after the knockdown of NAMPT in both wild-type and IDH mutant cells (Figure 6B). The observation that GMX1778 treatment also affected NAMPT knockdown cells strongly indicated nonspecific effects besides NAMPT inhibition.

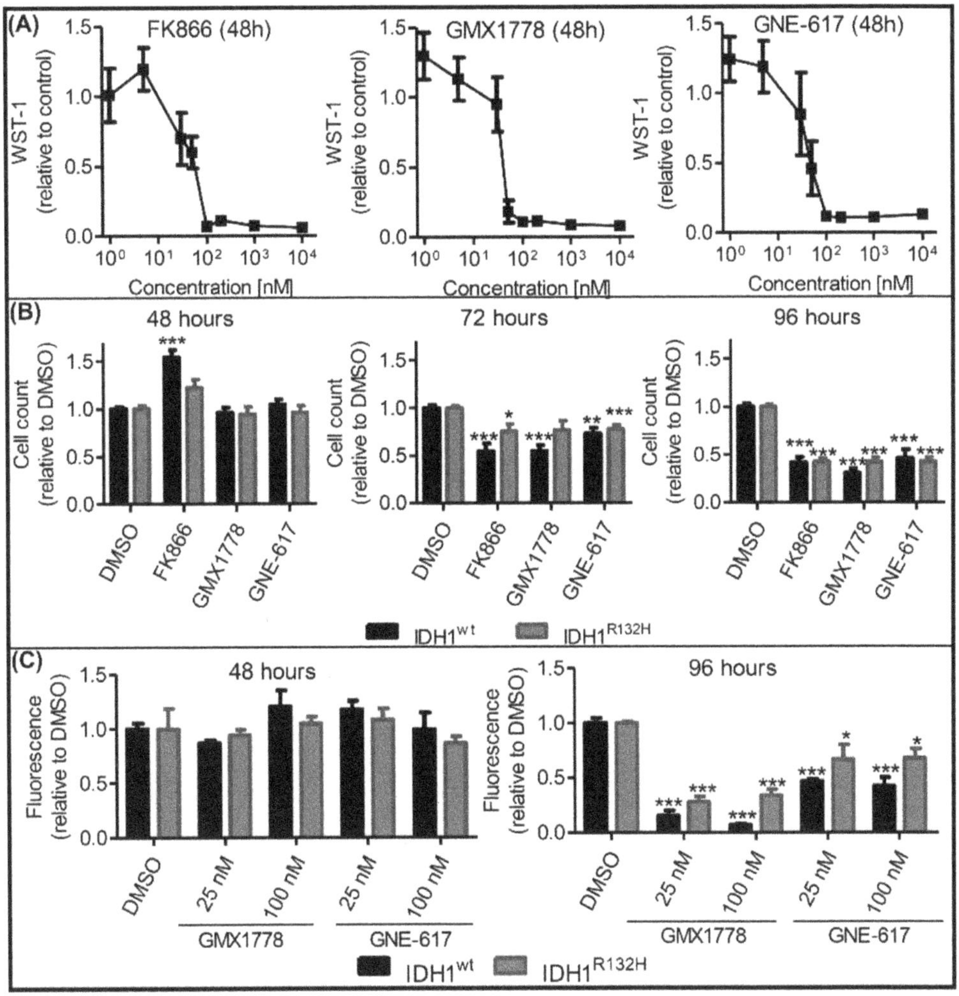

Figure 5. Effects of NAMPT inhibitors on cell viability were independent of IDH1 status. (**A**) Metabolic activity of IDH1^{wt} cells (HT7606) after treatment with different concentrations of the NAMPT inhibitors FK866, GMX1778 and GNE-617 for 48 h. Data are presented as a ratio to treatment with DMSO (nb = 3). (**B**) Cell count after treatment with 100 nM of FK866, GMX1778 and GNE-617 for 48, 72 or 96 h. Data are presented as a ratio to treatment with DMSO (nb = 3 per group). (**C**) Cell viability after treatment with 25 nM or 100 nM of GMX1778 or GNE-617 for 48 or 96 h. Data are presented as a ratio to treatment with DMSO (nb = 2 per group). Each condition was normalized to the cell count of each sample (nb = 2 per group) (nt = 3; * $p \leq 0.05$, ** $p \leq 0.01$, *** $p \leq 0.001$ compared with DMSO unless indicated otherwise).

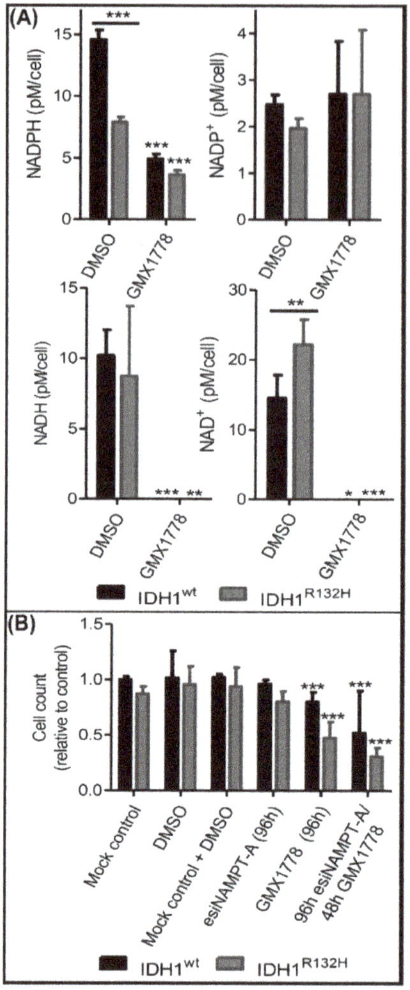

Figure 6. Combinational esiNAMPT knockdown and GMX1778 treatment. (**A**) NADH and NADPH levels after 48 h of treatment with DMSO or 25 nM GMX1778. Each condition was normalized to the cell count of each sample (nb = 2 per group) (* $p \leq 0.05$, ** $p \leq 0.01$, *** $p \leq 0.001$ compared with DMSO unless indicated otherwise). (**B**) The cells were treated for 96 h with a mock control, DMSO, mock control and DMSO, esiNAMPT-A or 25 nM GMX1778. The esiNAMPT-A + GMX1778 treatment was executed by treating cells for 48 h with 25 nM GMX1778 after 48 h of esiNAMPT-A pretreatment (nb = 2 per group). All data present the ratio to control cells in a medium without any treatment (not shown) (nt = 3; * $p \leq 0.05$, *** $p \leq 0.001$ compared with the mock control).

3. Discussion

IDH1 mutations were previously shown to alter redox metabolism in glioma cells. Besides differences in NADPH levels that may be directly attributed to the neoenzymatic function of mutated IDH1, changes in NAD^+ levels were also described [17,19]. Here, we confirmed the alteration of NADPH and NADH levels in a newly created $IDH1^{R132H}$ mutated glioma cell model. The silencing of various enzymes involved in NAD^+ synthesis and phosphorylation revealed a striking susceptibility of $IDH1^{R132H}$ cells to those treatments compared with $IDH1^{wt}$ cells, confirming increased dependence on NADPH synthesis. In contrast, small-molecule inhibitors of NAMPT—a central NAD^+ synthesis enzyme—could not reproduce the distinction between $IDH1^{R132H}$ cells and $IDH1^{wt}$ cells, indicating different responses of the cells to NAMPT depletion on mRNA and pharmacological inhibition.

We found that the $NADPH/NADP^+$ ratios were significantly decreased in $IDH1^{R132H}$ cells compared with $IDH1^{wt}$ cells, which was attributable to lower NADPH levels in those cells. This observation was in line with previous findings of decreased $NADP^+$-dependent IDH activity [18,33] and confirmed observations from us and others of decreased $NADPH/NADP^+$ ratios in stably transduced $IDH1^{R132H}$ glioma cells [7,16,17]. As $NADP^+$-dependent wild-type IDH is the main generator of NADPH in glioblastoma [18], the loss of NADPH was most likely a direct consequence of the NADPH-consuming neoenzymatic activity of mutated IDH1 [8]. Intriguingly, the $NADH/NAD^+$ ratios were also decreased in our $IDH1^{R132H}$ cells due to increased NAD^+ levels. We hypothesize that the higher NAD^+ levels are the result of the compensatory upregulation of NAD^+ synthesis as a reaction to the above described NADPH loss. The NADPH pool, which is instrumental as an antioxidant for ROS scavenging, can be replenished via phosphorylation of NAD^+ via NADK to $NADP^+$ and a subsequent reduction to NADPH.

Treatment with AGI-5198 [34], a selective inhibitor of mutant IDH1, normalized both the $NADPH/NADP^+$ and $NADH/NAD^+$ ratios in $IDH1^{R132H}$ to the level of $IDH1^{wt}$ cells. Previous studies showed increased NADPH levels in $IDH1^{R132H}$ cells after AGI-5198 treatment [35]; however, NADH and NAD^+ levels were not investigated. The normalization of the $NADPH/NADP^+$ ratio could be explained by the restoration of NADPH production capacity of $IDH1^{R132H}$ cells after inhibition of $IDH1^{R132H}$ [33]. Consequently, the proposed compensatory maintenance of high NAD^+ levels would be expendable. Here, we show for the first time that $IDH1^{R132H}$ inhibition indeed also normalized $NADH/NAD^+$ ratios, which underlined the linkage of NAD^+ and $NADP^+$ homeostasis, hence supporting our hypothesis of compensatory upregulation of NAD^+ synthesis in $IDH1^{R132H}$ cells.

There are four known NAD^+ synthesis pathways in mammalian cells starting from the substrates nicotinamide, nicotinamide riboside, nicotinic acid and tryptophan [36]. Previous analysis of key enzymes of those pathways identified NAMPT as the only one expressed in all investigated glioma cell lines and patient-derived glioma cell models [17]. Along with the reported overexpression of NAMPT in glioma cells, as well as its proposed association with oncogenic effects and poor prognosis in glioma [37,38], this made NAMPT an intriguing target. We found that $IDH1^{R132H}$ leads to the upregulation of NAMPT expression on mRNA level in our glioma cell model, in line with NAD^+ synthesis upregulation in those cells. However, the NAMPT protein levels were similar in the $IDH1^{R132H}$ and $IDH1^{wt}$ cells. These results confirmed our previous findings of lower NAMPT protein levels in $IDH1^{R132H}$ cells compared with $IDH1^{wt}$ cells in vitro and in vivo, as well as a discrepancy between the NAMPT mRNA and protein levels [17]. This indicated distinct post-transcriptional regulation of NAMPT expression in $IDH1^{R132H}$ and IDH^{wt} cells. Apart from the well-described effects of $IDH1^{R132H}$-produced 2-HG on DNA methylation, the oncometabolite also seems to impact the regulation of mRNA translation [39]. However, those mechanisms and their clinical implications remain to be investigated.

The knockdown of NAMPT expression led to reduced metabolic activity (shown by the WST1 level) independently of the IDH1 status, but only in $IDH1^{R132H}$ cells it also resulted in reduced viability. The reduced metabolic activity in both $IDH1^{R132H}$ and $IDH1^{wt}$ cells could be explained as a direct consequence of the deficiency of NAD^+ and its derivatives following the loss of the NAD^+ salvage pathway from nicotinamide. The observed $IDH1^{R132H}$ cell-specific reduction in cell viability after the NAMPT knockdown therefore indicated an enhanced dependence on NAD^+ synthesis in $IDH1^{R132H}$ cells compared with $IDH1^{wt}$. This phenomenon might be explained by lower basal NADPH levels in $IDH1^{R132H}$ cells, rendering them more vulnerable to the depletion of NAD^+, which, according to our hypothesis, is used for NADPH replenishment. Unexpectedly, NAMPT knockdown only slightly reduced the NAD^+ level in $IDH1^{R132H}$ cells directly but instead reduced the NADH level in $IDH1^{R132H}$ cells significantly. According to our observations, the $IDH1^{R132H}$ cells might unavoidably replenish the NADPH level via phosphorylation of NADH by NADK2. Therefore, a lack in NAD^+ regeneration could result in diminished NADH level marking NADK2 as another therapy target.

NADPH synthesis from NAD^+ requires the enzyme NADK for phosphorylation of NAD^+ to $NADP^+$. If NAD^+ is indeed the precursor for the regeneration of scarce NADPH in $IDH1^{R132H}$ cells, one would expect an increased dependence of those cells, not only on NAMPT but also on NADK, as well as other NAD^+ synthesis enzymes. We previously found that NAD^+ restoration in glioblastoma cells utilizes the salvage pathways via NAMPT and NMRK1, while nicotinic acid phosphoribosyltransferase (NAPRT) and de novo synthesis from tryptophan via quinolinic acid phosphoribosyltransferase (QPRT) are not involved [17]. Therefore, to examine the dependence on NAD^+ synthesis and phosphorylation, we expanded our investigation to NADK, NMRK1 and NMNAT1—the enzyme downstream of NAMPT, NMRK1, NAPRT and QPRT—in addition to NAMPT. The selective vulnerability of $IDH1^{R132H}$ cells to knockdown of NAMPT and NMRK1 indicated that these cells used both pathways to synthesize NAD^+, making both enzymes possible targets for the selective treatment of IDH1 mutated tumor cells. However, NMRK1 expression varied substantially between different glioma cells [17], thus limiting the clinical practicality of NMRK1-inhibiting approaches. The cytotoxicity of NMNAT1 knockdown was not only seen in $IDH1^{R132H}$ cells but also, to a smaller extent, in $IDH1^{wt}$ cells. This observation, as well as the greater effect of NMNAT1 knockdown on $IDH1^{R132H}$ cells compared to NAMPT or NMRK1 knockdown, seems plausible considering the involvement of NMNAT1 in all available NAD^+ synthesis pathways. $IDH1^{wt}$ cells seem to be able to compensate for the loss of only one pathway, making NMNAT1 a less attractive target for selective therapeutic approaches. Furthermore, we found a selective susceptibility of $IDH1^{R132H}$ cells to NADK knockdown, confirming the increased dependence of those cells on NADPH regeneration via NAD^+ phosphorylation.

As a possible clinical approach to make use of the increased dependence of $IDH1^{R132H}$ cells on NAD^+ synthesis and regeneration, we examined small molecular NAMPT inhibitors FK866, GMX1778 and GNE-617, which were shown to induce cell death in a variety of tumor cells and have, in part, completed phase I trials [40]. The decrease in metabolic activity we found after treatment with these agents complies with our data after esiRNA-mediated NAMPT inhibition, as well as reports of decreased levels of NAD^+, NADH, $NADP^+$ and NADPH in glioma cells with and without IDH1 mutation [19,41–44]. Other studies revealed that the reduction in metabolic activity preceded the cytotoxic effects of NAMPT inhibitors by a few hours [22,45,46]. Accordingly, we found that cytotoxic effects did not arise until 24 h after metabolic impairment in our cells. Contrary to the NAMPT esiRNA treatment, those effects were not limited to $IDH1^{R132H}$ cells but occurred to a similar extent in $IDH1^{wt}$ cells. A possible explanation could be the off-target effects of the NAMPT inhibitors, yet several studies demonstrated their specific activity toward NAMPT inhibition [22,45]. Using a combinatorial treatment of NAMPT knockdown and subsequent GMX1778 addition, we confirmed the cytotoxic effect of this inhibitor in the absence of NAMPT expression, confirming that GMX1778 induced off-target mediated cytotoxicity. Hasmann et al. concluded that FK866 has very low nonspecific cytotoxicity by revealing a lack of acceleration of apoptosis induction when using 100-fold IC50 concentrations of FK866. Furthermore, they ascribed the NAD^+ depleting effect of FK866 to the inhibition of NAMPT by showing that NAD^+ synthesis from nicotinic acid was not impaired by FK866. Subsequently, a reduction in NAMPT activity resulting from FK866 treatment was confirmed via measurement of the radioactive nicotinamide mononucleotide formed from the ^{14}C-labeled substrate nicotinamide [45]. Watson et al. found NAD^+ to be the metabolite that was most profoundly changed in cells exposed to GMX1778. They also showed that GMX1778 had no effect on the NAD^+ synthesis from nicotinic acid and identified GMX1778 as an inhibitor of NAMPT by measuring NAMPT enzyme activity and the binding affinity of NAMPT for GMX1778 [22]. In conclusion, both studies demonstrated the specificity of FK866 or GMX17 toward NAMPT inhibition regarding the NAD^+-depleting effects of those inhibitors. However, possible off-target effects concerning other metabolic pathways have not been investigated thoroughly.

The described discrepancy between NAMPT mRNA and protein expression might also contribute to the different responses to NAMPT knockdown on the mRNA level and NAMPT protein inhibition. Glioma cells overexpressing NAMPT were shown to be more sensitive to its inhibition [38]. Our NAMPT mRNA expression and esiRNA knockdown data revealed a similar correlation. However, NAMPT protein levels affected by possible post-transcriptional regulation of NAMPT expression in IDH1^{R132H} cells could result in low on-target effects of the tested inhibitors. This observation might prevent selective cellular impact and a call for concentrations of NAMPT inhibitors at which systemic cellular cytotoxicity predominates and the survival advantage of IDH1^{wt} cells is lost. Accordingly, low concentrations of NAMPT inhibitors did not induce cytotoxicity in any of the cells.

Contrasting with our results, Tateishi et al. described the selective cytotoxicity of the NAMPT inhibitors FK866 and GMX1778 in IDH1 mutant cancer cells [19]. They found that the downregulation of NAPRT, which is the rate-limiting enzyme of another NAD^+ salvage pathway [36], causes a drop in NAD^+ levels in IDH1^{R132H} cells and, thus, a susceptibility of those cells to NAMPT inhibition. Since we previously showed that the cell model we used lacked NAPRT, independent of the IDH1 status [17], this did not explain the observed effects in this study. Our data suggest that the cytotoxic effects of NAMPT inhibitors are, in part, independent of IDH1 status, thus possibly limiting their suitability as a selective treatment option for IDH1 mutated glioma. However, we note that this finding was based on NAMPT inhibitor treatment of different clones of one patient-derived glioblastoma cell line. Glioma arising from different cellular backgrounds may respond differently to NAMPT inhibition according to varying metabolic properties, such as variations in the expression of NAD^+ synthesis enzymes [17]. Therefore, future studies are required to validate our findings regarding pharmacological NAMPT inhibition, ideally in patient-derived IDH mutant glioma models.

In conclusion, our data underline that targeting the NAMPT NAD^+ regeneration pathway as a promising therapeutic option for gliomas with IDH1^{R132H} mutation. However, the method of treatment should be carefully considered since NAMPT inhibition with small molecules might not be effective, depending on the individual molecular background of a tumor. New efforts in therapeutic methods, such as the targeted delivery of siRNA [47,48] or the use of extracellular vesicles as drug delivery systems [49], may allow us to selectively reduce NAMPT expression in IDH1^{R132H} glioma and might be worth investigating in future studies.

4. Materials and Methods

4.1. Cell Culture and Reagents

HT7606 is a previously established primary glioblastoma cell line obtained from a patient who underwent surgery at the Klinik und Poliklinik für Neurochirurgie, University Hospital Carl Gustav Carus, TU Dresden, after informed written consent and with approval of the local ethics committee [31]. The cells and all derived cell lines were cultured in Dulbecco's Modified Eagle Medium containing 4.5 g/L glucose, GlutaMAX™ and pyruvate, which was supplemented with a 10 mM Hepes Buffer, 4× Non-Essential Amino Acids, 100 U/mL penicillin/streptomycin (all from Gibco, Waltham, MA, USA) and 20% fetal bovine serum (Biochrom AG, Berlin, Germany). The cells were cultured in a humidified incubator at 37 °C containing 5% CO_2.

4.2. Genome Editing Using CRISPR/Cas9

The IDH1 c.395G > A point mutation was introduced in the HT7606 cell line using CRISPR/Cas9. Two different protocols were applied in three independent experiments using either the Cas9-plasmid pX458 (#48138 Addgene, Watertown, MA, USA) according to the protocol of Ran et al. 2013 [50] or the Cas9-NLS-tagRFP (Eupheria Biotech, Dresden, Germany) following the manufacturer's protocol. The cells were transfected using Lipofectamine 3000 (Thermo Fisher Scientific, Waltham, MA, USA). The target-specific gRNA 5′-GGGGATCAAGTAAGTCATGT-3′ was designed on the crispr.mit.edu platform. It was used for all experiments and either transfected after ligation into the pX458 plasmid or as RNA molecules. Additionally, a single-strand oligodeoxynucleotide was transfected as a DNA-repair template. Successfully transfected clones were selected using FACS and seeded as single cells. Edited clones were screened with allele-specific PCR using two different sets of primers. In the three experiments, 142, 243 and 88 clones were analyzed. In each of the experiments, the IDH1^{R132H} mutation was found in one clone, resulting in editing efficiencies of 0.4% to 1.1%. IDH1 c.395G > A point mutation validation and gRNA off-target screening were accomplished using Sanger sequencing (primers are listed in Supplementary Table S1, off-target regions are listed in Supplementary Table S2). IDH1^{R132H} expression was confirmed with cDNA Sanger sequencing and Western blotting.

4.3. Quantification of NAD$^+$/NADH and NADP$^+$/NADPH

NAD$^+$, NADH, NADP$^+$ and NADPH levels were quantified using the bioluminescent NAD/NADH-Glo™ Assay and the NADP/NADPH-Glo™ Assay (both Promega, Madison, WI, USA) following the manufacturer's protocol for measuring NAD(P)$^+$ and NAD(P)H individually. Per well, 8.000 cells were seeded in a 96-well plate. After 24 h, either the IDH1^{R132H} small molecule inhibitor AGI-5198 (1 μM; Merck Millipore GmbH, Darmstadt, Germany) or DMSO (0.001%) was added. After 48 h, cells were treated with 0.2 N NaOH with 1% dodecyltrimethylammonium bromide for lysis. Half of the lysed cell sample was then incubated for 15 min at 60 °C for the NAD(P)H measurement. The other half was treated with 0.4 N HCl and then incubated for 15 min at 60 °C for the NAD(P)$^+$ measurement. The samples were again split and incubated with either the NAD/NADH-Glo Detection reagent or the NADP/NADPH-Glo Detection reagent at room temperature in the dark for 35 min. Luminescence was measured with a Mithras LB940 microplate reader. Calculation of the NAD$^+$, NADH, NADP$^+$ and NADPH concentrations was done with a standard curve using the corresponding metabolites.

4.4. RNA Extraction, Protein Extraction and Western Blotting

GeneMATRIX Universal DNA/RNA/Protein Purification Kit (Roboklon, Berlin, Germany) was used to extract RNA and proteins from the same samples following the manufacturer's instructions. The concentration of RNA was determined with a Qubit® RNA HS Assay Kit and a Qubit® 2.0 Fluorometer (Life Technologies, Carlsbad, CA, USA). The extracted proteins were quantified using a Pierce™ BCA Protein Assay Kit (Pierce Biotechnology, Pittsburgh, PA, USA) and an Infinite M200 with Magellan™ Data Analysis Software (TECAN Group AG, Mannedorf, Switzerland). Proteins were separated on NuPAGE 4–12% Bis-Tris gels (Invitrogen, Waltham, MA USA) for 45 min at 200 V (15–30 μg protein per lane) and transferred onto Hybond ECL Membranes (GE Healthcare, Chicago, IL, USA) for 1 h at 30 V using a NuPAGE® Transfer Buffer (Thermo Fisher Scientific, Waltham, MA, USA).

The membranes were washed twice with phosphate-buffered saline with 0.05% Tween 20 and blocked in 5% (*w*/*v*) nonfat milk for 60 min at room temperature before overnight (4 °C) incubation with primary antibodies against NAMPT (P4D5AT, Enzo Life Science, Inc., Farmingdale, NY, USA; 1:1500), IDH1 (ab117976, Abcam, Cambridge, UK; 1:1000), IDH1^{R132H} (DIA-H09, Dianova, Hamburg, Germany; 1:250), nicotinamide riboside kinase (NMRK1; PA5-26654, Thermo Fisher Scientific, Waltham, MA, USA; 1:1000), nicotinamide mononucleotide adenylyltransferase (NMNAT1; HPA059447, Sigma-Aldrich Corp., Burlington, MA, USA; 1:1000) and NAD$^+$ kinase (NADK; HPA048909, Sigma-Aldrich Corp., Burlington, MA, USA; 1:500). Antibodies against glyceraldehyde 3-phosphate dehydrogenase (GAPDH; H86045M, Meridian Life Science, Memphis, TN, USA; 1:2,000,000) and Actin (A2228, Sigma-Aldrich Corp., Burlington, MA, USA; 1:1000) were used as a loading control. All antibodies were diluted in 5% (*w*/*v*) nonfat milk. Human recombinant NAMPT protein (Abnova, Taipei, Taiwan) served as a positive control. Subsequently, the membranes were probed with an anti-mouse secondary antibody (AP127P, Merck Millipore GmbH,

Darmstadt, Germany; 1:10,000) for 1 h at room temperature. The Lumi-LightPLUS Western Blotting Substrate (Roche, Basel, Switzerland) was used for band detection with a Gel iX20 Imager (INTAS Science Imaging Instruments GmbH, Göttingen, Germany). The intensity of the Western blot bands was adjusted to the GAPDH control bands and quantified using the ImageJ freeware (http://rsb.info.nih.gov/ij/index.html, accessed on 25 March 2020).

4.5. Real-Time Quantitative PCR (RT-qPCR)

cDNA was synthesized from RNA extracts using the SuperScript™ VILO cDNA Synthesis Kit (Thermo Fisher Scientific, Waltham, MA, USA). RNA from normal brain tissue (Human adult normal tissue: Brain and Human Adult Normal Tissue 5 Donor Pool: Brain, both from Biochain Institute Inc., Newark, CA, USA) was used as the control. Quantitative analysis was performed with the SYBR™ Green PCR Master Mix (Applied Biosystems, Bedford, MA, USA) and a CFX96 Touch™ Real-Time PCR Detection System (Bio-Rad Laboratories, Hercules, CA, USA) according to the manufacturer's instructions. Primers are listed in Supplementary Table S1. The relative gene expression was calculated via normalization to the reference genes GAPDH and ARF1 according to the comparative Ct method [51].

4.6. esiRNA Knockdown

Endoribonuclease-prepared small interfering RNA (esiRNA) targeting NAMPT (esiNAMPT-A Catalog No. HU-05878-1, esiNAMPT-B Catalog No. esiSEC), nicotinamide mononucleotide adenylyltransferase (esiNMNAT1, Catalog No. HU-01617-1), NMRK1 (esiNMRK1 Catalog No. HU-09786-1) and NAD^+ kinase (esiNADK Catalog No. HU-08503-1) were purchased from Eupheria Biotech GmbH (Dresden, Germany). esiRNA targeting the sea pansy enzyme Renilla luciferase (mock control, Catalog No. RLUC) was used as a non-targeting control. Cells were transfected with esiRNA using Lipofectamine™ 3000 (Thermo Fisher Scientific, Waltham, MA, USA). The components were diluted in Opti-MEM™ Reduced Serum Medium (Gibco, Waltham, MA, USA). Proper concentrations were confirmed in standard transfection efficiency experiments. Knockdown efficiencies were determined by RT-qPCR and Western blot analysis.

4.7. NAMPT Inhibition Using Small-Molecule Inhibitors

For the NAMPT small-molecule inhibitor treatment, cells were seeded in 96-well plates. After 48 h, the cell culture medium was replaced with a medium containing different concentrations of one of the NAMPT small-molecule inhibitors FK866, GMX1778 (both from Sigma-Aldrich Corp., Burlington, MA, USA) and GNE-617 (ApexBio, Houston, TX, USA) which were previously diluted in DMSO (final concentration 0.001%).

4.8. WST-1 Assay

Cells were seeded with 2000 cells per well in transparent 96-well plates and 10 μL of WST-1 reagent (Sigma-Aldrich Corp., Burlington, MA, USA) were added to each well. After incubation for 4 h at 37 °C in 5% CO_2, absorption was measured at 440 nm with 690 nm as a reference wavelength using an Infinite M200 with Magellan™ Data Analysis Software v.7.2 (TECAN Group AG, Mannedorf, Switzerland).

4.9. Combinatorial esiNAMPT and GMX1778 Treatment

Cells were seeded in Corning® 384 well black/clear bottom plates with 500 cells per well. esiNAMPT-A was applied in reverse transfection with Lipofectamine™ RNAiMAX (Thermo Fisher Scientific, Waltham, MA, USA) on the day of cell seeding. GMX1778 was applied 48 h after the esiRNA treatment or 24 h after cell seeding for a GMX1778 single treatment. Cell viability was measured with automated picture analysis as described below.

4.10. Measurement of Cell Viability with CyQUANT® Direct Cell Proliferation Assay

Cell viability was analyzed with the CyQUANT® Direct Cell Proliferation Assay (Molecular Probes Inc., Thermo Fisher Scientific, Waltham, MA, USA) following the manufacturer's instructions. Fluorescence was measured at an excitation wavelength of 480 nm and emission wavelength of 535 nm using an Infinite M200 with Magellan™ Data Analysis Software (TECAN Group AG, Mannedorf, Switzerland).

4.11. Measurement of Cell Viability with Automated Picture Analysis

A viability assay after the described treatments was performed in black 384-well plates (Greiner, Frickenhausen, Germany). Cells were stained with Hoechst 33,342 and propidium iodide (Invitrogen™, Waltham, MA, USA). Plates were incubated at 37 °C for 10 min and then analyzed on an Operetta High-Content Imaging System (PerkinElmer, Waltham, MA, USA). Pictures were evaluated using Harmony™ software (PerkinElmer, Waltham, MA, USA). Living cells were calculated by subtracting the number of propidium iodide stained cells from the number of Hoechst-33,342-stained cells in each well.

4.12. Statistics

Statistical analysis was performed using GraphPad Prism 5 (GraphPad Software, Inc., San Diego, CA, USA). All data are presented as mean ± SEM. The number of cell lines (biological replicates, nb) and technical replicates (nt) for specific experiments are indicated in the figure captions. Single groups were compared using an unpaired t-test. Multiple groups were analyzed with one-way analysis of variance (ANOVA) followed by Dunnett's post hoc test. Any p-values < 0.05 were considered statistically significant.

Supplementary Materials: The following supporting information can be downloaded at: https://www.mdpi.com/article/10.3390/ijms23105787/s1.

Author Contributions: Conceptualization: M.C., M.P. and B.K.; methodology: S.R., F.B., D.W., K.G. and A.T.; investigation and experiments: M.C., M.P., J.B. and S.R.; data curation and formal analysis: M.C., D.W. and M.P.; writing: M.C., D.W. and M.P.; review and editing: M.C., D.W. and B.K.; supervision: F.B., A.T., E.S. and B.K.; project administration: E.S. and B.K.; funding acquisition: M.C., E.S. and B.K. All authors have read and agreed to the published version of the manuscript.

Funding: This work was supported by the Else Kröner-Fresenius-Stiftung (EKFS) (Else Kröner-Promotionskolleg Dresden) and by the Deutsche Krebshilfe (#70112014).

Institutional Review Board Statement: Not applicable.

Informed Consent Statement: Not applicable.

Data Availability Statement: The data used for this study are contained within the article or Supplementary Materials.

Acknowledgments: We are grateful to A. Krüger for providing excellent technical assistance. We thank G. Eisenhofer for their constructive criticism of the manuscript.

Conflicts of Interest: The authors declare no conflict of interest.

References

1. Cohen, A.L.; Colman, H. Glioma biology and molecular markers. *Cancer Res. Treat.* **2015**, *163*, 15–30.
2. Rath, B.H.; Fair, J.M.; Jamal, M.; Camphausen, K.; Tofilon, P.J. Astrocytes Enhance the Invasion Potential of Glioblastoma Stem-Like Cells. *PLoS ONE* **2013**, *8*, e54752. [CrossRef] [PubMed]
3. Tykocki, T.; Eltayeb, M. Ten-year survival in glioblastoma. A systematic review. *J. Clin. Neurosci. Off. J. Neurosurg. Soc. Australas.* **2018**, *54*, 7–13. [CrossRef] [PubMed]
4. Chien, L.-N.; Gittleman, H.; Ostrom, Q.T.; Hung, K.; Sloan, A.E.; Hsieh, Y.; Kruchko, C.; Rogers, L.R.; Wang, Y.G.; Chiou, H.-Y.; et al. Comparative Brain and Central Nervous System Tumor Incidence and Survival between the United States and Taiwan Based on Population-Based Registry. *Front. Public Health* **2016**, *4*, 151. [CrossRef] [PubMed]
5. Cohen, A.L.; Holmen, S.L.; Colman, H. IDH1 and IDH2 mutations in gliomas. *Curr. Neurol. Neurosci. Rep.* **2013**, *13*, 345. [CrossRef] [PubMed]

6. Louis, D.N.; Perry, A.; Reifenberger, G.; von Deimling, A.; Figarella-Branger, D.; Cavenee, W.K.; Ohgaki, H.; Wiestler, O.D.; Kleihues, P.; Ellison, D.W. The 2016 World Health Organization Classification of Tumors of the Central Nervous System: A summary. *Acta Neuropathol.* **2016**, *131*, 803–820. [CrossRef]
7. Yan, H.; Parsons, W.; Jin, G. IDH1 and IDH2 Mutations in Gliomas. *N. Engl. J. Med.* **2009**, *360*, 765–773. [CrossRef] [PubMed]
8. Dang, L.; White, D.W.; Gross, S.; Bennett, B.D.; Bittinger, M.A.; Driggers, E.M.; Fantin, V.R.; Jang, H.G.; Jin, S.; Keenan, M.C.; et al. Cancer-associated IDH1 mutations produce 2-hydroxyglutarate. *Nature* **2009**, *462*, 739–744. [CrossRef]
9. Hartong, D.T.; Dange, M.; McGee, T.L.; Berson, E.L.; Dryja, T.P.; Colman, R.F. Insights from retinitis pigmentosa into the roles of isocitrate dehydrogenases in the Krebs cycle. *Nat. Genet.* **2008**, *40*, 1230–1234. [CrossRef]
10. Waitkus, M.S.; Diplas, B.H.; Yan, H. Isocitrate dehydrogenase mutations in gliomas. *Neuro. Oncol.* **2016**, *18*, 16–26. [CrossRef]
11. Xu, W.; Yang, H.; Liu, Y.; Yang, Y.; Wang, P.; Kim, S.-H.; Ito, S.; Yang, C.; Wang, P.; Xiao, M.-T.; et al. Oncometabolite 2-Hydroxyglutarate Is a Competitive Inhibitor of α-Ketoglutarate-Dependent Dioxygenases. *Cancer Cell* **2011**, *19*, 17–30. [CrossRef] [PubMed]
12. Figueroa, M.E.; Wahab, O.A.; Lu, C.; Ward, P.S.; Patel, J.; Shih, A.; Li, Y.; Bhagwat, N.; Vasanthakumar, A.; Fernandez, H.F.; et al. Leukemic IDH1 and IDH2 mutations result in a hypermethylation phenotype, disrupt TET2 function, and impair hematopoietic differentiation. *Cancer Cell* **2010**, *18*, 553–567. [CrossRef] [PubMed]
13. Koivunen, P.; Lee, S.; Duncan, C.G.; Lopez, G.; Lu, G.; Ramkissoon, S.; Losman, J.A.; Joensuu, P.; Bergmann, U.; Gross, S.; et al. Transformation by the R Enantiomer of 2-Hydroxyglutarate Linked to EglN Activation. *Nature* **2013**, *483*, 484–488. [CrossRef]
14. Shi, J.; Sun, B.; Shi, W.; Zuo, H.; Cui, D.; Ni, L.; Chen, J. Decreasing GSH and increasing ROS in chemosensitivity gliomas with IDH1 mutation. *Tumour Biol.* **2015**, *36*, 655–662. [CrossRef] [PubMed]
15. Ohka, F.; Ito, M.; Ranjit, M.; Senga, T.; Motomura, A.; Motomura, K.; Saito, K.; Kato, K.; Kato, Y.; Wakabayashi, T.; et al. Quantitative metabolome analysis profiles activation of glutaminolysis in glioma with IDH1 mutation. *Tumour Biol.* **2014**, *35*, 5911–5920. [CrossRef] [PubMed]
16. Fack, F.; Tardito, S.; Hochart, G.; Oudin, A.; Zheng, L.; Fritah, S.; Golebiewska, A.; Nazarov, P.; Bernard, A.; Hau, A.; et al. Altered metabolic landscape in IDH-mutant gliomas affects phospholipid, energy, and oxidative stress pathways. *EMBO Mol. Med.* **2017**, *9*, 1681–1695. [CrossRef]
17. Biedermann, J.; Preussler, M.; Conde, M.; Peitzsch, M.; Richter, S.; Wiedemuth, R.; Abou-El-Ardat, K.; Krüger, A.; Meinhardt, M.; Schackert, G.; et al. Mutant IDH1 Differently Affects Redox State and Metabolism in Glial Cells of Normal and Tumor Origin. *Cancers* **2019**, *11*, 2028. [CrossRef]
18. Bleeker, F.E.; Atai, N.A.; Lamba, S.; Jonker, A.; Rijkeboer, D.; Bosch, K.S.; Tigchelaar, W.; Troost, D.; Vandertop, W.P.; Bardelli, A.; et al. The prognostic IDH1R132mutation is associated with reduced NADP+-dependent IDH activity in glioblastoma. *Acta Neuropathol.* **2010**, *119*, 487–494. [CrossRef]
19. Tateishi, K.; Wakimoto, H.; Iafrate, A.J.; Tanaka, S.; Loebel, F.; Lelic, N.; Wiederschain, D.; Bedel, O.; Deng, G.; Zhang, B.; et al. Extreme Vulnerability of IDH1 Mutant Cancers to NAD+ Depletion. *Cancer Cell* **2015**, *28*, 773–784. [CrossRef]
20. Revollo, J.R.; Grimm, A.A.; Imai, S. The regulation of nicotinamide adenine dinucleotide biosynthesis by Nampt/PBEF/visfatin in mammals. *Curr. Opin. Gastroenterol.* **2007**, *23*, 164–170. [CrossRef]
21. O'Brien, T.; Oeh, J.; Xiao, Y.; Liang, X.; Vanderbilt, A.; Qin, A.; Yang, L.; Lee, L.B.; Ly, J.; Cosino, E.; et al. Supplementation of nicotinic acid with NAMPT inhibitors results in loss of in vivo efficacy in NAPRT1-deficient tumor models. *Neoplasia* **2013**, *15*, 1314–1329. [CrossRef] [PubMed]
22. Watson, M.; Roulston, A.; Belec, L.; Billot, X.; Marcellus, R.; Bédard, D.; Bernier, C.; Branchaud, S.; Chan, H.; Dairi, K.; et al. The small molecule GMX1778 is a potent inhibitor of NAD+ biosynthesis: Strategy for enhanced therapy in nicotinic acid phosphoribosyltransferase 1-deficient tumors. *Mol. Cell. Biol.* **2009**, *29*, 5872–5888. [CrossRef] [PubMed]
23. Xiao, Y.; Elkins, K.; Durieux, J.K.; Lee, L.; Oeh, J.; Yang, L.X.; Liang, X.; DelNagro, C.; Tremayne, J.; Kwong, M.; et al. Dependence of Tumor Cell Lines and Patient-Derived Tumors on the NAD Salvage Pathway Renders Them Sensitive to NAMPT Inhibition with GNE-618. *Neoplasia* **2013**, *15*, 1151–1160. [CrossRef] [PubMed]
24. Reddy, P.S.; Umesh, S.; Thota, B.; Tandon, A.; Pandey, P.; Hegde, A.S.; Balasubramaniam, A.; Chandramouli, B.A.; Santosh, V.; Rao, M.R.S.; et al. PBEF1/NAmPRTase/Visfatin: A potential malignant astrocytoma/glioblastoma serum marker with prognostic value. *Cancer Biol. Ther.* **2008**, *7*, 663–668. [CrossRef] [PubMed]
25. Shackelford, R.E.; Mayhall, K.; Maxwell, N.M.; Kandil, E.; Coppola, D. Nicotinamide Phosphoribosyltransferase in Malignancy: A Review. *Genes Cancer* **2013**, *4*, 447–456. [CrossRef]
26. Sampath, D.; Zabka, T.S.; Misner, D.L.; O'Brien, T.; Dragovich, P.S. Inhibition of nicotinamide phosphoribosyltransferase (NAMPT) as a therapeutic strategy in cancer. *Pharmacol. Ther.* **2015**, *151*, 16–31. [CrossRef]
27. Lenting, K.; Verhaak, R.; Ter Laan, M.; Wesseling, P.; Leenders, W. Glioma: Experimental models and reality. *Acta Neuropathol.* **2017**, *133*, 263–282. [CrossRef]
28. Piaskowski, S.; Bienkowski, M.; Stoczynska-Fidelus, E.; Stawski, R.; Sieruta, M.; Szybka, M.; Papierz, W.; Wolanczyk, M.; Jaskolski, D.J.; Liberski, P.P.; et al. Glioma cells showing IDH1 mutation cannot be propagated in standard cell culture conditions. *Br. J. Cancer* **2011**, *104*, 968–970. [CrossRef]
29. Golebiewska, A.; Hau, A.C.; Oudin, A.; Stieber, D.; Yabo, Y.A.; Baus, V.; Barthelemy, V.; Klein, E.; Bougnaud, S.; Keunen, O.; et al. Patient-derived organoids and orthotopic xenografts of primary and recurrent gliomas represent relevant patient avatars for precision oncology. *Acta Neuropathol.* **2020**, *140*, 919–949. [CrossRef]

30. Klink, B.; Miletic, H.; Stieber, D.; Huszthy, P.C.; Valenzuela, J.A.C.; Balss, J.; Wang, J.; Schubert, M.; Sakariassen, P.Ø.; Sundstrøm, T.; et al. A Novel, Diffusely Infiltrative Xenograft Model of Human Anaplastic Oligodendroglioma with Mutations in FUBP1, CIC, and IDH1. *PLoS ONE* **2013**, *8*, e59773. [CrossRef]
31. Hendruschk, S.; Wiedemuth, R.; Aigner, A.; Töpfer, K.; Cartellieri, M.; Martin, D.; Kirsch, M.; Ikonomidou, C.; Schackert, G.; Temme, A. RNA interference targeting survivin exerts antitumoral effects in vitro and in established glioma xenografts in vivo. *Neuro. Oncol.* **2011**, *13*, 1074–1089. [CrossRef] [PubMed]
32. Zhu, H.; Zhang, Y.; Chen, J.; Qiu, J.; Huang, K.; Wu, M.; Xia, C. IDH1 R132H Mutation Enhances Cell Migration by Activating AKT-mTOR Signaling Pathway, but Sensitizes Cells to 5-FU Treatment as NADPH and GSH Are Reduced. *PLoS ONE* **2017**, *12*, e0169038. [CrossRef] [PubMed]
33. Molenaar, R.J.; Botman, D.; Smits, M.A.; Hira, V.V.; van Lith, S.A.; Stap, J.; Henneman, P.; Khurshed, M.; Lenting, K.; Mul, A.N.; et al. Radioprotection of IDH1-mutated cancer cells by the IDH1-mutant inhibitor AGI-5198. *Cancer Res.* **2015**, *75*, 4790–4802. [CrossRef] [PubMed]
34. Garrett-Bakelman, F.E.; Melnick, A.M. Differentiation therapy for IDH1/2 mutant malignancies. *Cell Res.* **2013**, *23*, 975–977. [CrossRef] [PubMed]
35. Tiburcio, P.D.B.; Gillespie, D.L.; Jensen, R.L.; Huang, L.E. Extracellular glutamate and IDH1(R132H) inhibitor promote glioma growth by boosting redox potential. *J. Neurooncol.* **2020**, *146*, 427–437. [CrossRef]
36. Chiarugi, A.; Dölle, C.; Felici, R.; Ziegler, M. The NAD metabolome—A key determinant of cancer cell biology. *Nat. Rev. Cancer* **2012**, *12*, 741–752. [CrossRef]
37. Gujar, A.D.; Le, S.; Mao, D.D.; Dadey, D.Y.; Turski, A.; Sasaki, Y.; Aum, D.; Luo, Y.; Dahiya, S.; Yuan, L.; et al. An NAD+-dependent transcriptional program governs self-renewal and radiation resistance in glioblastoma. *Proc. Natl. Acad. Sci. USA* **2016**, *113*, E8247–E8256. [CrossRef]
38. Lucena-Cacace, A.; Otero-Albiol, D.; Jiménez-García, M.P.; Peinado-Serrano, J.; Carnero, A. NAMPT overexpression induces cancer stemness and defines a novel tumor signature for glioma prognosis. *Oncotarget* **2017**, *8*, 99514–99530. [CrossRef]
39. Karpel-Massler, G.; Ishida, C.T.; Bianchetti, E.; Zhang, Y.; Shu, C.; Tsujiuchi, T.; Banu, M.A.; Garcia, F.; Roth, K.A.; Bruce, J.N.; et al. Induction of synthetic lethality in IDH1-mutated gliomas through inhibition of Bcl-xL. *Nat. Commun.* **2017**, *8*, 1067. [CrossRef]
40. Dalamaga, M.; Christodoulatos, G.S.; Mantzoros, C.S. The role of extracellular and intracellular Nicotinamide phosphoribosyltransferase in cancer: Diagnostic and therapeutic perspectives and challenges. *Metabolism* **2018**, *82*, 72–87. [CrossRef]
41. Bowlby, S.C.; Thomas, M.J.; D'Agostino, R.B.; Kridel, S.J. Nicotinamide Phosphoribosyl Transferase (Nampt) Is Required for De Novo Lipogenesis in Tumor Cells. *PLoS ONE* **2012**, *7*, e40195. [CrossRef] [PubMed]
42. Cerna, D.; Li, H.; Flaherty, S.; Takebe, N.; Coleman, C.N.; Yoo, S.S. Inhibition of nicotinamide phosphoribosyltransferase (NAMPT) activity by small molecule GMX1778 regulates Reactive Oxygen Species (ROS)-mediated cytotoxicity in a p53- and nicotinic acid phosphoribosyltransferase1 (NAPRT1)-dependent manner. *J. Biol. Chem.* **2012**, *287*, 22408–22417. [CrossRef] [PubMed]
43. Feng, J.; Yan, P.-F.; Zhao, H.-Y.; Zhang, F.-C.; Zhao, W.-H.; Feng, M. Inhibitor of Nicotinamide Phosphoribosyltransferase Sensitizes Glioblastoma Cells to Temozolomide via Activating ROS/JNK Signaling Pathway. *BioMed. Res. Int.* **2016**, *2016*, 1450843. [CrossRef] [PubMed]
44. van Horssen, R.; Willemse, M.; Haeger, A.; Attanasio, F.; Guneri, T.; Schwab, A.; Stock, C.M.; Buccione, R.; Fransen, J.A.M.; Wieringa, B. Intracellular NAD(H) levels control motility and invasion of glioma cells. *Cell. Mol. Life Sci.* **2013**, *70*, 2175–2190. [CrossRef]
45. Hasmann, M.; Schemainda, I. FK866, a Highly Specific Noncompetitive Inhibitor of Nicotinamide Phosphoribosyltransferase, Represents a Novel Mechanism for Induction of Tumor Cell Apoptosis. *Cancer Res.* **2003**, *63*, 7436–7442.
46. Tan, B.; Young, D.A.; Lu, Z.H.; Wang, T.; Meier, T.I.; Shepard, R.L.; Roth, K.; Zhai, Y.; Huss, K.; Kuo, M.-S.; et al. Pharmacological inhibition of nicotinamide phosphoribosyltransferase (NAMPT), an enzyme essential for NAD+biosynthesis, in human cancer cells: Metabolic basis and potential clinical implications. *J. Biol. Chem.* **2013**, *288*, 3500–3511. [CrossRef]
47. Lozada-Delgado, E.L.; Grafals-Ruiz, N.; Vivas-Mejía, P.E. RNA interference for glioblastoma therapy: Innovation ladder from the bench to clinical trials. *Life Sci.* **2017**, *188*, 26–36. [CrossRef]
48. Tietze, S.; Schau, I.; Michen, S.; Ennen, F.; Janke, A.; Schackert, G.; Aigner, A.; Appelhans, D.; Temme, A. A Poly(Propyleneimine) Dendrimer-Based Polyplex-System for Single-Chain Antibody-Mediated Targeted Delivery and Cellular Uptake of SiRNA. *Small* **2017**, *13*, 1700072. [CrossRef]
49. Rufino-Ramos, D.; Albuquerque, P.R.; Carmona, V.; Perfeito, R.; Nobre, R.J.; Pereira de Almeida, L. Extracellular vesicles: Novel promising delivery systems for therapy of brain diseases. *J. Contr. Release* **2017**, *262*, 247–258. [CrossRef]
50. Ran, F.A.; Hsu, P.P.D.; Wright, J.; Agarwala, V.; Scott, D.A.; Zhang, F. Genome engineering using the CRISPR-Cas9 system. *Nat. Protoc.* **2013**, *8*, 2281–2308. [CrossRef]
51. Schmittgen, T.D.; Livak, K.J. Analyzing real-time PCR data by the comparative C T method. *Nat. Protoc.* **2008**, *3*, 1101–1108. [CrossRef] [PubMed]

International Journal of
Molecular Sciences

MDPI

Article

Circadian Gene *cry* Controls Tumorigenesis through Modulation of Myc Accumulation in Glioblastoma Cells

Patricia Jarabo *,†, Carmen de Pablo †, Amanda González-Blanco and Sergio Casas-Tintó *

Instituto Cajal, CSIC, 28002 Madrid, Spain; cdpcarreras@gmail.com (C.d.P.); amanda11_4@hotmail.com (A.G.-B.)
* Correspondence: pjarabo@cajal.csic.es (P.J.); scasas@cajal.csic.es (S.C.-T.)
† These authors contributed equally to this work.

Abstract: Glioblastoma (GB) is the most frequent malignant brain tumor among adults and currently there is no effective treatment. This aggressive tumor grows fast and spreads through the brain causing death in 15 months. GB cells display a high mutation rate and generate a heterogeneous population of tumoral cells that are genetically distinct. Thus, the contribution of genes and signaling pathways relevant for GB progression is of great relevance. We used a *Drosophila* model of GB that reproduces the features of human GB and describe the upregulation of the circadian gene *cry* in GB patients and in a *Drosophila* GB model. We studied the contribution of *cry* to the expansion of GB cells and the neurodegeneration and premature death caused by GB, and we determined that *cry* is required for GB progression. Moreover, we determined that the PI3K pathway regulates *cry* expression in GB cells, and in turn, *cry* is necessary and sufficient to promote Myc accumulation in GB. These results contribute to understanding the mechanisms underlying GB malignancy and lethality, and describe a novel role of Cry in GB cells.

Keywords: cancer; neurodegeneration; glioma; *Drosophila*; disease model; PI3K; EGFR; genetics

Citation: Jarabo, P.; de Pablo, C.; González-Blanco, A.; Casas-Tintó, S. Circadian Gene *cry* Controls Tumorigenesis through Modulation of Myc Accumulation in Glioblastoma Cells. *Int. J. Mol. Sci.* **2022**, 23, 2043. https://doi.org/10.3390/ijms23042043

Academic Editors: Nives Pećina-Šlaus and Ivana Jovčevska

Received: 20 January 2022
Accepted: 8 February 2022
Published: 12 February 2022

Publisher's Note: MDPI stays neutral with regard to jurisdictional claims in published maps and institutional affiliations.

1. Introduction

Glioblastoma (GB) is the most common and aggressive type of glioma among all brain tumors, and it accounts for 57.3% of all gliomas [1]. It was classified in 2016 as a WHO grade IV diffuse oligodendroglial and astrocytic brain tumor, but the most recent classification (2021) includes these type of tumors in the "Gliomas, glioneuronal tumors, and neuronal tumors" group, termed as Glioblastoma, IDH-wildtype [2]. Despite current treatments, the median survival of GB patients is 15 months [3], and it is estimated that only 6.8% of patients survive five years after diagnosis [1]. To understand the genetic, molecular and cellular bases of gliomagenesis is fundamental for the development of effective therapies. In terms of histopathology and genetic expression, GB is a very heterogeneous type of tumor, even within the same patient [4]. However, there are common mutations in GB affecting different pathways that show mutual exclusion: the p53 pathway, the Rb pathway and components of the PI3K pathway [5].

Previous studies from our lab used a GB model in *Drosophila*, developed by Read and collaborators in 2009, that recapitulates key aspects of the disease both genetically and phenotypically [6–13]. This model is based on the expression of constitutively active forms of the *epidermal growth factor receptor* ($EGFR^{\lambda}$) and *phosphatidyl inositol 3 kinase* (*PI3K*) catalytic subunit ($dp110^{CAAX}$) (orthologues of EGFR and PI3K catalytic subunit in *Drosophila*, respectively). We used the binary expression system Gal4/UAS [14] to express $EGFR^{\lambda}$ and *PI3K* $dp110^{CAAX}$ specifically in glial cells under the control of *repo-Gal4* driver [6]. The co-activation of EGFR and PI3K signaling pathways in *Drosophila* glial cells reproduces the cascade of signaling events that occurs in GB patients [6]. In consequence, GB cells upregulate *myc* expression, which is essential for tumoral transformation, and the glial tumor cell numbers increase along with the expansion of the glial membrane. As a result,

GB progression causes a reduction in the number of synapses in neighboring neurons and premature death [6,9,15]. Furthermore, EGFR and PI3K pathway co-activation regulates processes such as progression and entry into the cell cycle and protein synthesis [6,7].

c-myc is one of the oncogenes most amplified in human cancer, including GB. About 60%–80% of human GB cases show elevated Myc levels [16]. Myc regulates cell proliferation, transcription, differentiation, apoptosis and cell migration. It is the point where EGFR and PI3K pathways converge; thus, Myc is considered essential for GB transformation [6,16–18]. Furthermore, in vitro and in vivo studies have shown that *myc* inhibition prevents glioma formation, inhibits cell proliferation and survival and even induces disease regression [16,19]. These features are conserved in *Drosophila* [6].

In the recent years, the study of alterations in circadian rhythm genes has emerged in different types of cancer, including GB [20]. Previous reports suggested that circadian rhythm genes play essential roles in different aspects of tumor progression. The central clock organizes the oscillations and rhythmicity of the physiological processes and modulates the expression of genes related to cell proliferation or differentiation, such as cell cycle components [21], proto-oncogenes and tumor suppressors [22].

In mammals, the structure responsible for coordinating circadian behavior throughout the body is the suprachiasmatic nucleus (SCN), located in the anterior region of the hypothalamus and made up of about 50,000 neurons in humans [23]. All the neurons that compose the central clock express the core circadian genes that control the oscillations that organize the cycles of the whole organism in absence of environmental cues. Furthermore, synchronization of the internal clock with light/dark cycles relies on cryptochrome protein (Cry), a blue light photopigment expressed in certain subsets of clock neurons. Cry is a receptor of near-UV/blue light and a regulator of gene expression that belongs to the group of DNA photolyases. It was suggested that the last universal common ancestor (LUCA) had one or several photolyases, supporting the evolutionary conservation of cryptochrome genes [24]. However, the mammalian gene that plays the role of *Drosophila cry* remains unknown. Interestingly, *Drosophila* Cry also acts as the mammalian Cry when expressed in peripheral clocks [25]. Besides, *cry1* expression is androgen responsive, Cry1 regulates DNA repair and the G2/M transition and it is associated with poor outcome in prostate cancer and colorectal cancer.

Regarding GB, studies in patients with primary gliomas show an association between a specific *per1* variant with overall glioma risk. Several circadian genes, including *cry1*, exhibited differential expression in GB samples compared to control brains as described in the literature [26,27], and in human cancer gene expression databases (https://www.proteinatlas.org, accessed on 1 February 2022; https://cancer.sanger.ac.uk/, accessed on 10 January 2022). Besides, the expression of the circadian gene *clk* is significantly enhanced in high-grade gliomas and correlates with tumor progression [28]. Moreover, *per1* and *per2* expression increases the efficacy of radiotherapy also in GB cells [29].

Furthermore, high levels of *cry1* inversely correlate with median survival in GB patients, acting as signal of poor prognosis (http://gepia.cancer-pku.cn/detail.php?gene=CRY1, accessed on 1 February 2022). Still, the functional mechanism of Cry in cancer susceptibility and carcinogenesis remains unsolved.

Different studies show a relationship between Cry and Myc [30]; c-Myc levels decrease in *cry1*/*cry2* null mutant mice [31]. Besides, *cry1* expression is induced by Myc in GB cells in culture [32].

Taking into account the deregulation in the expression of circadian genes in tumor tissues and the pre-established relationship between Cry and *myc*, which is a key player in GB, here we show that *cry* is regulated by PI3K pathway, *cry* expression enhances Myc accumulation in GB cells and it is necessary for GB progression.

2. Results

2.1. Cry Expression in Glioblastoma

To determine if *cry* expression was affected in glioma samples, we extracted RNA from the heads of 7-day-old adult control and glioma flies. Quantitative RT-PCR results (see Table 1 in Materials and Methods) indicate that *cry* mRNA levels are 50 times higher in glioma samples as compared to controls (Figure 1A). This result goes in line with the data retrieved from TCGA-GBM dataset (at http://gliovis.bioinfo.cnio.es/, accessed on 1 February 2022) that indicate a significant increase of *cry1* mRNA levels (RNA-seq) in GB samples, as compared to non-tumor tissue.

Next, to determine if *cry* upregulation occurs in GB cells, we used a specific reporter line that generates a green fluorescent protein (GFP) tagged form of Cry (GFP-Cry) and visualized adult brains in confocal microscopy. The images show the GFP signal (Cry) and glial membrane marked in red with myristoylated red fluorescent protein (mRFP) (Figure 1B–E,B′–E′). The quantification of GFP-Cry and mRFP co-localization is higher in glioma samples than in controls (Figure 1B,C,F) suggesting an accumulation of Cry in glioma cells. This signal is restored to control levels upon *cry* knockdown by means of RNAi expression in glial or glioma cells (Figure 1D–F).

Next, we analyzed human mRNA expression databases for Glioblastoma multiforme (http://gliovis.bioinfo.cnio.es/, accessed on 10 January 2022). The results indicate that *cry* in GB patients is transcriptionally upregulated in primary tumors (Figure 1G) and *cry1* upregulation correlates with worse prognosis (Figure 1H). Moreover, *cry1* is also upregulated in secondary GB (Figure 1I) and correlates with poor prognosis in secondary GB patients (Figure 1J). All together, these results indicate that *cry* is transcriptionally upregulated in GB cells in *Drosophila* and patients and suggest a role in GB malignancy and aggressiveness.

2.2. Cry Mediates GB Progression and Neurodegeneration

To determine the contribution of *cry* to GB progression, we used a previously validated protocol to quantify tumor progression and the associated neurodegeneration in *Drosophila* [7,9,11]. We stained adult control brains and compared them with GB, GB + *cryRNAi* and wt brains expressing *cryRNAi* in glial cells. We used a specific antibody against repo to visualize the nuclei of all glial cells and quantified the number of glial cells in the confocal images (Figure 2A–E). The results indicate that GB samples have a significant increase in the number of glial cells compared to control samples, but this increase depends on *cry* expression (Figure 2A–C,E). Besides, knockdown of *cry* in normal glia does not alter the number of glial cells (Figure 2D,E). In addition, we quantified the volume of glial membrane. We used Imaris software to measure the volume of the red signal that corresponds to a myristoilated form of RFP (mRFP) expressed in glial cells under the control of *repo-Gal4*. The quantification of the volume show a significant expansion of glial membrane in GB compared to control samples, but this increase depends on *cry* expression (Figure 2A′–C′,F). Again, knockdown of *cry* in normal glia does not alter the volume of glial membrane (Figure 2D′,F). These results suggest that *cry* expression is required for GB progression, but not for normal glia development.

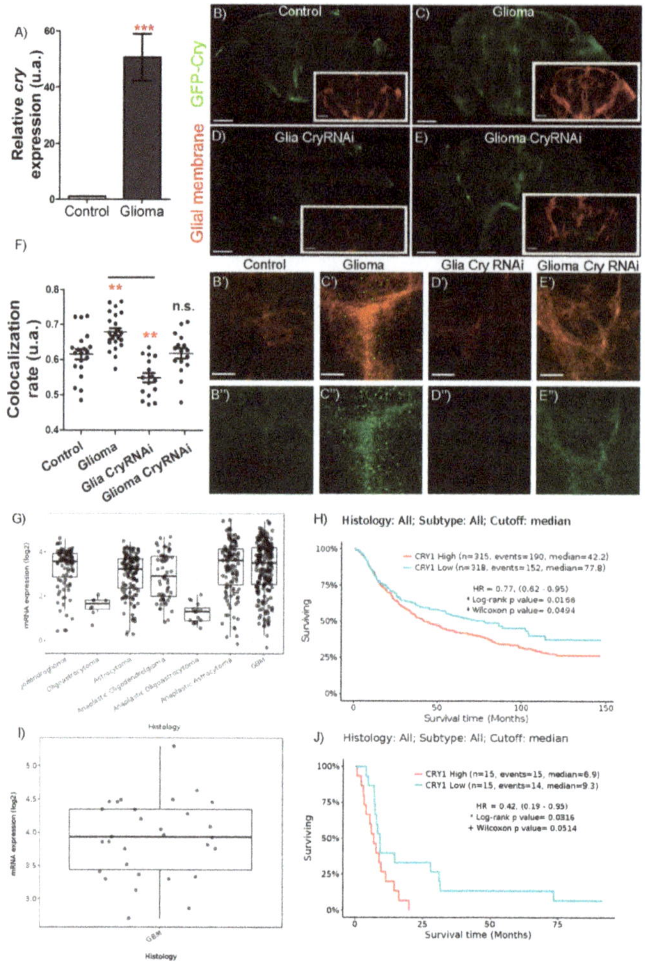

Figure 1. Circadian gene *cry* is upregulated in both human GB samples and GB *Drosophila* model. (**A**) RT-qPCR analysis of complete brains of 7-day-old adult flies from *repo-Gal4 > UAS-LacZ* (Control) and *repo-Gal4 > UAS-dEGFR*$^{\lambda}$, *UAS-dp110* ^{CAAX} (Glioma) genotypes in LD conditions at ZT6 for the circadian gene *cry* (*t*-test) in n = 90. (**B–E**) Confocal microscopy images of brains of 7-day-old adult flies from (**B**) *repo-Gal4 > UAS-LacZ* (Control), (**C**) *repo-Gal4 > UAS-dEGFR*$^{\lambda}$, *UAS-dp110* ^{CAAX} (Glioma), (**D**) *repo-Gal4 > UAS-dEGFR*$^{\lambda}$, *UAS-dp110* ^{CAAX}, *UAS-cryRNAi* (Glioma CryRNAi) and (**E**) *repo-Gal4 > UAS-cryRNAi* (Glia CryRNAi) after using (**B′–E″**) magnifications of the brain lobe central region, the reporter GFP-Cry in green and the glial membrane are marked in red. (**F**) Co-localization between GFP-Cry signal and the glial membrane (mRFP). Statistical analysis in at least n = 16 (ANOVA, post-hoc Bonferroni). (**G**) Data on overexpression of *cry1* in human primary gliomas and GB against normal tissue. (**H**) Graph showing a lower life expectancy in those patients with primary GB and *cry1* overexpressed compared to patients with primary GB with low expression of *cry1*. (**I**) Data on *cry1* expression in human secondary GB compared to normal tissue. (**J**) Graph showing a lower life expectancy in those patients with secondary GB and *cry1* overexpressed compared to patients with secondary GB with low expression of *cry1*. Images obtained from gliovis.bioinfo.cnio.es based on the 2016 classification of brain tumors (scale bar, 100 μm in (**B–E**) and 20 μm in (**B′–E′**) (n.s. not significant, ** p-value < 0.01, *** p-value < 0.001).

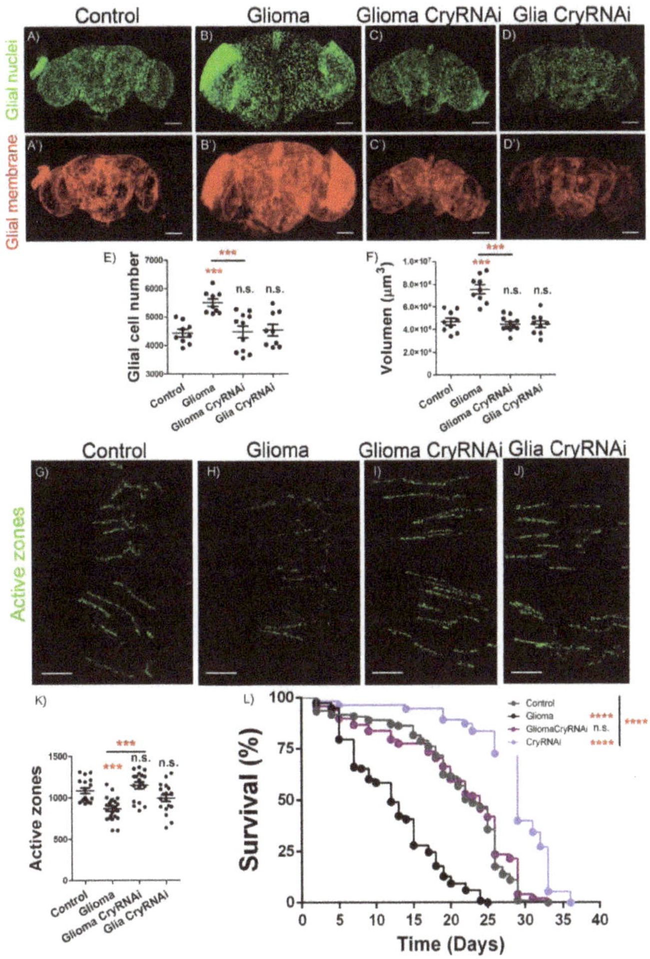

Figure 2. Ectopic downregulation of *cry* prevents GB tumorigenesis and effects. (**A–D**) Confocal microscopy images of brains from 7-day-old adult flies with the following genotypes: (**A**) *repo-Gal4 > UAS-LacZ* (Control), (**B**) *repo-Gal4 > UAS-dEGFR*$^{\lambda}$, *UAS-dp110* ^{CAAX} (Glioma), (**C**) *repo-Gal4 > UAS-dEGFR*$^{\lambda}$, *UAS-dp110* ^{CAAX}, *UAS-cryRNAi* (Glioma CryRNAi) and (**D**) *repo-Gal4 > UAS-cryRNAi* (Glia CryRNAi) with glial nuclei marked in green with anti-repo (scale bar, 100 µm). (**A′–D′**) Glial membrane is shown in red by the expression of mRFP. (**E**) Quantification of glial cells number and (**F**) quantification of glial membrane volume. Statistical analysis for at least $n = 11$ per genotype (ANOVA, post-hoc Bonferroni). (**G–J**) Confocal images of adult NMJ of 7-day-old flies from (**G**) *repo-Gal4 > UAS-LacZ* (Control) (**H**) *repo-Gal4 > UAS-dEGFR*$^{\lambda}$, *UAS-dp110*^{CAAX} (Glioma), (**I**) *repo-Gal4 > UAS-dEGFR*$^{\lambda}$, *UAS-dp110* ^{CAAX}, *UAS-cryRNAi* (Glioma CryRNAi) and (**J**) *repo-Gal4 > UAS-cryRNAi* (Glia CryRNAi) genotypes. Active zones are visualized by nc82 (anti-Brp) antibody and marked in green (scale bar, 25 µm). (**K**) Quantification and statistical analysis of active zones in at least $n = 17$ per genotype (ANOVA, post-hoc Bonferroni). (**L**) Graph shows a survival assay of *repo-Gal4 > UAS-LacZ* (Control, grey), *repo-Gal4 > UAS-dEGFR*$^{\lambda}$, *UAS-dp110*^{CAAX} (Glioma, black), *repo-Gal4 > UAS-dEGFR*$^{\lambda}$, *UAS-cryRNAi,* (Glioma *CryRNAi*, dark green) and *repo-Gal4 > UAS-cryRNAi,* (Glia *CryRNAi*, light green) flies and statistical analysis in $n = 90$ (Mantel-Cox test) (n.s. not significant, *** p-value < 0.001, **** p-value < 0.0001).

Next, we studied the impact of GB progression and *cry* expression in neighboring neurons. We counted the number of synapses in motor neurons of adult neuromuscular junction (NMJ), a standardized tissue to study neurodegeneration [9,11,33]. To visualize synapses, we used an anti-Brp antibody (nc82) to detect active zones in the neurons, and counted the number of synapses in control samples, GB, GB + *cryRNAi* and normal glia + *cryRNAi* (Figure 2G–J). The quantification of synapse number (Figure 2K) shows that GB induction provokes a significant reduction in the number of synapses as compared to control samples, compatible with a neurodegenerative process. This effect was previously described [7,9,11] as a consequence of GB progression. Moreover, *cry* knockdown in GB prevents the reduction in the number of synapses, and *cryRNAi* expression in normal glial cells does not cause any detectable change in the number of synapses. Finally, we aimed to determine the systemic effect of *cry*. We expressed *cryRNAi* in glia or GB cells, and we analyzed the life span of adult flies. The results show that GB causes a significant reduction of life span and a premature death, which is prevented by *cryRNAi* expression in GB cells. Moreover, *cryRNAi* expression in normal glial cells does not reduce lifespan but causes a significant increase in the average lifespan (Figure 2L).

2.3. Signaling Pathway to Control Cry Upregulation

To decipher the specific signaling pathway responsible for *cry* transcriptional activation in GB cells, we analyzed the contribution of the two main pathways activated in this model of GB, EGFR and PI3K. Both pathways converge in the expression of the gene *myc* (see Figure 3A for detailed genetic epistasis in GB). Thus, we analyzed the contribution of *PI3K*, *EGFR* and *myc* to *cry* upregulation. We measured the fluorescent signal of GFP-*cry* reporter in control adult brains (Figure 3B–B") and compared it with adult brains upon expression of the constitutively active forms of *PI3K* (Figure 3C–C") or *EGFR* (Figure 3D–D") in glial cells (under the control of *repo-Gal4*). In addition, we analyzed the GFP-cry signal in glial cells upon *myc* upregulation (Figure 3E–E"). We quantified in the confocal images the signal of GFP that co-localizes with glial membranes (mRFP) (Figure 3F). The results indicate that *PI3K* expression is sufficient to increase *GFP-cry* signal but not *EGFR* or *myc* overexpression. These results suggest that *PI3K* upregulation induces *cry* transcription, and *EGFR* or *myc* expression do not induce *cry* expression in glial cells.

2.4. Cry Regulates Myc Expression in Glial Cells

Next, to determine the epistatic relation between *cry* and *myc*, we analyzed Myc protein accumulation in glial cells upon *cry* expression. First, to analyze if Cry is sufficient to cause an increase in Myc protein levels, we used a specific antibody against Myc and analyzed Myc signal levels upon *cry* overexpression, *myc* overexpression or *cry* + *myc* overexpression in glial cells (Figure 4A–D'). The quantification of Myc surface signal that coincides with glial cells (anti-repo) showed that *cry* expression in glia is sufficient to increase Myc protein signal in glial cells, comparable to *myc* upregulation. In addition, *cry* + *myc* upregulation show a summation effect on the increase of Myc protein levels (Figure 4E). To conclude if *cry* is required for *myc* expression in GB, we quantified glial Myc signal in the control, *cry RNAi*, GB, GB + *cryRNAi* and *cry* upregulation (Figure 4F–J'). The quantifications indicate that *cryRNAi* in glial cells does not reduce the amount of Myc in glial cells (Figure 4K). In addition, GB condition triggers the number of Myc positive glial cells, as well as *cry* upregulation in glial cells (Figure 4K). Finally, *cryRNAi* expression in GB cells prevents the accumulation of Myc in GB cells. Taking all these results together, we conclude that *cry* is sufficient to trigger Myc accumulation in glial cells, and *cry* expression is necessary for Myc accumulation in GB condition.

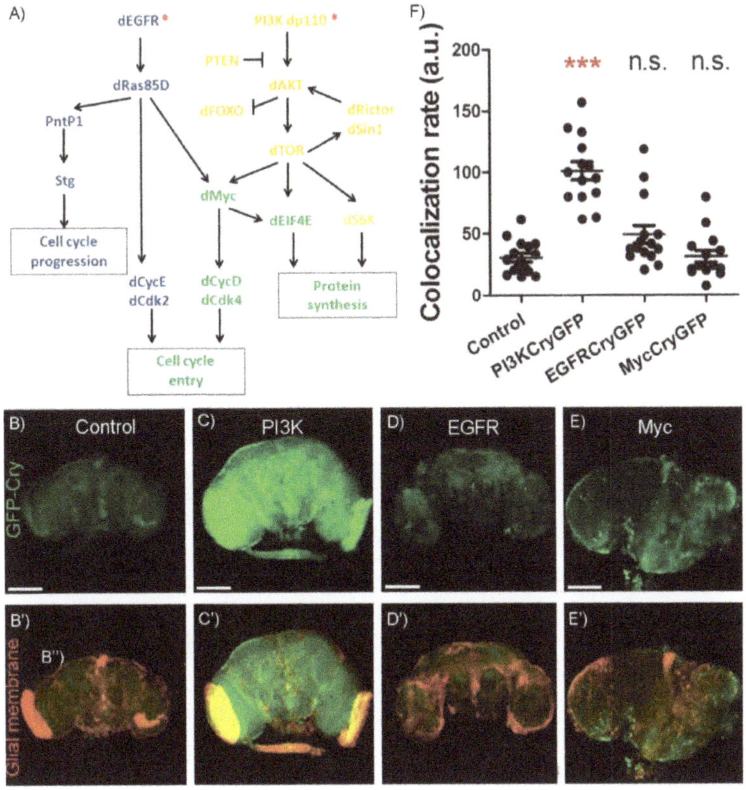

Figure 3. PI3K upregulates the levels of *cry*. (**A**) Scheme of EGFR (blue) and PI3K (yellow) signaling pathways involved in GB tumoral transformation with Myc as convergence point (green) (modified from [6]). (**B–E**) Confocal microscopy images of brains from 7-day-old adult flies with the following genotypes: (**B**) *repo-Gal4 > UAS-LacZ* (Control), (**C**) *repo-Gal4 > UAS-dp110*^{CAAX} (PI3K), (**C**) *repo-Gal4 > UAS-dEGFR*$^{\lambda}$ (EGFR) and (**D**) *repo-Gal4 > UAS-dmyc* (Myc) using the reporter *GFP-cry* visualized in green (scale bar, 100 µm) and **B′–E′**) the glial membrane is marked in red by the expression of mRFP. Green and red signal merge produces the yellow signal. (**F**) Co-localization between *GFP-cry* and the glial membrane (mRFP) and statistical analysis in at least $n = 16$ (ANOVA, post-hoc Bonferroni) (n.s. not significant, * *p*-value < 0.05, *** *p*-value < 0.001).

2.5. Cry Contribution to GB Progression

To investigate the contribution of Cry to glioma progression, we determined the number of glial cells and volume of glial membrane network in control adult brain, GB (*PI3K* + *EGFR*), *PI3K* + *cry*, *EGFR* + *cry* or *myc* + *cry* expressed in glial cells (Figure 5A–E′). The quantification showed that all these genetic combinations cause an increase in the number of glial cells as compared to control brains (Figure 5F). However, only the GB condition provoked an expansion of the glial membrane volume, and the combination of *PI3K* + *cry*, *EGFR* + *cry* or *myc* + *cry* showed a volume of glial membrane comparable to control brains (Figure 5G). To further determine the contribution of *cry* to GB expansion, we analyzed the contribution of single gene upregulation in glial cells for *cry* or *myc*, and the combination of *cry* + *myc* expression (Figure 5H–K′). The quantification of glial cell number showed that *cry* or *myc* expression alone, or in combination, is sufficient to increase the number of glial cells with respect to control samples (Figure 5L). Nevertheless, none of these genetic modifications is sufficient to expand glial membrane volume (Figure 5M).

These results suggest that *cry* or *myc* are sufficient to trigger glial cell number increase in adult brains, but not to expand the volume of glial membrane network.

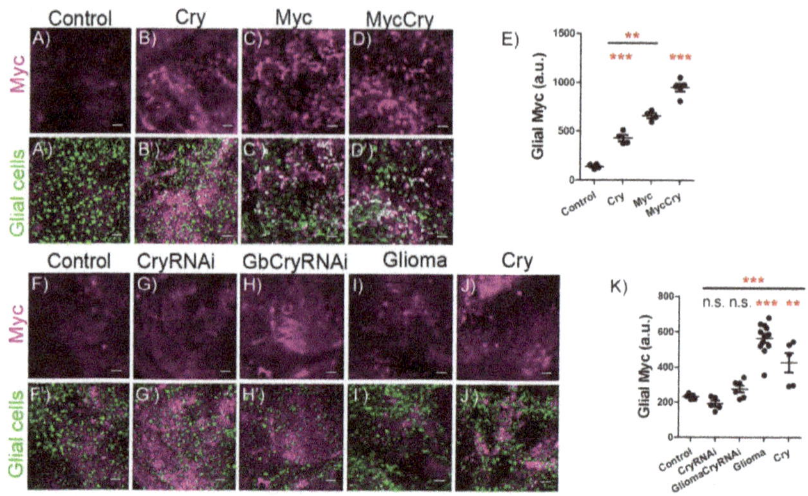

Figure 4. Cry increases glial Myc protein levels in physiological and GB conditions. (**A–D**) Confocal microscopy images of brains from 7-day-old adult flies with the following genotypes: (**A**) *repo-Gal4 > UAS-LacZ* (Control), (**B**) *repo-Gal4 > UAS-cry* (Cry), (**C**) *repo-Gal4 > UAS-dmyc* (Myc) and (**D**) *repo-Gal4 > UAS-dmyc, UAS-cry* (MycCry) with Myc marked in magenta (anti-Myc). (**A'–D'**) Glial nuclei marked in green (anti-Repo) (scale bar, 25 μm). (**E**) Glial Myc quantification and statistical analysis for at least $n = 9$ per genotype (ANOVA, post-hoc Bonferroni). (**F–J**) Confocal microscopy images of brains 7-day-old adult flies from (**F**) *repo-Gal4 > UAS-LacZ* (Control), (**G**) *repo-Gal4 > UAS-cryRNAi* (CryRNAi), (**H**) *repo-Gal4 > UAS-dEGFR*$^{\lambda}$, *UAS-dp110* ^{CAAX}, *UAS-cryRNAi* (Glioma CryRNAi), (**I**) *repo-Gal4 > UAS-dEGFR*$^{\lambda}$, *UAS-dp110* ^{CAAX} (Glioma) and (**J**) *repo-Gal4 > UAS-cry* (Cry); Myc is marked in magenta (anti-Myc) (**F'–J'**) and glial nuclei are marked in green (anti-Repo) (scale bar, 25 μm). (**K**) Glial Myc quantification and statistical analysis for at least $n = 12$ per genotype (ANOVA, post-hoc Bonferroni) (n.s. not significant, ** p-value < 0.01, *** p-value <0.001).

2.6. Cry Upregulation in Glial Cells Causes Synapse Loss and Premature Death

It was previously described that GB progression induces synapse loss, an early symptom of neurodegeneration. To determine the contribution of *cry* to synapse loss, we counted the number of active zones in motor neurons of adult neuromuscular junction in the control, GB (*PI3K* + *EGFR*), *PI3K* + *cry*, *EGFR* + *cry* or *myc* + *cry* samples (Figure 5N–R). The quantification of the number of active zones showed that the expression in glial cells of GB (*PI3K* + *EGFR*), *PI3K* + *cry*, *EGFR* + *cry* or *myc* + *cry* is sufficient to reduce the number of synapses in NMJ neurons (Figure 5S).

Finally, to evaluate the systemic effect of GB and glial expression of *PI3K* + *cry*, *EGFR* + *cry* or *my c* + *cry*, we analyzed the lifespan of adult individuals. The results show that GB causes a premature death, as previously described in *Drosophila* and mice Xenografts [8,9,11], glial upregulation of *EGFR* + *cry* or *myc* + *cry* causes a significant reduction of lifespan but less aggressive than GB, and *PI3K* + *cry* upregulation in glial cells does not reduce lifespan (Figure 5T).

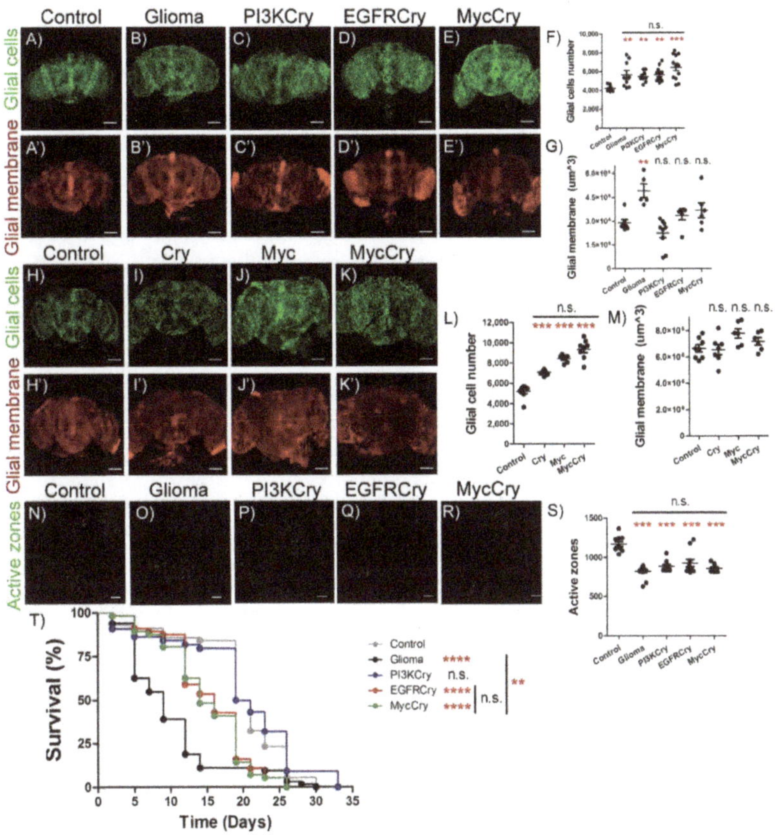

Figure 5. *EGFR-cry* co-expression induces glial cells number increase, synapse number and survival reduction. (**A–E**) Confocal microscopy images of brains from 7-day-old adult flies with the following genotypes: (**A**) *repo-Gal4* > *UAS-LacZ* (Control), (**B**) *repo-Gal4* > *UAS-dEGFR*$^{\lambda}$, *UAS-dp110* ^{CAAX} (Glioma), (**C**) *repo-Gal4* > *UAS-dp110* ^{CAAX}, *UAS-cry* (PI3KCry), (**D**) *repo-Gal4* > *UAS-dEGFR*$^{\lambda}$, *UAS-cry* (EGFRCry) and (**E**) *repo-Gal4* > *UAS-dmyc, UAS-cry* (MycCry) with glial nuclei marked in green (anti-Repo) (scale bar, 100 µm). (**A′–E′**) Glial membrane is visualized in red by the expression. (**F**) Glial cells number and (**G**) glial membrane volume quantification and statistical analysis for at least $n = 12$ per genotype (ANOVA, post-hoc Bonferroni). (**H–K**) Confocal images of adult brains of 7-day-old flies from (**H**) *repo-Gal4* > *UAS-LacZ* (Control) (**I**) *repo-Gal4* > *UAS-cry* (Cry), (**J**) *repo-Gal4* > *UAS-dmyc* (Myc) and (**K**) *repo-Gal4* > *UAS-cry, UAS-dmyc* (MycCry) genotypes with glial nuclei marked in green (anti-Repo) (scale bar, 100 µm) (**H′–K′**) and glial membrane shown in red (mRFP). (**L**) Glial cells number and (**M**) glial membrane volume quantification and statistical analysis for at least $n = 9$ per genotype (ANOVA, post-hoc Bonferroni). (**N–R**) Confocal microscopy images of NMJ of 7-day-old adult flies from (**N**) *repo-Gal4* > *UAS-LacZ* (Control), (**O**) *repo-Gal4* > *UAS-dEGFR*$^{\lambda}$, *UAS-dp110* ^{CAAX} (Glioma), (**P**) *repo-Gal4* > *UAS-dp110* ^{CAAX}, *UAS-cry* (PI3KCry), (**Q**) *repo-Gal4* > *UAS-dEGFR*$^{\lambda}$, *UAS-cry* (EGFRCry) and (**R**) *repo-Gal4* > *UAS-dmyc, UAS-cry* (MycCry). Active zones are marked with anti-Brp (nc-82) visualized in green (nc82, anti-Brp) (scale bar, 25µm). (**S**) Quantification and statistical analysis of active zones in at least $n = 13$ per genotype (ANOVA, post-hoc Bonferroni). (**T**) Graph shows a survival assay of *repo-Gal4* > *UAS-LacZ* (Control, grey), *repo-Gal4* > *UAS-dEGFR*$^{\lambda}$, *UAS-dp110*^{CAAX} (Glioma, black), *repo-Gal4* > *UAS-dp110*^{CAAX}, *UAS-cry* (PI3KCry, blue), *repo-Gal4* > *UAS-dEGFR*$^{\lambda}$, *UAS-cry,* (EGFRCry, red) and *repo-Gal4* > *UAS-cry, UAS-dmyc* (MycCry, green) flies and statistical analysis in $n = 90$ (Mantel-Cox test) (n.s. not significant, ** p-value < 0.01, *** p-value < 0.001, p-value < 0.0001).

3. Discussion

Different studies have established a relation between alterations in circadian rhythm genes and cancer [32,34]. Specifically, one of the genes associated with different types of cancer is *cry* [35–37]. Thus, this study aims to investigate the role of *cry* in a *Drosophila* GB model.

The previous work of Luo et al. 2012 [38] describes a reduction of the number of glial cells positive for *cry1/2* expression in glioma tissue compared to normal tissue. However, the authors show that glioma cells that are positive for Cry1/2 show an increase in the amount of Cry1/2 with respect to non-tumoral tissue. Moreover, both Madden et al. 2014 [26] (with a sample 10 times larger than that of Luo et al. 2012) and Wang et al. 2021 [27] (using data from three different databases) analyzed the expression of circadian genes in glioma tissue compared to healthy tissue and conclude that *cry1* is overexpressed in glioma tissue. We also found this result in other databases such as https://www.proteinatlas.org/ and http://gliovis.bioinfo.cnio.es/, accessed on 10 January 2022, [39], which in turn is compatible with the observations in the *Drosophila* model of GB.

Nonetheless, Fan et al. [40] investigated the role of Cry2 in rat glioma cells and observed that *cry2* mRNA and protein levels showed aberrant rhythmic periodicity of 8 h, compared to 24 h in normal tissue. Thus, futures studies on the contribution of circadian rhythms genes should take into consideration the variations of expression.

On the contrary, Dong et al. [41] state that glioblastoma stem cells (GSCs) displayed robust circadian rhythms dependent on core clock transcription factors. The use of Cry1/2 agonists induced anti-tumor effects suggesting that GSCs are sensitive to *cry1/2* activity. Taken the different conclusions into consideration, most of the literature and our data suggest that *cry* is upregulated in glioma cells and promotes glioma progression; however, the role of *cry* expression in *Drosophila*, or *cry1/2* expression in mammals, may differ according to the glioma subtype, the specific mutations in glioma cells and the cell population of study within the glioma and the hour of the day.

We described an increase in *cry1 mRNA* levels in human GB samples and in a well-studied *Drosophila* model of GB. However, we cannot conclude that GB cells show an increase in *cry* transcription, or an enhancement of *cry* mRNA stability. The *Drosophila* GB model is based on the activation of the two most frequently mutated pathways in GB, PI3K and EGFR, which converge in Myc as a coincidence point. These pathways are of great relevance to promote GB cells expansion, GB progression and, in consequence, the deterioration of neighboring neurons and a premature death. The results indicate that *cry* upregulation in *Drosophila* GB cells depends on *PI3K* expression, and it is required for GB cells number increase and synapse loss (Figure 6). In addition, *cry* expression in glial cells is sufficient to increase the number of glial cells. However, *cry* expression is expendable for normal wt glial growth during development. Taking into consideration that *cry* is upregulated in GB cells and promotes glia cells number increase, we did not observe any contribution to normal glia development, which makes Cry a potential target for GB treatment.

Besides, we show that Cry is necessary and sufficient to induce *myc* expression in GB cells. This agrees with in vitro studies that revealed an increase in Myc levels as a result of *cry* upregulation [32]. Therefore, we propose that *cry* is part of the PI3K-Myc signaling pathway in GB, where *cry* upregulation would be associated with glial cells number increase. However, PI3K is a highly promiscuous enzyme that participates in numerous signaling pathways, and the results suggest that Cry contribution is restricted to the malignant features of GB dependent on *myc*, such as GB cell number increase and neurodegeneration. However, *cry* expression is independent of glial membrane expansion characteristic of GB progression. Besides, *cry* expression in glial cells partially reduces lifespan, but is less aggressive than GB. This result suggests that Cry plays a central role in GB and is required for GB formation, and *cry* mutations might be responsible for several features of GB. The human gene expression databases indicate that *cry1* expression levels correlate negatively with lifespan, and it is associated with a poor prognosis. In conclusion, these results suggest that further studies on the contribution of Cry1 to human GB progression could lead to novel strategies to treat GB patients.

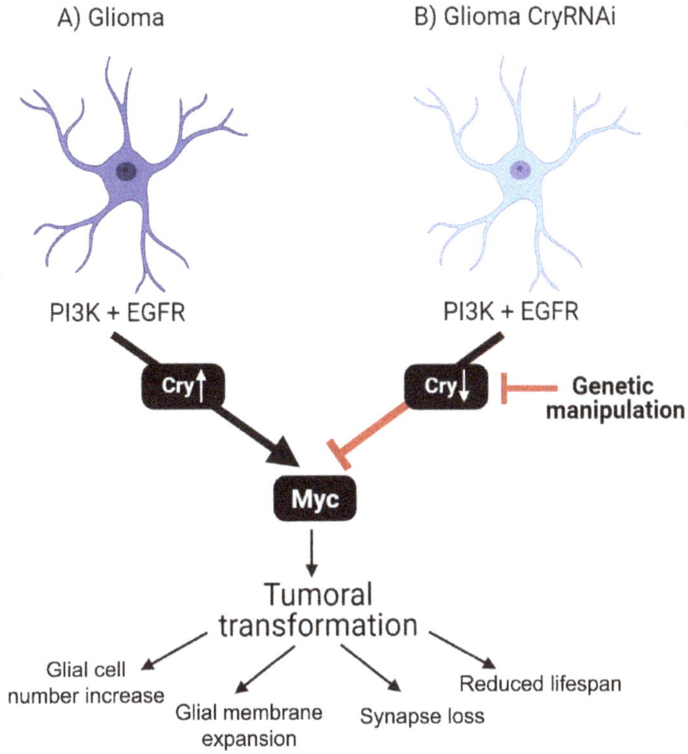

Figure 6. Schematic representation of Cry contribution to GB progression. (**A**) Glia cells are transformed to glioma by the co-activation of PI3K and EGFR pathways. These signals take to the accumulation of Myc in glioma cells mediated by *cry* expression. In consequence, glioma cells number increases, expand glial membrane and cause a reduction in synapse number of neurons (neurodegeneration). This process triggers premature death and reduced lifespan of the fly. (**B**) Glioma cells are transformed by PI3K and EGFR pathways activation, but upon *cry* knockdown (*cry RNAi*), they do not accumulate Myc, do not display cellular features of glioma and prevent lifespan reduction.

Recent publications describe the communication between GB cells and neurons in human GB cells and mice xenografts based on the establishment of electrical and chemical synapses, which are essential for tumor progression [42,43]. A possible explanation for GB prevention by *cry* downregulation arises from the non-circadian function of Cry as a regulator of synapse number through the genetic and physical association with the key presynaptic protein Bruchpilot (Brp) [44,45]. In *Drosophila*, *cry* mutants show reduced *brp* expression levels. Actually, Cry interacts physically with Brp to modulate its stability, and triggers its degradation activated by light. Therefore, it is possible that *cry* overexpression in glial cells promotes the establishment of synapses with neurons. Moreover, the absence of light input impairs Brp degradation in glial cells, thus promoting tumoral progression. In conclusion, further experiments are required to unveil the molecular interactions of Cry and Brp proteins, including the putative formation of abnormal synapses between glial and neurons under the GB condition.

The studies of other groups describe the beneficial effects of haloperidol on *cry1* expression in GB cells, but these results obtained in cell culture suggest that the doses required to treat patients might be toxic; in consequence, specific delivery strategies combined with haloperidol are worth of study. In addition, we observed significant effects of *cry* knockdown in normal glial cells, in line with Bolukbasiet al., who recently described the extension of lifespan by *foxo* upregulation in glial cells [46]. We observed an effect of *cry* upregulation in the number of glial cells (Figure 5L). Given that *cry* and *foxo* respond to PI3K pathway, it is tempting to speculate that *cry* expression is relevant for lifespan extension by PI3K pathway, and associated behaviors such as diet restriction.

The classical definition of Cry as a regulator of circadian rhythms can now be expanded to the biology of glial cells, GB progression and the expansion of lifespan in *Drosophila*. This plethora of different phenotypes associated with one gene is now a common feature previously described for Troponin I [47–49], Caspases [50–52] or even other circadian genes as *per1* [53,54] and contributes to the explanation of the multiple phenotypes observed in patients.

This study describes the epistatic relationship between *PI3K*, *cry* and *myc* and the relevance for GB progression. The strengths of this study rely on the importance of understanding the mechanisms underlying the progression of a fatal tumor as GB, and the reliability of *Drosophila* as an animal model useful to study human disease. However, it is important to take into consideration the limitations of the results to put them in perspective. We used a model based on the activation of PI3K and EGFR pathways that reproduces the key features of human disease progression, but the contribution of additional mutations such as IDH or TP53 in GB require further studies; thus, new models in flies or other animal models of study will contribute to validate and narrow down our findings.

4. Materials and Methods

4.1. Fly Stocks and Genetics

All fly stocks were maintained at 25 °C (unless otherwise specified) on a 12/12 h light/dark cycles at constant humidity in a standard medium. The stocks used from Bloomington Stock Center were *tub-Gal80^{TS}* (BL-7019), *Repo-Gal4* (BL-7415) and *UAS-LacZ* (BL-8529). Other fly stocks used were *UAS-dEGFR$^{\lambda}$*, *UAS-dp110^{CAAX}* (gift from R. Read [6]), *UAS-cry* (gift from F. Royer [55]), *GFP-cry* (BDSC_76317, gift from P.E. Hardin), *UAS-PI3K* (gift from J. Botas [56]), *UAS-cryRNAi* (gift from F. Royer [57]) and UAS-dMyc (gift from E. Moreno [58]).

The stock containing *UAS-cryRNAi* was previously generated and validated [57]; this construct produces a double-stranded RNA that corresponds to the 300–799 region of *cryRA* mRNA.

The glioma-inducing line contains the *UAS-dEGFR$^{\lambda}$* and *UAS-dp110^{CAAX}* transgenes that encode for the constitutively active forms of the human orthologues PI3K and EGFR, respectively [6]. The *Repo-Gal4* line drives the *Gal4* expression to glial cells and precursors [59,60] combined with the *UAS-dEGFR$^{\lambda}$* and *UAS-dp110^{CAAX}* line allow us to generate a glioma thanks to the Gal4 system [14]. To visualize glial or GB cells membrane, we induced the expression of a myristoylated form of red fluorescent protein (UAS-mRFP, described in [9]) under the control of the specific glial promoter *repo-Gal4*.

Gal80^{TS} is a repressor of the Gal4 activity at 18 °C, although at 29 °C is inactivated [61]. The *tub-Gal80^{TS}* construct was used in all the crosses to avoid the lethality caused by the glioma development during the larval stage. The crosses were kept at 17 °C until the adult flies emerged. To inactivate the Gal80^{TS} protein and activate the Gal4/UAS system to allow for the expression of our genes of interest; the adult flies were maintained at 29 °C for 7 days except in the survival assay (flies were at 29 °C until death).

4.2. Immunostaining and Image Acquisition

All tissues were treated in simultaneously for each experiment. Adult brains were dissected and fixed with 4% formaldehyde in phosphate-buffered saline for 20 min, whereas adult NMJ were fixed for 10 min; in both cases, samples were washed 3 × 15 min with PBS + 0.4% triton, blocked for 1 h with PBS + 0.4% triton + BSA 5%, incubated overnight with primary antibodies, washed 3 × 15 min, incubated with secondary antibodies for 2 h and mounted in Vectashield mounting medium with DAPI in the case of the brains. The primary antibodies used were anti-Repo mouse (1/200; DSHB, Iowa City, IA, USA) to recognize glial nuclei, anti-Bruchpilot-nc82-mouse (1/50; DSHB, Iowa City, IA, USA) to recognize the presynaptic protein Bruchpilot, anti-HRP rabbit (1/400; Cell Signaling, Danvers, MA, USA) to recognize neuronal membranes, anti-GFP rabbit (1:500; DSHB, Iowa City, IA, USA) and anti-Myc guinea pig (1/100; DSHB, Iowa City, IA, USA) to recognize the nuclear protein Myc. The secondary antibodies used were anti-mouse, -rabbit or -guinea pig Alexa 488 or 647 (1/500; Life Technologies, Carlsbad, CA, USA). Images were taken by a Leica SP5 confocal microscopy applying same conditions for each experiment.

4.3. qRT-PCR

The mRNA for all samples was extracted from adult brains and processed in parallel. For this, 1- to 4-day-old male adult mice were maintained at 29 °C for 7 days and collected on dry ice at ZT6. Total RNA was extracted by triplicate from 30 heads. RNA was extracted with TRIzol and phenol chloroform. cDNA was synthetized from 1 µg of RNA and cDNA samples from 1:5 dilutions were used for real-time PCR reactions. Transcription levels were determined in a 14 µL volume in duplicate using SYBR Green (Applied Biosystems, Waltham, MA, USA) and 7500 qPCR (Thermo Fisher Scientific, Waltham, MA, USA). We analyzed transcription levels of *cry* using Rp49 as a housekeeping gene reference.

Sequences of primers were as follows.

Table 1. Sequences of primers used to detect *cry* expression. *Rp49* is used as housekeeping gene.

Primer Name	5′-3′ Sequence
Rp49 F	GCATACAGGCCCAAGATCGT
Rp49 R	AACCGATGTTGGGCATCAGA
cry F	TTCTTCCCATCAAAACTGG
cry R	AAACGCATCCGATTGTAACC

After completing each real-time PCR run, with cycling conditions of 95 °C for 10 min, 40 cycles of 95 °C for 15 s and 55 °C for 1 min, and outlier data were analyzed using 7500 software (Applied Biosystems, Waltham, MA, USA). Ct values by triplicate of duplicates from three biological samples were analyzed calculating 2DDCt.

4.4. Survival Assays

Lifespan was determined under 12:12 h LD cycles at 29 °C conditions. Three replicates of 30 1- to 4-day-old male adults were collected in vials containing standard *Drosophila* media and transferred every 2–3 days to fresh *Drosophila* media.

4.5. Quantification

Fluorescent reporter-relative *cry* signals within brains were determined from images taken at the same confocal settings avoiding saturation. For the analysis of co-localization rates, "co-localization" tool from LAS AF Lite software (Leica, Wetzlar, Germany), was used taking the co-localization rate data for the statistics analyzing the co-localization between the green signal (both cases) and signal coming from glial tissue from three slices per brain in similar positions of the *z* axis.

Glial network was marked by a *UAS-myristoylated-RFP* reporter (mRFP) specifically expressed under the control of *repo-Gal4*. The total volume was quantified using the Imaris surface tool (Imaris 6.3.1 software, Oxford Instruments, Abingdon, UK). Glial nuclei were marked by staining with the anti-Repo (DSHB). The number of Repo + cells and number of synapses (anti-nc82; DSHB) were quantified by using the spots tool in Imaris 6.3.1 software (Oxford Instruments, Abingdon, UK). We selected a minimum size and threshold for the spot in the control samples of each experiment: 0.5 µm for active zones and 2 µm for glial cell nuclei. Myc glial signal was quantified using the Imaris surface tool (Imaris 6.3.1 software, Oxford Instruments, Abingdon, UK) creating a mask for the glial nuclei signal and exclusively selecting the myc signal corresponding to glial nuclei. Then we applied the same conditions to the analysis of the corresponding experimental sample.

4.6. Statistics

The results were analyzed using the GraphPad Prism 5 software. Quantitative parameters were divided into parametric and nonparametric using the D'Agostino and Pearson omnibus normality test, and the variances were analyzed with F test. The *t*-test and ANOVA test with Bonferroni's post hoc were used in parametric parameters, using Welch's correction when necessary. The survival assays were analyzed with Mantel–Cox test. The p limit value for rejecting the null hypothesis and considering the differences between cases as statistically significant was $p < 0.05$ (*). Other p-values are indicated as ** when $p < 0.01$ and *** when $p < 0.001$.

4.7. Human GB Databases

We used a public open access database (http://gliovis.bioinfo.cnio.es/, accessed on 1 February 2022) to analyze the expression of human *Cry1* gene in GB samples. We used the "Adult" samples in CGGA Dataset and included the data from primary and secondary tumor types. The data shown in Figure 1 correspond to the "expression" and "survival" tabs. Please note that nomenclature corresponds to the 2016 classification. GBM—Glioblastoma multiforme.

Author Contributions: Conceptualization, P.J. and S.C.-T.; methodology, P.J., C.d.P., A.G.-B. and S.C.-T.; software, P.J., C.d.P., A.G.-B. and S.C.-T.; validation, P.J., C.d.P., A.G.-B. and S.C.-T.; formal analysis, P.J., C.d.P., A.G.-B. and S.C.-T.; investigation, P.J., C.d.P., A.G.-B. and S.C.-T.; resources, P.J. and S.C.-T.; data curation, P.J. and S.C.-T.; writing original draft preparation, P.J. and S.C.-T.; writing, review and editing, P.J., C.d.P., A.G.-B. and S.C.-T.; visualization, P.J., C.d.P., A.G.-B. and S.C.-T.; supervision, P.J. and S.C.-T.; project administration, S.C.-T.; funding acquisition, P.J. and S.C.-T. All authors have read and agreed to the published version of the manuscript.

Funding: Research was funded by grant PID2019-110116GB-100 from the Spanish Ministerio de Ciencia e Innovación to S.C.-T. and Khalifa Capital donation to P.J.

Institutional Review Board Statement: Not applicable.

Informed Consent Statement: Not applicable.

Acknowledgments: We thank Alberto Ferrús and Paco Martín for critiques of the manuscript and for helpful discussions, and Esther Seco for fly stocks maintenance. We want to thank the Vienna *Drosophila* Resource Centre, the Bloomington *Drosophila* Stock Centre and the Developmental Studies Hybridoma Bank for supplying fly stocks and antibodies, and FlyBase for its wealth of information. We acknowledge the support of the Confocal Microscopy Unit and Molecular Biology Unit at the Cajal Institute and the *Drosophila* Transgenesis Unit. The results shown here are in part based upon data generated by the TCGA Research Network: https://www.cancer.gov/tcga (accessed on 1 February 2022).

Conflicts of Interest: The authors declare no conflict of interest. The funders had no role in the design of the study; in the collection, analyses, or interpretation of data; in the writing of the manuscript, or in the decision to publish the results.

References

1. Ostrom, Q.T.; Truitt, G.; Gittleman, H.; Brat, D.J.; Kruchko, C.; Wilson, R.; Barnholtz-Sloan, J.S. Relative survival after diagnosis with a primary brain or other central nervous system tumor in the National Program of Cancer Registries, 2004 to 2014. *Neuro. Oncol. Pract.* **2020**, *7*, 306–312. [CrossRef] [PubMed]
2. Louis, D.N.; Perry, A.; Wesseling, P.; Brat, D.J.; Cree, I.A.; Figarella-Branger, D.; Hawkins, C.; Ng, H.K.; Pfister, S.M.; Reifenberger, G.; et al. The 2021 WHO Classification of Tumors of the Central Nervous System: A summary. *Neuro. Oncol.* **2021**, *23*, 1231–1251. [CrossRef] [PubMed]
3. Wirsching, H.G.; Weller, M. The Role of Molecular Diagnostics in the Management of Patients with Gliomas. *Curr. Treat. Options Oncol.* **2016**, *17*, 51. [CrossRef] [PubMed]
4. Rich, J.N.; Bigner, D.D. Development of novel targeted therapies in the treatment of malignant glioma. *Nat. Rev. Drug Discov.* **2004**, *3*, 430–446. [CrossRef]
5. Brennan, C.W.; Verhaak, R.G.W.; McKenna, A.; Campos, B.; Noushmehr, H.; Salama, S.R.; Zheng, S.; Chakravarty, D.; Sanborn, J.Z.; Berman, S.H.; et al. The somatic genomic landscape of glioblastoma. *Cell* **2013**, *155*, 462. [CrossRef]
6. Read, R.D.; Cavenee, W.K.; Furnari, F.B.; Thomas, J.B. A Drosophila model for EGFR-Ras and PI3K-dependent human glioma. *PLoS Genet.* **2009**, *5*, e1000374. [CrossRef]

7. Portela, M.; Mitchell, T.; Casas-Tintó, S. Cell-to-cell communication mediates glioblastoma progression in Drosophila. *Biol. Open* **2020**, *9*, bio053405. [CrossRef]
8. Portela, M.; Segura-Collar, B.; Argudo, I.; Sáiz, A.; Gargini, R.; Sánchez-Gómez, P.; Casas-Tintó, S. Oncogenic dependence of glioma cells on kish/TMEM167A regulation of vesicular trafficking. *Glia* **2019**, *67*, 404–417. [CrossRef]
9. Portela, M.; Venkataramani, V.; Fahey-Lozano, N.; Seco, E.; Losada-Perez, M.; Winkler, F.; Casas-Tintó, S. Glioblastoma cells vampirize WNT from neurons and trigger a JNK/MMP signaling loop that enhances glioblastoma progression and neurodegeneration. *PLoS Biol.* **2019**, *17*, e3000545. [CrossRef]
10. Portela, M.; Casas-Tintó, S. New Cellular Dimensions on Glioblastoma Progression. *Neurosci. Insights* **2020**, *15*, 2633105520923076. [CrossRef]
11. Jarabo, P.; de Pablo, C.; Herranz, H.; Martín, F.A.; Casas-Tintó, S. Insulin signaling mediates neurodegeneration in glioma. *Life Sci. Alliance* **2021**, *4*. [CrossRef] [PubMed]
12. Formica, M.; Storaci, A.M.; Bertolini, I.; Carminati, F.; Knævelsrud, H.; Vaira, V.; Vaccari, T. V-ATPase controls tumor growth and autophagy in a Drosophila model of gliomagenesis. *Autophagy* **2021**, *17*, 4442–4452. [CrossRef] [PubMed]
13. Vigneswaran, K.; Boyd, N.H.; Oh, S.Y.; Lallani, S.; Boucher, A.; Neill, S.G.; Olson, J.J.; Read, R.D. YAP/TAZ Transcriptional Coactivators Create Therapeutic Vulnerability to Verteporfin in EGFR-mutant Glioblastoma. *Clin. Cancer Res.* **2021**, *27*, 1553–1569. [CrossRef] [PubMed]
14. Brand, A.H.; Perrimon, N. Targeted gene expression as a means of altering cell fates and generating dominant phenotypes. *Development* **1993**, *118*, 401–415. [CrossRef]
15. Casas-Tintó, S.; Portela, M. Cytonemes, their formation, regulation, and roles in signaling and communication in tumorigenesis. *Int. J. Mol. Sci.* **2019**, *20*, 5641. [CrossRef]
16. Annibali, D.; Whitfield, J.R.; Favuzzi, E.; Jauset, T.; Serrano, E.; Cuartas, I.; Redondo-Campos, S.; Folch, G.; Gonzàlez-Juncà, A.; Sodir, N.M.; et al. Myc inhibition is effective against glioma and reveals a role for Myc in proficient mitosis. *Nat. Commun.* **2014**, *5*, 4632. [CrossRef]
17. Kozono, D.; Li, J.; Nitta, M.; Sampetrean, O.; Gonda, D.; Kushwaha, D.S.; Merzon, D.; Ramakrishnan, V.; Zhu, S.; Zhu, K.; et al. Dynamic epigenetic regulation of glioblastoma tumorigenicity through LSD1 modulation of MYC expression. *Proc. Natl. Acad. Sci. USA* **2015**, *112*, E4055–E4064. [CrossRef]
18. Oktay, Y.; Ülgen, E.; Can, Ö.; Akyerli, C.B.; Yüksel, Ş.; Erdemgil, Y.; Durasl, I.M.; Henegariu, O.I.; Nanni, E.P.; Selevsek, N.; et al. IDH-mutant glioma specific association of rs55705857 located at 8q24.21 involves MYC deregulation. *Sci. Rep.* **2016**, *6*, 27569. [CrossRef]
19. Tateishi, K.; Iafrate, A.J.; Ho, Q.; Curry, W.T.; Batchelor, T.T.; Flaherty, K.T.; Onozato, M.L.; Lelic, N.; Sundaram, S.; Cahill, D.P.; et al. Myc-Driven Glycolysis Is a Therapeutic Target in Glioblastoma. *Clin. Cancer Res.* **2016**, *22*, 4452–4465. [CrossRef]
20. Wagner, P.M.; Prucca, C.G.; Caputto, B.L.; Guido, M.E. Adjusting the Molecular Clock: The Importance of Circadian Rhythms in the Development of Glioblastomas and Its Intervention as a Therapeutic Strategy. *Int. J. Mol. Sci.* **2021**, *22*, 8289. [CrossRef]
21. Lahti, T.; Merikanto, I.; Partonen, T. Circadian clock disruptions and the risk of cancer. *Ann. Med.* **2012**, *44*, 847–853. [CrossRef]
22. Kettner, N.M.; Katchy, C.A.; Fu, L. Circadian gene variants in cancer. *Ann. Med.* **2014**, *46*, 208. [CrossRef]
23. Jarabo, P.; Martin, F.A. Neurogenetics of Drosophila circadian clock: Expect the unexpected. *J. Neurogenet.* **2017**, *31*, 250–265. [CrossRef] [PubMed]
24. Vechtomova, Y.L.; Telegina, T.A.; Kritsky, M.S. Evolution of Proteins of the DNA Photolyase/Cryptochrome Family. *Biochemistry (Moscow)* **2020**, *85*, S131–S153. [CrossRef] [PubMed]
25. Collins, B.; Mazzoni, E.O.; Stanewsky, R.; Blau, J. Drosophila CRYPTOCHROME is a circadian transcriptional repressor. *Curr. Biol.* **2006**, *16*, 441–449. [CrossRef] [PubMed]
26. Madden, M.H.; Anic, G.M.; Thompson, R.C.; Nabors, L.B.; Olson, J.J.; Browning, J.E.; Monteiro, A.N.; Egan, K.M. Circadian pathway genes in relation to glioma risk and outcome. *Cancer Causes Control* **2014**, *25*, 25–32. [CrossRef]
27. Wang, Z.; Su, G.; Dai, Z.; Meng, M.; Zhang, H.; Fan, F.; Liu, Z.; Zhang, L.; Weygant, N.; He, F.; et al. Circadian clock genes promote glioma progression by affecting tumour immune infiltration and tumour cell proliferation. *Cell Prolif.* **2021**, *54*, e12988. [CrossRef]
28. Chen, Z.; Liu, P.; Li, C.; Luo, Y.; Chen, I.; Liang, W.; Chen, X.; Feng, Y.; Xia, H.; Wang, F. Deregulated expression of the clock genes in gliomas. *Technol. Cancer Res. Treat.* **2013**, *12*, 91–97. [CrossRef]
29. Zhanfeng, N.; Yanhui, L.; Zhou, F.; Shaocai, H.; Guangxing, L.; Hechun, X. Circadian genes Per1 and Per2 increase radiosensitivity of glioma in vivo. *Oncotarget* **2015**, *6*, 9951–9958. [CrossRef]
30. Burchett, J.B.; Knudsen-clark, A.M.; Altman, B.J. MYC Ran Up the Clock: The Complex Interplay between MYC and the Molecular Circadian Clock in Cancer. *Int. J. Mol. Sci.* **2021**, *22*, 7761. [CrossRef]
31. Liu, Z.; Selby, C.P.; Yang, Y.; Lindsey-Boltz, L.A.; Cao, X.; Eynullazada, K.; Sancar, A. Circadian regulation of c-MYC in mice. *Proc. Natl. Acad. Sci. USA* **2020**, *117*, 21609–21617. [CrossRef] [PubMed]
32. Altman, B.J.; Hsieh, A.L.; Sengupta, A.; Krishnanaiah, S.Y.; Stine, Z.E.; Walton, Z.E.; Gouw, A.M.; Venkataraman, A.; Li, B.; Goraksha-Hicks, P.; et al. MYC Disrupts the Circadian Clock and Metabolism in Cancer Cells. *Cell Metab.* **2015**, *22*, 1009–1019. [CrossRef] [PubMed]
33. Arnés, M.; Romero, N.; Casas-Tintó, S.; Acebes, Á.; Ferrús, A. PI3K activation prevents Aβ42-induced synapse loss and favors insoluble amyloid deposit formation. *Mol. Biol. Cell* **2020**, *31*, 244–260. [CrossRef] [PubMed]
34. Kelleher, F.C.; Rao, A.; Maguire, A. Circadian molecular clocks and cancer. *Cancer Lett.* **2014**, *342*, 9–18. [CrossRef]

35. Shafi, A.A.; McNair, C.M.; McCann, J.J.; Alshalalfa, M.; Shostak, A.; Severson, T.M.; Zhu, Y.; Bergman, A.; Gordon, N.; Mandigo, A.C.; et al. The circadian cryptochrome, CRY1, is a pro-tumorigenic factor that rhythmically modulates DNA repair. *Nat. Commun.* **2021**, *12*, 401. [CrossRef] [PubMed]
36. Mampay, M.; Flint, M.S.; Sheridan, G.K. Tumour brain: Pretreatment cognitive and affective disorders caused by peripheral cancers. *Br. J. Pharmacol.* **2021**, *178*, 3977–3996. [CrossRef]
37. Yang, Y.; Lindsey-Boltz, L.A.; Vaughn, C.M.; Selby, C.P.; Cao, X.; Liu, Z.; Hsu, D.S.; Sancar, A. Circadian clock, carcinogenesis, chronochemotherapy connections. *J. Biol. Chem.* **2021**, *297*, 101068. [CrossRef]
38. Luo, Y.; Wang, F.; Chen, L.A.; Chen, X.W.; Chen, Z.J.; Liu, P.F.; Li, F.F.; Li, C.Y.; Liang, W. Deregulated expression of cry1 and cry2 in human gliomas. *Asian Pac. J. Cancer Prev.* **2012**, *13*, 5725–5728. [CrossRef]
39. Bowman, R.L.; Wang, Q.; Carro, A.; Verhaak, R.G.W.; Squatrito, M. GlioVis data portal for visualization and analysis of brain tumor expression datasets. *Neuro. Oncol.* **2017**, *19*, 139. [CrossRef]
40. Fan, W.; Caiyan, L.; Ling, Z.; Jiayun, Z. Aberrant rhythmic expression of cryptochrome2 regulates the radiosensitivity of rat gliomas. *Oncotarget* **2017**, *8*, 77809. [CrossRef]
41. Dong, Z.; Zhang, G.; Qu, M.; Gimple, R.C.; Wu, Q.; Qiu, Z.; Prager, B.C.; Wang, X.; Kim, L.J.Y.; Morton, A.R.; et al. Targeting Glioblastoma Stem Cells through Disruption of the Circadian Clock. *Cancer Discov.* **2019**, *9*, 1556. [CrossRef] [PubMed]
42. Venkataramani, V.; Tanev, D.I.; Strahle, C.; Studier-Fischer, A.; Fankhauser, L.; Kessler, T.; Körber, C.; Kardorff, M.; Ratliff, M.; Xie, R.; et al. Glutamatergic synaptic input to glioma cells drives brain tumour progression. *Nature* **2019**, *573*, 532–538. [CrossRef] [PubMed]
43. Venkatesh, H.S.; Morishita, W.; Geraghty, A.C.; Silverbush, D.; Gillespie, S.M.; Arzt, M.; Tam, L.T.; Espenel, C.; Ponnuswami, A.; Ni, L.; et al. Electrical and synaptic integration of glioma into neural circuits. *Nature* **2019**, *573*, 539–545. [CrossRef] [PubMed]
44. Górska-Andrzejak, J.; Makuch, R.; Stefan, J.; Görlich, A.; Semik, D.; Pyza, E. Circadian expression of the presynaptic active zone protein bruchpilot in the lamina of Drosophila melanogaster. *Dev. Neurobiol.* **2013**, *73*, 14–26. [CrossRef] [PubMed]
45. Damulewicz, M.; Mazzotta, G.M.; Sartori, E.; Rosato, E.; Costa, R.; Pyza, E.M. Cryptochrome is a regulator of synaptic plasticity in the visual system of drosophila melanogaster. *Front. Mol. Neurosci.* **2017**, *10*, 165. [CrossRef] [PubMed]
46. Bolukbasi, E.; Woodling, N.S.; Ivanov, D.K.; Adcott, J.; Foley, A.; Rajasingam, A.; Gittings, L.M.; Aleyakpo, B.; Niccoli, T.; Thornton, J.M.; et al. Cell type-specific modulation of healthspan by Forkhead family transcription factors in the nervous system. *Proc. Natl. Acad. Sci. USA* **2021**, *118*, e2011491118. [CrossRef]
47. Casas-Tintó, S.; Ferrús, A. Troponin-I mediates the localization of selected apico-basal cell polarity signaling proteins. *J. Cell Sci.* **2019**, *132*, jcs225243. [CrossRef]
48. Casas-Tintó, S.; Portela, M.; Ferrús, A. Widening the concept of oncogene. *Aging (Albany NY)* **2016**, *8*, 2262. [CrossRef]
49. Casas-Tintó, S.; Maraver, A.; Serrano, M.; Ferrús, A. Troponin-I enhances and is required for oncogenic overgrowth. *Oncotarget* **2016**, *7*, 52631–52642. [CrossRef]
50. Arthurton, L.; Nahotko, D.A.; Alonso, J.; Wendler, F.; Baena-Lopez, L.A. Non-apoptotic caspase activation preserves Drosophila intestinal progenitor cells in quiescence. *EMBO Rep.* **2020**, *21*, e48892. [CrossRef]
51. Baena-Lopez, L.A. All about the caspase-dependent functions without cell death. *Semin. Cell Dev. Biol.* **2018**, *82*, 77–78. [CrossRef] [PubMed]
52. Baena-Lopez, L.A.; Arthurton, L.; Xu, D.C.; Galasso, A. Non-apoptotic Caspase regulation of stem cell properties. *Semin. Cell Dev. Biol.* **2018**, *82*, 118–126. [CrossRef] [PubMed]
53. Kwapis, J.L.; Alaghband, Y.; Kramár, E.A.; López, A.J.; Vogel Ciernia, A.; White, A.O.; Shu, G.; Rhee, D.; Michael, C.M.; Montellier, E.; et al. Epigenetic regulation of the circadian gene Per1 contributes to age-related changes in hippocampal memory. *Nat. Commun.* **2018**, *9*, 3323. [CrossRef]
54. Zhang, L.; Ptáček, L.J.; Fu, Y.H. Diversity of human clock genotypes and consequences. *Prog. Mol. Biol. Transl. Sci.* **2013**, *119*, 51–81. [CrossRef]
55. Emery, P.; Stanewsky, R.; Helfrich-Förster, C.; Emery-Le, M.; Hall, J.C.; Rosbash, M. Drosophila CRY is a deep brain circadian photoreceptor. *Neuron* **2000**, *26*, 493–504. [CrossRef]
56. Orme, M.H.; Alrubaie, S.; Bradley, G.L.; Walker, C.D.; Leevers, S.J. Input from Ras is required for maximal PI(3)K signalling in Drosophila. *Nat. Cell Biol.* **2006**, *8*, 1298–1302. [CrossRef] [PubMed]
57. Picot, M.; Cusumano, P.; Klarsfeld, A.; Ueda, R.; Rouyer, F. Light Activates Output from Evening Neurons and Inhibits Output from Morning Neurons in the Drosophila Circadian Clock. *PLoS Biol.* **2007**, *5*, e315. [CrossRef]
58. Moreno, E.; Basler, K. dMyc transforms cells into super-competitors. *Cell* **2004**, *117*, 117–129. [CrossRef]
59. Lee, B.P.; Jones, B.W. Transcriptional regulation of the Drosophila glial gene repo. *Mech. Dev.* **2005**, *122*, 849–862. [CrossRef] [PubMed]
60. Casas-Tintó, S.; Arnés, M.; Ferrús, A. Drosophila enhancer-Gal4 lines show ectopic expression during development. *R. Soc. Open Sci.* **2017**, *4*, 170039. [CrossRef]
61. McGuire, S.E.; Le, P.T.; Osborn, A.J.; Matsumoto, K.; Davis, R.L. Spatiotemporal Rescue of Memory Dysfunction in Drosophila. *Science* **2003**, *302*, 1765–1768. [CrossRef] [PubMed]

International Journal of
Molecular Sciences

MDPI

Article

The Different Temozolomide Effects on Tumorigenesis Mechanisms of Pediatric Glioblastoma PBT24 and SF8628 Cell Tumor in CAM Model and on Cells In Vitro

Eligija Damanskienė [1,*], Ingrida Balnytė [1], Angelija Valančiūtė [1], Marta Maria Alonso [2], Aidanas Preikšaitis [3] and Donatas Stakišaitis [1,4,*]

1 Department of Histology and Embryology, Medical Academy, Lithuanian University of Health Sciences, 44307 Kaunas, Lithuania; ingrida.balnyte@lsmuni.lt (I.B.); angelija.valanciute@lsmuni.lt (A.V.)
2 Department of Pediatrics, Clínica Universidad de Navarra, University of Navarra, 31008 Pamplona, Spain; mmalonso@unav.es
3 Centre of Neurosurgery, Clinic of Neurology and Neurosurgery, Faculty of Medicine, Vilnius University, 03101 Vilnius, Lithuania; danas911@gmail.com
4 Laboratory of Molecular Oncology, National Cancer Institute, 08660 Vilnius, Lithuania
* Correspondence: eligija.damanskiene@lsmuni.lt (E.D.); donatas.stakisaitis@lsmuni.lt (D.S.)

Abstract: It is necessary to elucidate the individual effects of temozolomide (TMZ) on carcinogenesis and tumor resistance to chemotherapy mechanisms. The study aimed to investigate the TMZ 50 and 100 μM dose effect difference between PBT24 and SF8628 cell line high-grade pediatric glioblastoma (phGBM) xenografts in a chicken chorioallantoic membrane (CAM) model, on PCNA and EZH2 immunohistochemical expression in the tumor and on the expression of NKCC1, KCC2, E- and N-cadherin genes in TMZ-treated and control cell groups in vitro. TMZ at a 100 μg dose reduced the incidence of PBT24 xenograft invasion into the CAM, CAM thickening and the number of blood vessels in the CAM ($p < 0.05$), but did not affect the SF8628 tumor in the CAM model. The TMZ impact on PBT24 and SF8628 tumor PCNA expression was similarly significantly effective but did not alter EZH2 expression in the studied tumors. The TMZ at 50 μM caused significantly increased RNA expression of the NKCC1 gene in both studied cell types compared with controls ($p < 0.05$). The expression of the KCC2 gene was increased in PBT24 TMZ-treated cells ($p < 0.05$), and no TMZ effect was found in SF8628-treated cells. The study supports the suggestion that individual sensitivity to TMZ should be assessed when starting treatment.

Keywords: pediatric glioblastoma; temozolomide; NKCC1; KCC2; EZH2; PCNA; CAM

Citation: Damanskienė, E.; Balnytė, I.; Valančiūtė, A.; Alonso, M.M.; Preikšaitis, A.; Stakišaitis, D. The Different Temozolomide Effects on Tumorigenesis Mechanisms of Pediatric Glioblastoma PBT24 and SF8628 Cell Tumor in CAM Model and on Cells In Vitro. *Int. J. Mol. Sci.* **2022**, *23*, 2001. https://doi.org/10.3390/ijms23042001

Academic Editors: Nives Pećina-Šlaus and Ivana Jovčevska

Received: 3 January 2022
Accepted: 6 February 2022
Published: 11 February 2022

Publisher's Note: MDPI stays neutral with regard to jurisdictional claims in published maps and institutional affiliations.

1. Introduction

Pediatric high-grade glioblastoma multiforme (phGBM) is a highly malignant brain tumor and the most common cause of death [1]. Five-year survival in phGBM patients is less than 20 percent [2]. The standard glioblastoma treatment includes surgical resection, radiotherapy and temozolomide (TMZ) chemotherapy [3]. Following therapy with TMZ in adults, the treatment has similarly employed TMZ for phGBM patients [2,4]. Notwithstanding, patients having an initial response to TMZ fail therapy: approximately 55% of glioblastoma patients develop resistance to TMZ chemotherapy [5,6]. phGBM is different from adult gliomas. The unique developmental origins and distinct biological factors of this heterogeneous group of tumors have highlighted the importance of avoiding treatment strategies based solely on adult glioblastoma, as this approach has not improved the outcome of phGBM [7]. Individual TMZ effectiveness depends on the resistance to TMZ, which would cause glioblastoma recurrence and a worse outcome [8]. It is essential to determine individual sensitivity to TMZ treatment and the personal effect of TMZ on cancerogenesis, which is critical for effective treatment.

The main antitumor effect of anticancer medicines is the inhibition of tumor cell proliferation and the promotion of apoptosis. In glioblastoma, markers of tumor progression include the expression of proliferating cell nuclear antigen (PCNA), enhancer of zeste homolog 2 (EZH2) and ionic transporters that regulate intracellular chloride levels [9–11]. PCNA is an important target for many aggressive tumors. The proliferation of tumor cells is correlated with a high degree of tumor malignancy, which can be evaluated by measuring the PCNA protein expression [12]. Thus, the potential for targeting PCNA in chemotherapy against aggressive tumors is actively pursued [13]. Researchers have reported Polycomb repressive complex 2 (PRC2) activity in phGBM [14]. PRC2 is usually accompanied by cancer progression [15]. PRC2 helps to support gene silencing and X chromosome inactivation through the enzymatic methylation of K27 on histone H3 by EZH2. EZH2 is a catalytic PRC2 subunit, a histone methyltransferase targeting lysine 27 of histone H3 [16]. This methylated H3-K27 consequence is associated with the silencing of different genes in phGBM [14]. Functional interconnections among EZH2-mediated histone methylation and DNA methylation indicate the gene silencing involved in the loss of tumor suppression [9].

High expression of the Na-K-2Cl co-transporter (NKCC1) in glioblastoma is involved in cell proliferation [11]. The sign of apoptosis is a reduction in cell volume arising from a loss of intracellular K^+([K^+]i) and chloride ([Cl^-]i) [10,17]. The increased [Cl^-]i level in glioblastoma cells is related to upregulated NKCC1 and downregulated K-Cl co-transporter (KCC2) [18,19]. Increased NKCC1 protein expression in human glioblastoma directly correlates with the tumor grade and cell migration; NKCC1 inhibition reduces glioblastoma cell migration and tumor invasion [20,21]. The repressing effect on glioma cell migration is expected to result from reduced [Cl^-]i [22]. Knockdown of NKCC1 in glioblastoma cells causes the formation of significantly more extensive focal adhesions and cell traction forces than in control cells [20]. Blockade of the NKCC1 protein function serves as a therapeutic strategy to overcome TMZ-mediated glioblastoma resistance [23].

In adult glioblastoma epileptic patients, a reduction in KCC2 neuropil staining [19] and a drop in [K^+]i and [Cl^-]i concentrations in the TMZ-treated glioma cells were described [24,25]; a loss of K^+ and Cl^- that occurred through apoptosis was confirmed [26]. Thus, it is necessary to investigate whether TMZ initiates KCC2 activity in glioblastoma cells, resulting in the parallel loss of K^+ and Cl^- ions.

High expression of NKCC1 is related to epithelial–mesenchymal transition (EMT) in gliomas, offering a new therapeutic strategy for inhibiting tumor progression [27]. The EMT process is linked with resistance to treatment, and EMT in glioblastoma cells may be complemented by enhanced N-cadherin (CDH2) expression, correlating with adverse prognosis [28]. The association of phGBM progression with cadherin-E (CDH1) and CDH2 is mainly unexplored, whereas the literature on the association of CDH1 and CDH2 expression with adult glioma progression is controversial [29–31]. TMZ generates a DNA O6-methylguanine lesion that triggers DNA restoration, drains the enzyme O6-methylguanine methyltransferase and begins glioblastoma cell apoptosis, produced by activating external apoptotic or mitochondrial-dependent pathways [32]. TMZ causes cell arrest in the G2/M cell cycle [33] and reduces cell proliferation during exposure to medicine [33,34]. Despite this, TMZ arouses NKCC1 expression and activity in glioblastoma cells [24].

The chicken chorioallantoic membrane (CAM) model clarifies investigational medicines for cancer treatment and is an alternative model to study tumor growth, invasion and angiogenesis [35,36]. The CAM model, being immunodeficient, allows transplantations from different tumor tissues and species without immune responses [37]. The CAM model has advantages over the rodent immunodeficiency models in that it is cheap, allows monitoring of the CAM epithelium basement membrane barrier disruption by the tumor, induces neoangiogenesis, permits the detection of drug effects and has a short-term duration, which can be helpful in the prediction of anticancer therapy's efficacy [38]. Our study demonstrated that the spheroid model did not reflect the treatment effect on tumor progression observed in the CAM model [39].

The present study aimed to investigate the differences between a 3-year-old girl's high-grade SF8628 cell line xenograft and a 13-year-old boy's high-grade PBT24 cell line [40] xenograft using the CAM model; to examine the tumor response to treatment with TMZ's effect on PCNA and EZH2 histological expression in cells of the tumor in the CAM; and to investigate the impact of TMZ on the NKCC1, KCC2, E- and N-cadherin gene expression in the studied phGBM cells.

2. Results

2.1. *The Biomicroscopy of PBT24 and SF8628 Xenograft on CAM*

Figure 1 shows stereomicroscopic images of the PBT24 and SF8628 control and TMZ-treated tumor on the CAM at 5 days of chick embryo development (EDD9—day 2, EDD12—day 5 post-transplantation), extracted EDD12 CAM with tumor (ex ovo) and their histological view (H–E). On EDD9, the PBT24 and SF8628 control tumors were larger than on EDD12, indicating that the tumor invaded into the CAM mesenchyme three days later, and only the superficial part of the tumor was visible above the membrane. The PBT24 tumor on EDD9 was dense with precise edges. On EDD12, the PBT24 control and 50 μM TMZ-treated tumors were without clear borders, appearing smaller due to a smaller tumor on the membrane, as part of the tumor invaded the CAM (Figure 1; EDD9, EDD12, ex ovo, and H–E), and the vascular network ("spoked-wheel") formed around the tumor (Figure 1). Figure 2 shows the vascular network formed around the control and 50 μM TMZ-treated tumor, obviously visible after the injection of fluorescent dextran into the CAM vessel. The PBT24-100 μM TMZ tumor EDD12 was dense, with a clear border, growing on the surface of the CAM, with muted neo-angiogenesis (Figures 1 and 2).

The stereomicroscopically visible SF8628 control EDD12 tumor was smaller than the TMZ-treated tumors, and the histological H–E images show the profoundly invasive nature of the tumor growth (Figure 1). Compared to EDD9, the SF8628 TMZ-treated EDD12 tumor size was less pronounced as the tumor grew and was predominantly exposed on the CAM surface or was shallowly invasive into the CAM mesenchyme. There was an apparent "spoked-wheel" vascular network around the SF8628 control EDD12 and 50 μM TMZ-treated tumor, which was less pronounced around the 100 μM TMZ-treated tumor (Figures 1 and 2).

2.2. *The PBT24 and SF8628 Growth, Invasion into CAM Frequency, the CAM Thickness and the Number of Blood Vessels in CAM under the Tumors of the Study Groups*

Compared to the control, 100 μM TMZ reduced the frequency of PBT24 tumor invasion into the CAM ($p = 0.007$), while 50 μM TMZ had no effect on invasion ($p > 0.05$). A significant difference was found when the PBT24-50 μM TMZ and PBT24-100 μM TMZ groups were compared ($p = 0.019$). Compared to the SF8628 control at EDD12, treatment of the SF8628 tumor with TMZ did not reduce the tumor invasion frequency into the CAM in the SF8628-treated groups ($p > 0.05$; Table 1; Figure 3).

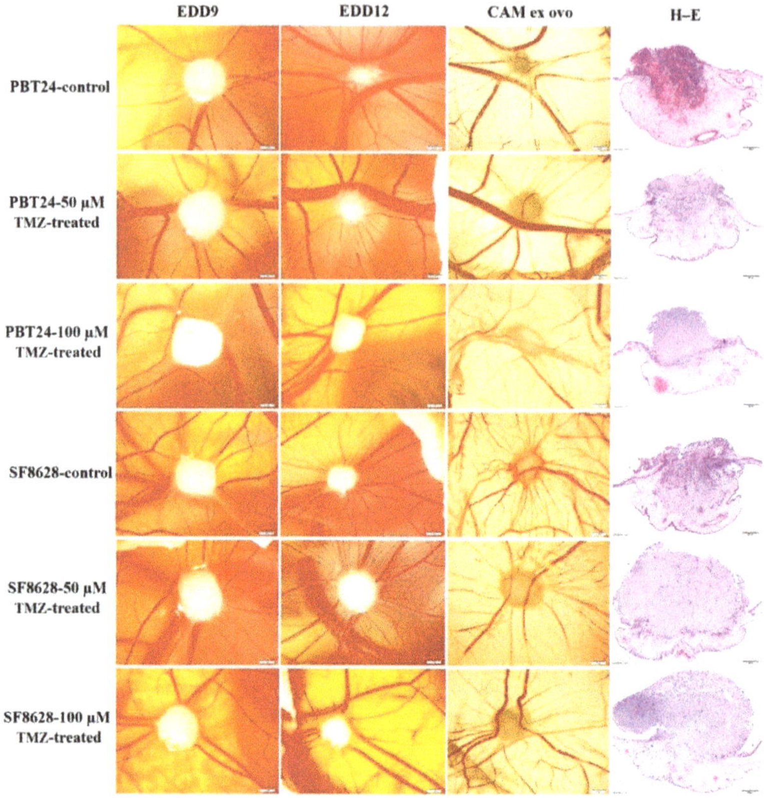

Figure 1. Stereomicroscopy of PBT24 and SF8628 tumors in vivo, a chorioallantoic membrane with tumor ex ovo and the histologic images of the study groups. EDD9, EDD12 and CAM ex ovo scale bar—1 mm; hematoxylin and eosin (H–E) stained preparations' scale bar—200 µm.

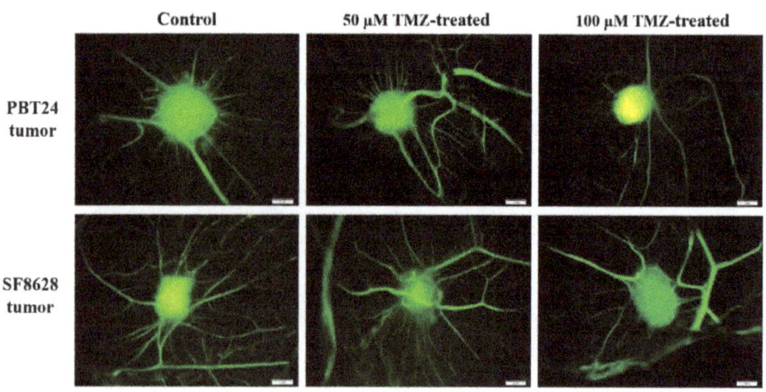

Figure 2. Fluorescent stereomicroscopy assay with fluorescent dextran of PBT24 and SF8628 tumors. Dextran highlighted the tumor and vascular network around it. Scale bar—1 mm.

Table 1. PBT24 and SF8628 tumor invasion into CAM frequency, the thickness of the CAM and the number of blood vessels in the CAM under the tumors of the study groups.

Study Group	*n*	Invasion Frequency (%)	CAM Thickness (μm)	Number of Blood Vessels
			Median (Range)	
PBT24-control	13	76.92	300.88 (65.23–700.87)	15 (6–28)
PBT24-50 μM TMZ	13	69.23	321.39 (67.02–516.85)	9 (3–14) [e]
PBT24-100 μM TMZ	10	20.0 [a,b]	55.48 (38.4–275.2) [c,d]	5.5 (3–13) [f]
SF8628-control	13	84.61	282.5 (47.85–539.7)	15 (5–21)
SF8628-50 μM TMZ	14	71.43	419.4 (84.49–683.7)	15 (5–29)
SF8628-100 μM TMZ	13	53.85	252.1 (55.51–529.1)	14 (7–19)

[a] $p = 0.007$, compared with PBT24-control; [b] $p = 0.0191$, compared with PBT24-50 μM TMZ; [c] $p = 0.0009$, compared with PBT24-50 μM TMZ; [d] $p = 0.0003$, compared with PBT24-control; [e] $p = 0.0012$, compared with PBT24-control; [f] $p = 0.0001$, compared with PBT24-control.

Compared with the PBT24 control on EDD12, the CAM thickness beneath the tumor was significantly lower in the PBT24-100 μM TMZ ($p = 0.0003$). The PBT24-100 μM TMZ CAM thickness was lower than in the PBT24-50 μM TMZ group ($p = 0.0009$). Treatment of the SF8628 tumor with both doses of TMZ did not affect the CAM thickness under the EDD12 tumor ($p > 0.05$; Table 1).

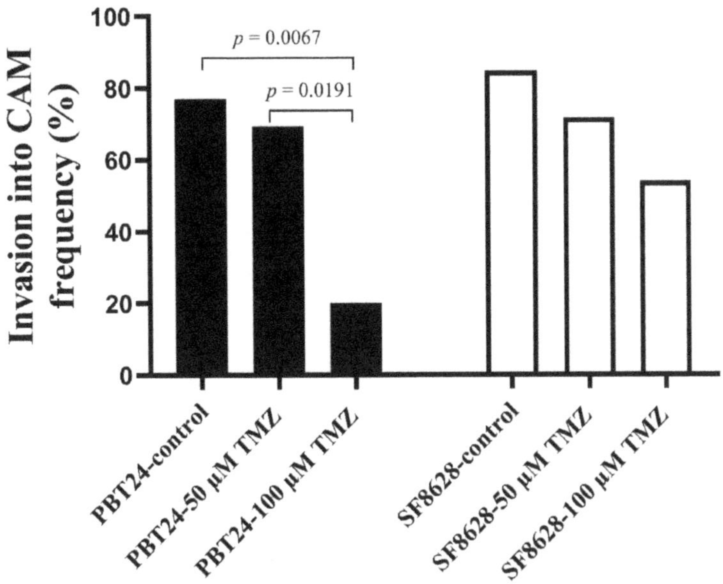

Figure 3. PBT24 and SF8628 tumor invasion into CAM frequency in control and TMZ-treated groups.

When comparing the neo-angiogenesis expression on EDD12 of the control in the CAM with that in the PBT24-treated TMZ groups, both doses of TMZ significantly inhibited angiogenesis in the CAM under the PBT24 tumor ($p < 0.002$), but there was no difference in this parameter between the TMZ-treated groups ($p > 0.05$). The treatment with TMZ had no suppressive effect on neo-angiogenesis in the SF8628 tumor groups ($p > 0.05$; Table 1).

2.3. The PCNA and EZH2 Expression in PBT24 and SF8628 Tumors

Table 2 and Figures 4 and 5 show the PCNA and EZH2 positively stained cell expression in the tissue of the studied tumors at EDD12.

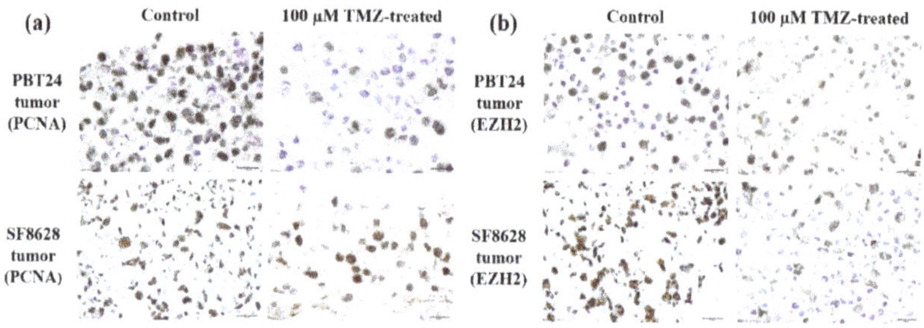

Figure 4. PCNA and EZH2 positively stained tumors of the PBT24 and SF8628 control and 100 μM TMZ-treated study groups. Dark brown nuclei indicate a PCNA-positive (**a**) and EZH2-positive cell (**b**). Scale bar—20 μm.

A significantly higher number of PCNA-positive cells was found in the SF8628 control than in PBT24 control tumors ($p < 0.05$; Figure 5a). The effect of TMZ treatment on the expression of the studied markers was similar in PBT24 and SF8628 tumors. Compared to the control, both doses of TMZ significantly reduced the number of PCNA-positive cells in PBT24 and SF8628 tumors ($p < 0.05$), but the 100 μM TMZ dose was significantly more effective compared to the 50 μM TMZ dose (Table 2; Figure 5a).

Table 2. The percentage of PCNA and EZH2 positively stained cells in PBT24 and SF8628 tumors of the study groups.

Study Group	PCNA-Positive Cells (%)		EZH2-Positive Cells (%)	
	n	Median (Range)	*n*	Median (Range)
PBT24-control	9	63.78 (34.87–80.95)	6	71.00 (42.63–78.70)
PBT24-50 μM TMZ	6	32.12 (21.78–42.42) [a]	7	69.15 (45.38–75.37)
PBT24-100 μM TMZ	6	18.15 (6.25–26.80) [b,c]	7	17.11 (2.38–95.06)
SF8628-control	8	90.81 (76.27–100) [d]	8	85.36 (72.04–97.45) [h]
SF8628-50 μM TMZ	6	76.17 (69.19–77.28) [e]	6	81.71 (72.13–94.08)
SF8628-100 μM TMZ	7	55.65 (12.45–83.57) [f,g]	6	42.55 (8.68–92.45)

[a] $p = 0.0028$, compared with PBT24-control; [b] $p = 0.0004$, compared with PBT24-control; [c] $p = 0.0152$, compared with PBT24-50 μM TMZ; [d] $p = 0.001$, compared with PBT24-control; [e] $p = 0.0080$, compared with SF8628-control; [f] $p = 0.0012$, compared with SF8628-control; [g] $p = 0.0350$, compared with SF8628-50 μM TMZ; [h] $p = 0.0127$, compared with PBT24-control.

A significantly higher level of EZH2-positive cells was detected in SF8628 control tumors than in PBT24 control tumors ($p < 0.05$). No significant reduction in EZH2-positive cells was observed in TMZ-treated tumor tissue compared to the matched control or when comparing the expression of EZH2-positive cells between the treated groups ($p > 0.05$; Table 2; Figure 5b).

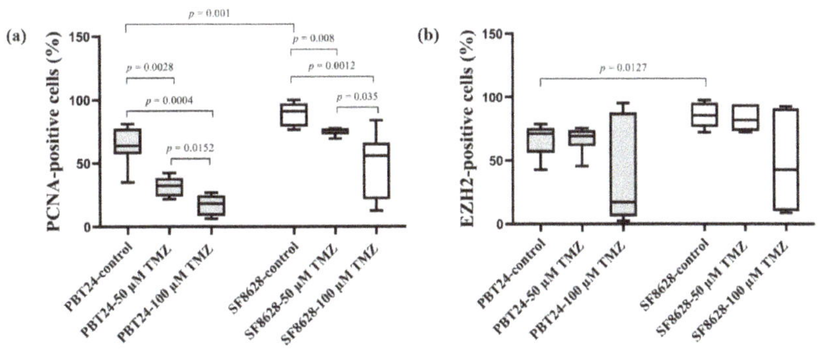

Figure 5. The percentage of PCNA-positive (**a**) and EZH2-positive (**b**) cells in PBT24 and SF8628 tumors.

2.4. The Expression of SLC12A2 (NKCC1 Co-Transporter) and SLC12A5 (KCC2 Co-Transporter) Gene in PBT24 and SF8628 Cell Study Groups

The expression of *SLC12A2* in PBT24 and SF8628 cell groups is shown in Table 3 and Figure 6. We found no difference in *SLC12A2* expression between the PBT24 and SF8628 control groups. Treatment with 50 μM TMZ significantly increased *SLC12A2* expression in PBT24 ($p = 0.0022$) and SF8628 cells ($p = 0.0022$). The effect of 50 μM TMZ on *SLC12A2* expression was significantly lower in SF8628-50 μM TMZ than in PBT24-50 μM TMZ cells (Figure 6a,c; Table 3). Compared to the control, the 50 μM TMZ dose increased *SLC12A2* expression in PBT24 cells by two-fold ($2^{-\Delta\Delta CT} = 2.04$) and in SF8628 cells by 1.5-fold ($2^{-\Delta\Delta CT} = 1.5$).

Table 3. RNA expression of *SLC12A2* and *SLC12A5* gene in PBT24 and SF8628 cell study groups.

Study Group	*n*	CT Mean		ΔCT Mean ± SD	ΔΔCT
		SLC12A2	***GAPDH***		
PBT24-control	6	22.951	19.372	3.579 ± 0.73	
PBT24-50 μM TMZ	6	21.766	19.214	2.552 ± 0.2 [a]	−1.027
SF8628-control	6	22.894	19.017	3.876 ± 0.21	
SF8628-50 μM TMZ	6	22.215	18.966	3.249 ± 0.15 [b,c]	−0.628
		SLC12A5	***GAPDH***	**ΔCT mean ± SD**	**ΔΔCT**
PBT24-control	6	32.564	19.372	13.191 ± 0.83	
PBT24-50 μM TMZ	6	31.047	19.214	11.833 ± 0.19 [d]	−1.359
SF8628-control	6	36.831	19.017	17.814 ± 0.43 [e]	
SF8628-50 μM TMZ	6	36.127	18.966	17.161 ± 0.29 [f]	−0.652

[a] $p = 0.0022$, compared with PBT24-control (*SLC12A2*); [b] $p = 0.0022$, compared with PBT24-50 μM TMZ (*SLC12A2*); [c] $p = 0.0022$, compared with SF8628-control (*SLC12A2*); [d] $p = 0.0022$, compared with PBT24-control (*SLC12A5*); [e] $p = 0.0022$, compared with PBT24-control (*SLC12A5*); [f] $p = 0.0022$, compared with PBT24-50 μM TMZ (*SLC12A5*).

The expression of the *SLC12A5* and *GAPDH* genes and the differences in the expression found when comparing the groups studied are shown in Table 3. The expression of *SLC12A5* in SF8628 control cells was significantly lower than in the PBT24 control ($p = 0.0022$). The TMZ dose of 50 μM increased *SLC12A5* expression 2.6-fold ($2^{-\Delta\Delta CT} = 2.6$) in PBT24 cells and 1.6-fold in SF8628 cells compared to the respective control.

The 50 μM dose of TMZ significantly increased *SLC12A5* expression in PBT24 cells ($p = 0.0022$), and the treatment had no significant effect on the gene expression in SF8628 cells. There was significantly lower expression of *SLC12A5* in the SF8628-50 µM TMZ group compared to the PBT24-50 µM TMZ group ($p = 0.0022$; Figure 6b,d; Table 3).

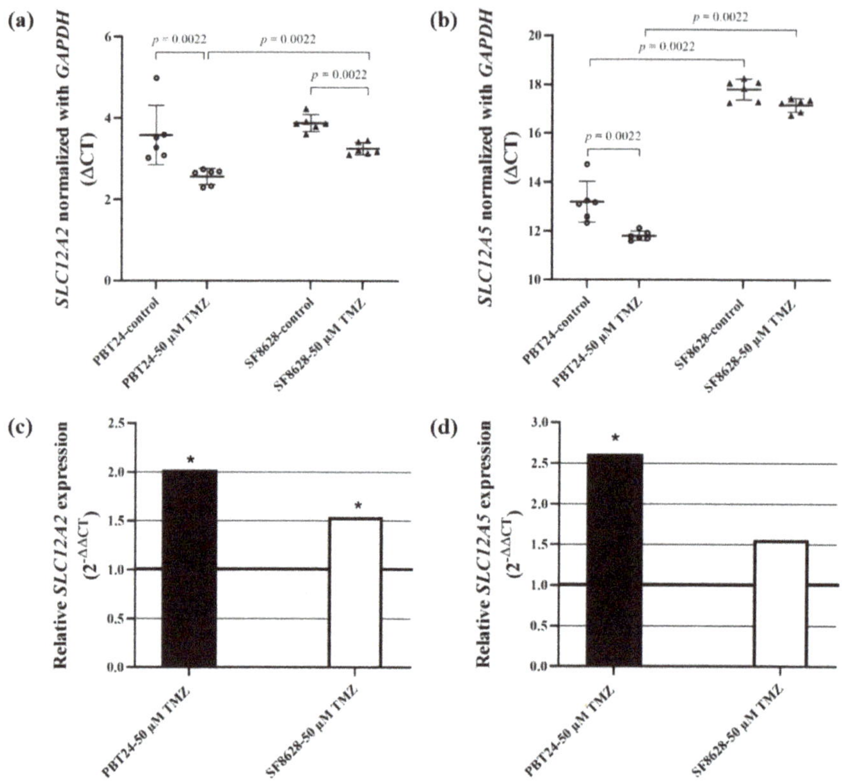

Figure 6. *SLC12A2* (**a**) and *SLC12A5* (**b**) expression in PBT24 and SF8628 control groups and 50 µM TMZ-treated groups. Data are after normalization with the *GAPDH* gene. Delta threshold cycle (ΔCT) values were used for the graph (the horizontal bars represent the mean; the short horizontal lines show standard deviation (SD) values). *SLC12A2* (**c**) and *SLC12A5* (**d**) relative expression in PBT24 and SF8628 50 µM TMZ-treated groups. The relative gene expression in TMZ-treated groups compared with respective controls. The 1.0 line shows the starting point of gene expression; * $p < 0.05$.

Comparison of the mean value of the ΔCT *SLC12A5*/ΔCT *SLC12A2* ratio of PBT24 and SF8628 cells showed the significantly higher ratio value of SF8628 (4.70 (4.27–4.81) compared to the PBT24 control (3.87 (2.95–4.17); $p < 0.002$). The treatment with TMZ increased the ΔCT *SLC12A5*/ΔCT *SLC12A2* value in PBT24 cells to 4.50 (4.3–5.1) ($p < 0.003$), and in SF8628 cells to 5.26 (5.05–5.58) ($p < 0.003$). When comparing the ratio value between PBT24- and SF8628-treated cell groups, it was significantly higher in the SF8628-50 µM TMZ than in the PBT24-treated group ($p < 0.009$; Figure 7).

The correlation (r) between *SLC12A2* and *SLC12A5* ΔCT values was 0.71 in control PBT24 cells, 0.14 in control SF8628 cells, 0.66 in treated PBT24 and 0.49 in treated SF8628 cells ($p > 0.05$ in all groups).

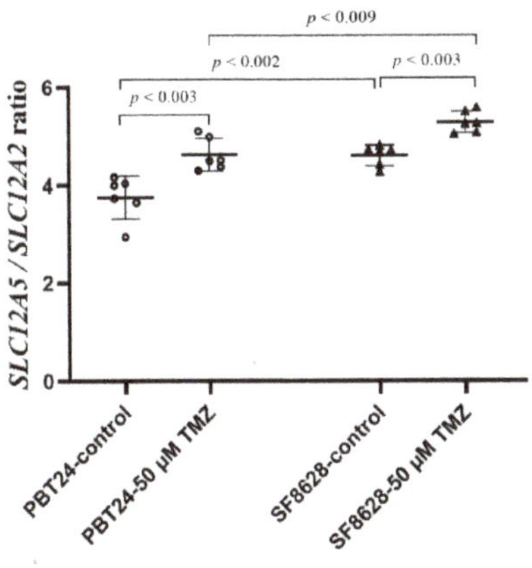

Figure 7. Comparison of the mean of ΔCT *SLC12A5*/ΔCT *SLC12A2* ratio value with SD among the PBT24 and SF8628 cell study groups.

2.5. The Expression of CDH1 (E-Cadherin) and CDH2 (N-Cadherin) Gene in PBT24 and SF8628 Cell Study Groups

The expression of *CDH1*, *CDH2* and *GAPDH* genes and the differences between the study groups are shown in Table 4. Significantly higher *CDH1* expression in the PBT24 control cells than in the SF8628 control group was found ($p = 0.0022$). This shows the more invasive phenotype of SF8628 cells. *CDH1* expression in the PBT24-50 µM TMZ group was also higher than in the SF8628-50 µM TMZ group ($p = 0.0022$). No differences in *CDH2* expression were found when comparing PBT24 cells with SF8628 control groups ($p > 0.05$), but *CDH2* expression in the PBT24-50 µM TMZ group was found to be significantly higher than that in SF8628 cells treated with 50 µM TMZ ($p = 0.0022$; Figure 8a,b; Table 4).

Table 4. RNA expression of E- and N-cadherin in PBT24 and SF8628 cell study groups.

Study Group	*n*	CT Mean		ΔCT Mean ± SD	ΔΔCT
		CDH1	*GAPDH*		
PBT24-control	6	33.476	19.372	14.104 ± 1.05	
PBT24-50 µM TMZ	6	32.294	19.214	13.079 ± 0.81	−1.024
SF8628-control	6	38.689	19.017	19.672 ± 0.51 [a]	
SF8628-50 µM TMZ	6	38.851	18.966	19.885 ± 0.22 [b]	0.213
		CDH2	*GAPDH*	ΔCT mean ± SD	ΔΔCT
PBT24-control	6	22.924	19.372	3.552 ± 0.98	
PBT24-50 µM TMZ	6	22.182	19.214	2.968 ± 0.11	−0.584
SF8628-control	6	23.449	19.017	4.432 ± 0.23	
SF8628-50 µM TMZ	6	23.177	18.966	4.211 ± 0.21 [c]	−0.221

[a] $p = 0.0022$, compared with PBT24-control (*CDH1*); [b] $p = 0.0022$, compared with PBT24-50 µM TMZ (*CDH1*); [c] $p = 0.0022$, compared with PBT24-50 µM TMZ (*CDH2*).

Treatment with 50 µM TMZ did not significantly affect *CDH1* and *CDH2* expression compared to controls in either treated group ($p > 0.05$; Table 4). The 50 µM dose of TMZ increased *CDH1* expression in PBT24 cells by two-fold ($2^{-\Delta\Delta CT} = 2.0$) and *CDH2* expression by 1.5-fold, while SF8628 cells had a 0.8-fold decrease in *CDH1* expression and a 1.16-fold increase in *CDH2* expression compared to their controls (Figure 8c,d).

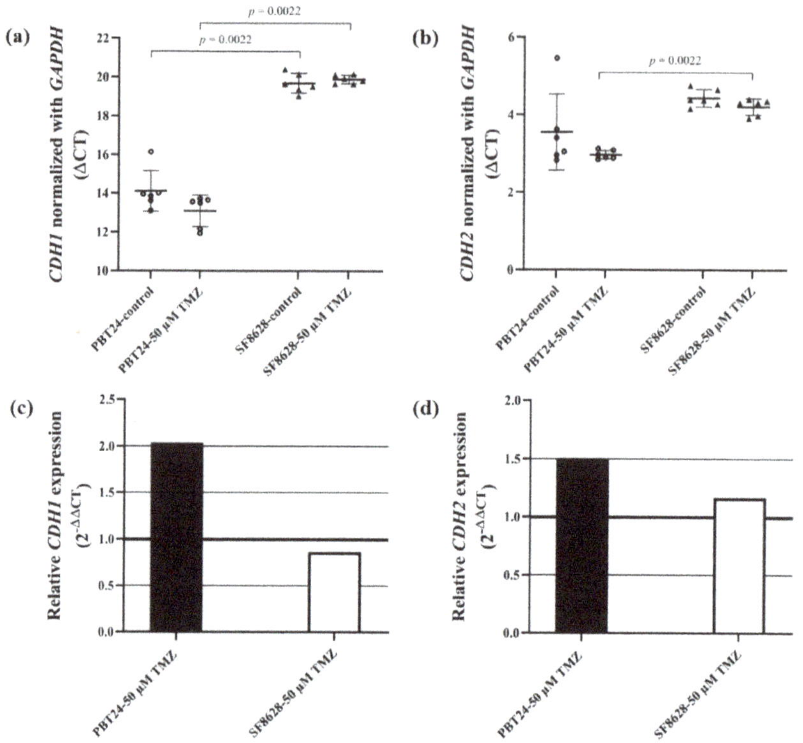

Figure 8. *CDH1* (**a**) and *CDH2* (**b**) expression in PBT24 and SF8628 control groups and 50 μM TMZ-treated groups. Data are after normalization with *GAPDH*. Delta threshold cycle (ΔCT) values were used for the graph (the horizontal bars represent the mean; the short horizontal lines show SD values). *CDH1* (**c**) and *CDH2* (**d**) gene expression in PBT24 and SF8628 50 μM TMZ-treated groups. The relative gene expression in TMZ-treated groups as compared with their control groups. The 1.0 line shows the starting point of gene expression.

3. Discussion

In a previous study, TMZ chemotherapy significantly improved overall survival in the elderly group but had a more limited effect in the younger group [41]. The ineffectiveness of glioblastoma treatment in the face of a high level of glioblastoma polymorphism has shown that targeting all patients with a single strategy is unrealistic to achieve treatment progress. Thus, a personalized pharmacological therapy for glioblastoma should be tailored to the individual patient's tumor pathophysiological, molecular, genetic and gender-related characteristics [42]. Our experimental study using a CAM model demonstrates the differences in the efficacy of TMZ therapy and the associated molecular mechanisms in the treatment of pediatric PBT24 and SF6828 tumors. The study found that TMZ at a dose of 100 μM significantly reduced the incidence of PBT24 tumor invasion into the CAM and the thickness of the CAM, and significantly inhibited neo-angiogenesis in the CAM beneath the PBT24 tumor, but had no effect on the SF8628 tumor growth and the corresponding parameters studied.

Cancer cell proliferation may involve a non-oncogenic structural protein, such as PCNA, which acts as a "hub" for large cellular complexes that is essential for tumor growth and cancer cell survival. Drugs reducing PCNA expression in tumor cells are expected to have a broader anticancer therapeutic spectrum than medicines targeting specific signal proteins [43]. The control SF8628 tumor had significantly more PCNA-positive cells than the PBT24 control. The effect of TMZ treatment on PCNA expression in PBT24 and SF8628 tumors was similar and dose-dependent. PCNA is involved in DNA metabolic processes, including DNA replication and repair, chromatin organization and transcription. PCNA is necessary for cell metabolic processes such as glycolysis [13].

The study found a significantly higher level of EZH2-positive cells in the SF8628 control tumor than in the PBT24 control. EZH2 is commonly overexpressed in glioblastoma and is firmly associated with tumor malignancy [44–46]. It is essential to assess the impact of the medicine integration on EZH2 blockade when explaining phGBM therapy strategies [15]. The inhibition of EZH2 reverses TMZ chemosensitivity in glioblastoma [47]. However, no significant reduction in EZH2-positive cells was observed in TMZ-treated PBT24 and SF8628 tumor tissue compared to their controls in our study.

Cancer progression is related to Cl^- and Na^+ in the tumor microenvironment [11,48,49]. The persistence of high neuronal levels of NKCC1 in pediatric glioblastoma supports the hypothesis of abnormal and immature neuronal cells in the phGBM. Strong NKCC1 immune reactivity in the aberrant neuronal component of glioblastoma and no upregulation of neuronal NKCC1 was observed in the perilesional area of tumor specimens [19]. Our study found no difference in *SLC12A2* expression between control PBT24 and SF8628 groups. Treatment with 50 μM TMZ significantly increased *SLC12A2* expression in PBT24 and SF8628 cells compared to controls in PBT24 cells by 2-fold and in SF8628 cells by 1.5-fold. Increased expression of NKCC1 protein and its elevated phosphorylation, with a concurrent increase in the phosphorylation of serine–threonine kinases WNK, in TMZ-treated glioblastoma was reported [24,50,51]. The researchers suggested that NKCC1 activity in TMZ-treated cells was stimulated via Cl^-/volume-sensitive regulatory kinases and the WNK-mediated signaling pathway, which is vital in protecting glioma from a loss of cell volume and Cl^- during TMZ treatment. The regulatory WNK kinases, a family of serine–threonine kinases, are activated by losing $[Cl^-]i$ and cell shrinkage [51,52]. The rapid upregulation of these proteins is likely due to de novo protein synthesis using mRNA reserves, allowing the glioblastoma cells to adapt instantaneously to the altered osmotic situation [24].

Therapeutic resistance has been proposed to emerge from the overexpression of the NKCC1 transporter, which intensifies DNA repair mechanisms against TMZ-induced apoptosis [24]. Inhibition of NKCC1 activity by bumetadine accelerates TMZ-treated glioblastoma cell apoptosis, and this suggests that NKCC1 activity remains functional and further regulates cell volume in TMZ-treated glioma, playing a role in $[Cl^-]i$ supplementation [24].

High-grade glioblastoma cells accumulate intracellular chloride ($[Cl^-]i$) to ~10-fold higher levels compared with the average in grade II glioma and the normal cortex [20]. It was proposed that some factors could dilute K^+ and Cl^- concentrations in TMZ-treated cells. Researchers reported that aquaporin 4 protein was downregulated in glioblastoma cells after chemotherapy and radiotherapy, with reduced peritumoral brain edema [53]. Silencing WNK kinase activity can promote Na-K-2Cl inhibition and K-Cl co-transporter activation via net transporter dephosphorylation, revealing WNKs' ability to modulate $[Cl^-]i$ [50]. Apoptosis requires persistent cell shrinkage and loss of cell volume via the reduction of $[K^+]i$ and $[Cl^-]i$, which occurs before any other detectable apoptosis features [25,26,54]. The study data show that SF8628 control cells have significantly lower KCC2 gene (*SLC12A5*) expression than PBT24 cells. TMZ treatment significantly increased *SLC12A5* expression in PBT24 cells, while treatment of SF8628 cells had no significant effect on gene expression. Moreover, *SLC12A5* expression was substantially lower in the SF8628-treated TMZ than in the PBT24-treated group.

Adult glioblastoma patients with epilepsy syndrome showed a decrease in KCC2 staining in tumor tissue [19] and a reduction in $[K^+]i$ and $[Cl^-]i$ levels in TMZ-treated glioma cells [24,25]. Loss of $[K^+]i$ and $[Cl^-]i$ in the glioma cell in parallel with expressed apoptosis was confirmed [26]. Therefore, it is important to carry out further research to determine whether the distinct effect of TMZ in stimulating KCC2 activity in glioblastoma cells is due to a patient-specific impact of TMZ in promoting KCC2 activity in cells and the relationship of this with the efficacy of tumor treatment.

The study data of the ΔCT *SLC12A5*/ΔCT *SLC12A2* ratio in PBT24 and SF8628 cells show that the ratio value of the SF8628 control cells is significantly higher than that of PBT24. Treatment with TMZ significantly increased the value of ΔCT *SLC12A5*/ΔCT *SLC12A2* in both TMZ-treated groups, and when comparing the value between the PBT24- and SF8628-treated groups, it was substantially higher in the SF8628 group. Additionally, these data suggest that the efficacy of TMZ treatment may be related to changes in $[Cl^-]$i, with a Cl^- concentration increase in SF8628 cells associated with increased NKCC1 gene expression and no modifications of KCC2 gene expression. In contrast, PBT24-treated cells showed an apparent rise in KCC2 gene expression, with a lower value of the co-transporter gene ratio, which may have led to a decrease in K^+ and Cl^- concentrations in TMZ-treated PBT24 cells. The reduction of the intracellular K^+ and Cl^- ion levels is related to the activation of caspases and triggers caspase cascade-related apoptosis mechanisms [17]. The decline of intracellular K^+, Na^+ and Cl^- results in an 80–85% loss of cell volume, DNA degradation and apoptotic body development in Jurkat cells [25].

Reactive astrocytes express NKCC1 in glioblastoma [19]. NKCC1 upregulation may lead to astrocyte swelling [55,56] and produce a GABAA receptor-mediated excitatory response, facilitating seizures [57–59]. The paradoxical excitatory action of GABAA depends on the relatively high $[Cl^-]$i content in the cell [58]. On the other hand, KCC2 is a neuron-specific Cl^- extruder that uses a K^+ gradient to maintain a low $[Cl^-]$i level to ensure the proper functioning of postsynaptic GABAA receptors. Studies over the last two decades have shown that low KCC2 activity results in excitatory GABAergic transmission associated with seizures. KCC2 expression and function are features of epileptic disorders in the developing and adult brain. The effect of drugs that activate KCC2 function in glioblastoma is important as a potential new therapeutic target for treating glioblastoma [60]. Future studies of the colocalization of Cl^- co-transporters with the GABAA receptor may shed light on the importance of the functional interaction of Cl^- transporters in glioblastoma cells. In our study, statistically significantly higher expression of *CDH1* was detected in PBT24 control cells compared to SF8628. Researchers have shown that a decrease in *CDH1* expression is associated with astrocytoma progression [29,30], while other studies showed that high E-cadherin expression is associated with a poorer prognosis of the disease [31]. The contribution of E-cadherin expression to adult glioblastoma and phGBM progression remains unclear. No differences were found when comparing the expression of *CDH2* in PBT24 cells with that in the SF8628 control. Treatment with TMZ had no statistically significant effect on *CDH1* and *CDH2* expression.

Treatment with TMZ was found to be effective in inhibiting PBT24 tumor growth on the CAM and its invasion into the CAM, and inhibiting neo-angiogenesis, but was ineffective on the SF8628 xenograft. This difference is possibly related to TMZ's differential effect on the carcinogenesis mechanisms regulating $[Cl^-]$i levels, where PBT24 cells showed initially higher KCC2 expression, and its activation by TMZ therapy. It cannot be excluded that the found differences among cell lines are also related to sex-specific disparities. Sex-specific analyses can improve accuracy in identifying the molecular subtype of glioblastoma, and patients can achieve a better outcome by personalizing treatment according to sex differences in molecular mechanisms [61].

4. Materials and Methods

4.1. Cell Lines and Cell Culture

A 13-year-old boy's high-grade glioblastoma PBT24 cell line cells were donated by Prof. M. M. Alonso (University of Navarra, Spain) [40] for the study. A 3-year-old girl's diffuse intrinsic pontine glioblastoma (DIPG) SF8628 cell line cells—harboring the histone H3.3 Lys 27-to-methionine (Sigma Aldrich, St. Louis, MO, USA)—were also studied [62,63]. The PBT24 cells were cultivated in Roswell Park Memorial Institute 1640 (RPMI) medium (Sigma Aldrich, St. Louis, MO, USA). The media were supplemented with 10% fetal bovine serum (FBS; Sigma Aldrich, St. Louis, MO, USA) containing 100 IU/mL of penicillin and 100 μg/mL of streptomycin (P/S; Sigma Aldrich, St. Louis, MO, USA). The SF8628 cells were cultivated in Dulbecco's Modified Eagle Medium (DMEM)–High-Glucose media (Sigma Aldrich, St. Louis, MO, USA). The media were supplemented with 10% fetal bovine serum (FBS; Sigma Aldrich, St. Louis, MO, USA) containing 100 IU/mL of penicillin and 100 μg/mL of streptomycin (P/S; Sigma Aldrich, St. Louis, MO, USA) and 2 mM L-Glutamine (Sigma Aldrich, St. Louis, MO, USA). Cells were incubated at 37 °C in a humidified 5% CO_2 atmosphere.

4.2. The CAM Model

According to the legislation in force in the EU and Lithuania, no approval for studies using the CAM model is needed from the Ethics Committee. Cobb 500 fertilized chicken eggs were obtained from a local hatchery (Rumšiškės, Lithuania) and kept in an incubator (Maino incubators, Oltrona di San Mamette, Italy) at 37 °C temperature and 60% relative air humidity. The eggs were rolled automatically once per hour until the third embryo development day (EDD3).

The CAM was detached from the eggshell at EDD3; the eggshell was cleaned with 70% ethanol, a small round hole was drilled in the location of the air chamber, and approximately 2 mL of the egg white was aspirated. A window of approximately 1 cm^2 in the eggshell was drilled and sealed with sterile transparent plastic tape. The eggs were kept in the incubator without rotation until GB cell tumor grafting on CAM at the seventh embryo development day (EDD7).

4.3. The PBT24 and SF8628 Tumor Study Groups

The growth and invasion into the CAM of the formatted PBT24 cell, as well as of SF8628 cell line xenografts, were investigated in the 6 groups. The study groups were as follows: PBT24-control (n = 13), PBT24-50 μM TMZ (n = 13), PBT24-100 μM TMZ (n = 10). The studied SF8628 tumor groups were the following: SF8628-control (n = 13), SF8628-50 μM TMZ (n = 14), SF8628-100 μM TMZ (n = 13).

Biomicroscopy in vivo and histological analyses of invasion, the thickness of the CAM and the number of vessels in the CAM under the tumor were performed.

The immunohistochemical (IHC) expression of PCNA in the tumor was studied in the following groups: PBT24-control (n = 9), PBT24-50 μM TMZ (n = 6), PBT24-100 μM TMZ (n = 6), SF8628-control (n = 8), SF8628-50 μM TMZ (n = 6), SF8628-100 μM TMZ (n = 7).

The expression of the EZH2 was investigated in the following: PBT24-control (n = 6), PBT24-50 μM TMZ (n = 7), PBT24-100 μM TMZ (n = 7), SF8628-control (n = 8), SF8628-50 μM TMZ (n = 6), SF8628-100 μM TMZ (n = 6). Efficacy studies of TMZ on GB in vivo and in vitro study at selected 100 and 50 μM doses were based on our and other investigators' data [64].

4.4. Biomicroscopy Data to Assess Tumor Growth and Drug Efficacy

The biomicroscopy of xenografts on CAM at embryo development from 9 to 12 days (EDD9–12) in vivo is suitable for evaluating the tumor growth characteristics and its malignancy, and detecting the disparities among different cell line tumors and the sensitivity to treatment. One sign of tumor malignancy and growth progression is the relatively rapid formation of vasculature around the tumor—a "spoked-wheel" consisting of tumor-attracted small blood vessels and formed by neo-angiogenesis new blood vessels. The tumor size, border clarity and changes in the "spoked-wheel" expression may serve as features of the drug effect on tumorigenesis.

4.5. Tumor Grafting on CAM In Vivo

An absorbable gelatin surgical sponge (Surgispon, Aegis Lifesciences, India) was cut manually with a blade into pieces of 9 mm^3 ($3 \times 3 \times 1$ mm). The 1×10^6 cells were resuspended in 20 μL of rat tail collagen, type I (Gibco, New York, NY, USA) (in the control group), and temozolomide (TMZ; Sigma Aldrich, St. Louis, MO, USA) dissolved in dimethyl sulfoxide (DMSO; Sigma Aldrich, St. Louis, MO, USA). A 20 μL liquid mixture of tumor cells was pipetted onto a piece of sponge. The 50 µM TMZ- and 100 µM TMZ-treated tumor groups and their controls were formed. At EDD7, the tumor was grafted onto the CAM among significant blood vessels. Its structural changes were observed with a stereomicroscope (SZX2-RFA16, Olympus, Tokyo, Japan) in vivo during the EDD9–12 period. The tumor images were acquired using a digital camera (DP92, Olympus, Tokyo, Japan) and CellSens Dimension 1.9 digital imaging software.

4.6. Histological Study of the Tumor

At EDD12, the specimens were harvested, fixed in a buffered 10% formalin solution for 24 h and embedded in paraffin wax. The tumor sample was cut with a microtome (Leica, Nussloch, Germany) into 3-µm-thick sections. The sections were stained with H–E and IHC methods. Visualization and photographing of H–E- and IHC-stained tumor slides were performed using a light microscope (BX40F4, Olympus, Tokyo, Japan) and a digital camera (XC30, Olympus, Tokyo, Japan) equipped with CellSens Dimension 1.9 software.

H–E-stained tumors were divided into two types: invasive and non-invasive. The tumor invasion into the CAM was categorized as the destruction of the chorionic epithelium (ChE) or/and tumor cell invasion into the CAM mesenchyme. The tumor not invaded into mesenchyme was located on the CAM surface, and the chorionic epithelium's integrity was not disrupted. The tumor invasion was examined in H–E slides at 20× and 40× magnifications.

4.7. Assessment of the CAM Thickness and the Number of Blood Vessels in CAM

The CAM thickness (width) was evaluated by photographing H–E-stained CAM at 4× magnification directly under the tumor. The thickness of CAM was measured (µm) in ten areas. The median CAM thickness was calculated in the area under the tumor.

The number of blood vessels was assessed by photographing the H–E-stained CAM at 4× magnification directly under the tumor. Blood vessels larger than 10 µm were counted.

4.8. Immunohistochemical Study

The expression of the PCNA and EZH2 markers was determined in tumor cells by immunohistochemistry. Primary antibodies to PCNA (PC10, Thermo Fisher Scientific, Branchburg, NJ, USA) and KMT6/EZH2 (phospho S21, ab84989, Abcam, Cambridge, UK) were used to detect PCNA and EZH2 positively stained tumor cells. Thin CAM sections of 3 µm were mounted onto adhesion slides (Thermo Fisher Scientific, Branchburg, NJ, USA), deparaffinized and rehydrated by standard techniques. Heat-induced antigen retrieval was performed using a Tris/EDTA buffer at pH 9 (K8002, Dako, Glostrup, Denmark) and a pressure cooker at 95 °C for 20 min (Thermo Fisher Scientific, Branchburg, NJ, USA). The Shandon CoverPlate System (Thermo Fisher Scientific, Branchburg, NJ, USA) was used for staining. Endogenous peroxidase was blocked with the Peroxidase Blocking Reagent (SM801, Dako, Glostrup, Denmark). The slides were treated with primary antibodies (1:100) for 30 min at room temperature. The primary antibody and antigen complex was determined using the horseradish peroxidase-labeled polymer dextran conjugated with a secondary mouse antibody and a linker (SM802 and SM804, respectively; Dako, Glostrup, Denmark) for 30 min at room temperature. Positive reactions were visualized using the 3,3′-diaminobenzidine-containing chromogen (DAB, DM827, Dako, Glostrup, Denmark), which gives a brown color to the site of the target antigen recognized by the primary antibody. After each step, a Tris-buffered saline solution containing Tween 20 (DM831, Dako, Glostrup, Denmark) was used as a wash buffer. Slides were counterstained with the Mayer hematoxylin solution (Sigma Aldrich, Taufkirchen, Germany), dehydrated, cleared and mounted.

For assessment of the tumor PCNA and EZH2 protein expression, two random vision fields (plot area 23,863.74 µm^2) of the immunohistochemically stained tumor were photographed at 40× magnification. All cells and the PCNA and EZH2 positively stained cells were calculated in selected vision fields, and the percentages of PCNA- and EZH2-positive cells were counted in each tumor.

4.9. Extraction of RNA from PBT24 and SF8628 Cell Line Cells

PBT24 and SF8628 cell line cells were treated with 50 μMTMZ for 24 h. The concentration of 50 μM was chosen because it corresponds to the mean plasma concentration of the drug in TMZ-treated patients [65]. Control groups were cultured in a cell culture medium depending on the cell line. According to the manufacturer's instructions, the total RNA was extracted using the TRIzol Plus RNA Purification Kit (Life Technologies, New York, NY, USA). The RNA quality and concentration were assessed using a NanoDrop2000 spectrophotometer (Thermo Scientific, Branchburg, NJ, USA). The total RNA integrity was analyzed using the Agilent 2100 Bioanalyzer system (Agilent Technologies, Santa Clara, CA, USA) with an Agilent RNA 6000 Nano Kit (Agilent Technologies, Santa Clara, CA, USA). The samples of RNA were stored at −80 °C until further analysis.

4.10. Determination of the SLC12A5, SLC12A2, CDH1 and CDH2 Gene Expression in PBT24 and SF8628 Cell Line Cells

RNA expression assays were performed for *SLC12A5* (Hs00221168_m1), SLC12A2(Hs0 0169032_m1), *CDH1* (Hs01023894_m1), *CDH2* (Hs00983056_m1) and *GAPDH* (Hs02786624_g1) genes. According to the manufacturer's instructions, reverse transcription was performed with the High-Capacity cDNA Reverse Transcription Kit with RNase Inhibitor (Applied Biosystems, Waltham, MA, USA) in a 20 μL reaction volume containing 50 ng RNA using the Biometra TAdvanced thermal cycler (Analytik Jena AG, Jena, Germany). The synthesized copy DNA (cDNA) was stored at 4 °C until use or at −80 °C for a longer time. Real-time polymerase chain reaction (PCR) was performed using an Applied Biosystems 7900 Fast Real-Time PCR System (Applied Biosystems, Waltham, MA, USA). The reactions were run in triplicate with 4 μL of cDNA template in a 20 μL reaction volume (10 μL of TaqMan Universal Master Mix II, no UNG (Applied Biosystems, Waltham, MA, USA), 1 μL of TaqMan Gene Expression Assay 20× (Applied Biosystems, Waltham, MA, USA), 5 μL of nuclease-free water (Invitrogen, Carlsbad, CA, USA)), with the program running at 95 °C for 10 min, followed by 45 cycles of 95 °C for 15 s, 50 °C for 2 min and 60 °C for 1 min.

The control and 24 h TMZ-treated groups ($n = 6$ per group) were tested for *SLC12A5, SLC12A2, CDH1* and *CDH2* expression.

4.11. Statistical Analysis

The statistical analysis was performed using the Statistical Package for Social Sciences, version 23.0 for Windows (IBM SPSS Statistics V23.0). The frequency of tumor invasion into the CAM was expressed as a percentage (%), and the chi-square test was used to compare tumor invasion into CAM frequency between the study groups. The Shapiro–Wilk test was used to verify the normality assumption. Data of PCNA and EZH2 positively stained cells, the number of blood vessels and the CAM thickness are expressed as median and range (minimum and maximum values). The difference between the two independent groups was evaluated using the nonparametric Mann–Whitney U test.

To investigate the KCC2, NKCC1, E-cadherin and N-cadherin genes RNA expression in the TMZ-treated and control groups, the threshold cycle (CT) value was normalized with the control *GAPDH*, and the ΔCT value was obtained. The Livak method (ΔΔCT) was used for calculating the relative fold change in expression levels [66]. The Spearman's rank correlation coefficient (r) was used to assess relationships between the *SLC12A5* and *SLC12A2* (ΔCT values were used). Differences at the value of $p < 0.05$ were considered significant. The figures were created using GraphPad Prism 7 and IBM SPSS Statistics 23.0 software.

5. Conclusions

PCNA and EZH2 marker assays of PBT24 and SF8628 glioblastoma tumors transplanted on CAM showed that the PBT24 tumor is less aggressive than the SF8628 tumor. TMZ treatment effectively decreased PBT24 xenograft growth but did not affect the SF8628 tumor. TMZ treatment reduced PCNA expression in PBT24 and SF8628 tumors and had no effect on EZH2 expression. TMZ activated Na-K-2Cl co-transporter gene expression in both tumors but increased K-Cl co-transporter gene expression only in PBT24 cells. The efficacy of the treatment may be related to changes in intracellular Cl^- levels induced by TMZ exposure. These data highlight the importance of studies on the activity of the K-Cl co-transporter in the context of personalized anticancer therapy efficacy.

Author Contributions: Conceptualization, D.S., I.B., A.V. and E.D.; Methodology, I.B. and A.V.; Validation, D.S., I.B. and A.V.; Formal analysis, E.D., D.S., I.B. and A.V.; Investigation, E.D.; Resources, D.S., I.B. and M.M.A.; Data curation, E.D.; Writing—Original Draft Preparation, E.D. and D.S.; Writing—Review and Editing, E.D., D.S., A.V., I.B., M.M.A. and A.P.; Visualization, E.D.; Supervision, D.S.; Project Administration, D.S. and I.B.; Funding Acquisition, D.S. and I.B. All authors have read and agreed to the published version of the manuscript.

Funding: This research was funded by the Research Council of Lithuania, grant number P-MIP-20-36.

Institutional Review Board Statement: Not applicable.

Informed Consent Statement: Not applicable.

Data Availability Statement: The data presented in this study are available on request from the corresponding author.

Conflicts of Interest: The authors declare no conflict of interest.

References

1. Durno, C.A.; Aronson, M.; Tabori, U.; Malkin, D.; Gallinger, S.; Chan, H.S.L. Oncologic surveillance for subjects with biallelic mismatch repair gene mutations: 10 year follow-up of a kindred. *Pediatr. Blood Cancer* **2012**, *59*, 652–656. [CrossRef] [PubMed]
2. Coleman, C.; Stoller, S.; Grotzer, M.; Stucklin, A.G.; Nazarian, J.; Mueller, S. Pediatric hemispheric high-grade glioma: Targeting the future. *Cancer Metastasis Rev.* **2020**, *39*, 245–260. [CrossRef] [PubMed]
3. Jihong, Z.; Malcolm, F.G.S.; Tracey, D.B. Temozolomide: Mechanisms of action, repair and resistance. *Curr. Mol. Pharmacol.* **2011**, *5*, 102–114.
4. Stupp, R.; Hegi, M.E.; Mason, W.P.; van den Bent, M.J.; Taphoorn, M.J.B.; Janzer, R.C.; Ludwin, S.K.; Allgeier, A.; Fisher, B.; Belanger, K.; et al. Effects of radiotherapy with concomitant and adjuvant temozolomide versus radiotherapy alone on survival in glioblastoma in a randomised phase III study: 5-year analysis of the EORTC-NCIC trial. *Lancet Oncol.* **2009**, *10*, 459–466. [CrossRef]
5. Karachi, A.; Dastmalchi, F.; Mitchell, D.A.; Rahman, M. Temozolomide for immunomodulation in the treatment of glioblastoma. *Neuro-Oncol.* **2018**, *20*, 1566–1572. [CrossRef] [PubMed]
6. Hegi, M.E.; Diserens, A.-C.; Gorlia, T.; Hamou, M.-F.; de Tribolet, N.; Weller, M.; Kros, J.M.; Hainfellner, J.A.; Mason, W.; Mariani, L.; et al. *MGMT* gene silencing and benefit from temozolomide in glioblastoma. *N. Engl. J. Med.* **2005**, *352*, 997–1003. [CrossRef]
7. Jones, C.; Karajannis, M.A.; Jones, D.T.W.; Kieran, M.W.; Monje, M.; Baker, S.J.; Becher, O.J.; Cho, Y.-J.; Gupta, N.; Hawkins, C.; et al. Pediatric high-grade glioma: Biologically and clinically in need of new thinking. *Neuro-Oncol.* **2017**, *19*, 153–161. [CrossRef]
8. Lee, S.Y. Temozolomide resistance in glioblastoma multiforme. *Genes Dis.* **2016**, *3*, 198–210. [CrossRef]
9. Margueron, R.; Reinberg, D. The polycomb complex PRC2 and its mark in life. *Nature* **2011**, *469*, 343–349. [CrossRef]
10. Maeno, E.; Ishizaki, Y.; Kanaseki, T.; Hazama, A.; Okada, Y. Normotonic cell shrinkage because of disordered volume regulation is an early prerequisite to apoptosis. *Proc. Natl. Acad. Sci. USA* **2000**, *97*, 9487–9492. [CrossRef]
11. Cong, D.; Zhu, W.; Kuo, J.S.; Hu, S.; Sun, D. Ion transporters in brain tumors. *Curr. Med. Chem.* **2015**, *22*, 1171–1181. [CrossRef] [PubMed]
12. Guzińska-Ustymowicz, K.; Pryczynicz, A.; Kemona, A.; Czyzewska, J. Correlation between proliferation markers: PCNA, Ki-67, MCM-2 and antiapoptotic protein Bcl-2 in colorectal cancer. *Anticancer Res.* **2009**, *29*, 3049–3052. [PubMed]
13. González-Magaña, A.; Blanco, F.J. Human PCNA structure, function, and interactions. *Biomolecules* **2020**, *10*, 570. [CrossRef] [PubMed]
14. Lewis, P.W.; Müller, M.M.; Koletsky, M.S.; Cordero, F.; Lin, S.; Banaszynski, L.A.; Garcia, B.A.; Muir, T.W.; Becher, O.J.; Allis, C.D. Inhibition of PRC2 activity by a gain-of-function H3 mutation found in pediatric glioblastoma. *Science* **2013**, *340*, 857–861. [CrossRef] [PubMed]

15. Simon, J.A.; Lange, C.A. Roles of the EZH2 histone methyltransferase in cancer epigenetics. *Mutat. Res.* **2008**, *647*, 21–29. [CrossRef]
16. Cao, R.; Wang, L.; Wang, H.; Xia, L.; Erdjument-Bromage, H.; Tempst, P.; Jones, R.S.; Zhang, Y. Role of histone H3 lysine 27 methylation in polycomb-group silencing. *Science* **2002**, *298*, 1039–1043. [CrossRef]
17. Bortner, C.D.; Cidlowski, J.A. Cell shrinkage and monovalent cation fluxes: Role in apoptosis. *Arch. Biochem. Biophys.* **2007**, *462*, 176–188. [CrossRef]
18. Pallud, J.; Le Van Quyen, M.; Bielle, F.; Pellegrino, C.; Varlet, P.; Cresto, N.; Baulac, M.; Duyckaerts, C.; Kourdougli, N.; Chazal, G.; et al. Cortical GABAergic excitation contributes to epileptic activities around human glioma. *Sci. Transl. Med.* **2014**, *6*, 244ra89. [CrossRef]
19. Aronica, E.; Boer, K.; Redeker, S.; Spliet, W.G.M.; van Rijen, P.C.; Troost, D.; Gorter, J.A. Differential expression patterns of chloride transporters, Na^+-K^+-$2Cl^-$-cotransporter and K^+-Cl^--cotransporter, in epilepsy-associated malformations of cortical development. *Neuroscience* **2007**, *145*, 185–196. [CrossRef]
20. Garzon-Muvdi, T.; Schiapparelli, P.; ap Rhys, C.; Guerrero-Cazares, H.; Smith, C.; Kim, D.-H.; Kone, L.; Farber, H.; Lee, D.Y.; An, S.S.; et al. Regulation of brain tumor dispersal by NKCC1 through a novel role in focal adhesion regulation. *PLoS Biol.* **2012**, *10*, e1001320. [CrossRef]
21. Haas, B.R.; Cuddapah, V.A.; Watkins, S.; Rohn, K.J.; Dy, T.E.; Sontheimer, H. With-No-Lysine Kinase 3 (WNK3) stimulates glioma invasion by regulating cell volume. *Am. J. Physiol. Cell Physiol.* **2011**, *301*, C1150–C1160. [CrossRef] [PubMed]
22. Haas, B.R.; Sontheimer, H. Inhibition of the sodium-potassium-chloride cotransporter isoform-1 reduces glioma invasion. *Cancer Res.* **2010**, *70*, 5597–5606. [CrossRef] [PubMed]
23. Luo, L.; Guan, X.; Begum, G.; Ding, D.; Gayden, J.; Hasan, M.N.; Fiesler, V.M.; Dodelson, J.; Kohanbash, G.; Hu, B.; et al. Blockade of cell volume regulatory protein NKCC1 increases TMZ-induced glioma apoptosis and reduces astrogliosis. *Mol. Cancer Ther.* **2020**, *19*, 1550–1561. [CrossRef] [PubMed]
24. Algharabil, J.; Kintner, D.B.; Wang, Q.; Begum, G.; Clark, P.A.; Yang, S.-S.; Lin, S.-H.; Kahle, K.T.; Kuo, J.S.; Sun, D. Inhibition of Na^+-K^+-$2Cl^-$ cotransporter isoform 1 accelerates temozolomide-mediated apoptosis in glioblastoma cancer cells. *Cell. Physiol. Biochem.* **2012**, *30*, 33–48. [CrossRef] [PubMed]
25. Bortner, C.D.; Sifre, M.I.; Cidlowski, J.A. Cationic gradient reversal and cytoskeleton-independent volume regulatory pathways define an early stage of apoptosis. *J. Biol. Chem.* **2008**, *283*, 7219–7229. [CrossRef]
26. Okada, Y.; Maeno, E. Apoptosis, cell volume regulation and volume-regulatory chloride channels. *Comp. Biochem. Physiol. Part A Mol. Integr. Physiol.* **2001**, *130*, 377–383. [CrossRef]
27. Sun, H.; Long, S.; Wu, B.; Liu, J.; Li, G. NKCC1 Involvement in the epithelial-to-mesenchymal transition is a prognostic biomarker in gliomas. *PeerJ* **2020**, *8*, e8787. [CrossRef]
28. Noh, M.-G.; Oh, S.-J.; Ahn, E.-J.; Kim, Y.-J.; Jung, T.-Y.; Jung, S.; Kim, K.-K.; Lee, J.-H.; Lee, K.-H.; Moon, K.-S. Prognostic significance of E-cadherin and N-cadherin expression in gliomas. *BMC Cancer* **2017**, *17*, 583. [CrossRef]
29. Bar, J.K.; Zub, L.; Lis-Nawara, A.; Noga, L.; Jeleń, M.; Paradowski, B. Expression and interactions between cell adhesion molecules CD44v6 and E-cadherin in human gliomas. *Adv. Clin. Exp. Med.* **2014**, *23*, 827–834. [CrossRef]
30. D'Urso, P.I.; D'Urso, O.F.; Storelli, C.; Catapano, G.; Gianfreda, C.D.; Montinaro, A.; Muscella, A.; Marsigliante, S. Retrospective protein expression and epigenetic inactivation studies of CDH1 in patients affected by low-grade glioma. *J. Neurooncol.* **2011**, *104*, 113–118. [CrossRef]
31. Lewis-Tuffin, L.J.; Rodriguez, F.; Giannini, C.; Scheithauer, B.; Necela, B.M.; Sarkaria, J.N.; Anastasiadis, P.Z. Misregulated E-cadherin expression associated with an aggressive brain tumor phenotype. *PLoS ONE* **2010**, *5*, e13665. [CrossRef] [PubMed]
32. Roos, W.P.; Batista, L.F.Z.; Naumann, S.C.; Wick, W.; Weller, M.; Menck, C.F.M.; Kaina, B. Apoptosis in malignant glioma cells triggered by the temozolomide-induced DNA lesion O6-methylguanine. *Oncogene* **2007**, *26*, 186–197. [CrossRef] [PubMed]
33. Hirose, Y.; Berger, M.S.; Pieper, R.O. P53 effects both the duration of G2/M arrest and the fate of temozolomide-treated human glioblastoma cells. *Cancer Res.* **2001**, *61*, 1957–1963. [PubMed]
34. Beier, D.; Röhrl, S.; Pillai, D.R.; Schwarz, S.; Kunz-Schughart, L.A.; Leukel, P.; Proescholdt, M.; Brawanski, A.; Bogdahn, U.; Trampe-Kieslich, A.; et al. Temozolomide preferentially depletes cancer stem cells in glioblastoma. *Cancer Res.* **2008**, *68*, 5706–5715. [CrossRef] [PubMed]
35. Ribatti, D. The chick embryo chorioallantoic membrane (CAM). A multifaceted experimental model. *Mech. Dev.* **2016**, *141*, 70–77. [CrossRef]
36. Kavaliauskaitė, D.; Stakišaitis, D.; Martinkutė, J.; Šlekienė, L.; Kazlauskas, A.; Balnytė, I.; Lesauskaitė, V.; Valančiūtė, A. The effect of sodium valproate on the glioblastoma U87 cell line tumor development on the chicken embryo chorioallantoic membrane and on EZH2 and P53 expression. *BioMed Res. Int.* **2017**, *2017*, 6326053. [CrossRef]
37. Ribatti, D. The chick embryo chorioallantoic membrane (CAM) assay. *Reprod. Toxicol.* **2017**, *70*, 97–101. [CrossRef]
38. Zhao, Z.; Bauer, N.; Aleksandrowicz, E.; Yin, L.; Gladkich, J.; Gross, W.; Kaiser, J.; Hackert, T.; Strobel, O.; Herr, I. Intraductal papillary mucinous neoplasm of the pancreas rapidly xenografts in chicken eggs and predicts aggressiveness. *Int. J. Cancer* **2018**, *142*, 1440–1452. [CrossRef]
39. Valiulytė, I.; Curkūnavičiūtė, R.; Ribokaitė, L.; Kazlauskas, A.; Vaitkevičiūtė, M.; Skauminas, K.; Valančiūtė, A. The anti-tumorigenic activity of Sema3C in the chick embryo chorioallantoic membrane model. *Int. J. Mol. Sci.* **2019**, *20*, 5672. [CrossRef]

40. Martinez-Velez, N.; Marigil, M.; García-Moure, M.; Gonzalez-Huarriz, M.; Aristu, J.J.; Ramos-García, L.-I.; Tejada, S.; Díez-Valle, R.; Patiño-García, A.; Becher, O.J.; et al. Delta-24-RGD combined with radiotherapy exerts a potent antitumor effect in diffuse intrinsic pontine glioma and pediatric high grade glioma models. *Acta Neuropathol. Commun.* **2019**, *7*, 64. [CrossRef]
41. Efremov, L.; Abera, S.F.; Bedir, A.; Vordermark, D.; Medenwald, D. Patterns of glioblastoma treatment and survival over a 16-years period: Pooled data from the german cancer registries. *J. Cancer Res. Clin. Oncol.* **2021**, *147*, 3381–3390. [CrossRef] [PubMed]
42. Ene, C.I.; Holland, E.C. Personalized medicine for gliomas. *Surg. Neurol. Int.* **2015**, *6*, S89–S95. [CrossRef] [PubMed]
43. Li, C.M.; Haratipour, P.; Lingeman, R.G.; Perry, J.J.P.; Gu, L.; Hickey, R.J.; Malkas, L.H. Novel peptide therapeutic approaches for cancer treatment. *Cells* **2021**, *10*, 2908. [CrossRef] [PubMed]
44. Mohammad, F.; Weissmann, S.; Leblanc, B.; Pandey, D.P.; Højfeldt, J.W.; Comet, I.; Zheng, C.; Johansen, J.V.; Rapin, N.; Porse, B.T.; et al. EZH2 is a potential therapeutic target for H3K27M-mutant pediatric gliomas. *Nat. Med.* **2017**, *23*, 483–492. [CrossRef] [PubMed]
45. Lee, J.; Son, M.J.; Woolard, K.; Donin, N.M.; Li, A.; Cheng, C.H.; Kotliarova, S.; Kotliarov, Y.; Walling, J.; Ahn, S.; et al. Epigenetic-mediated dysfunction of the bone morphogenetic protein pathway inhibits differentiation of glioblastoma-initiating cells. *Cancer Cell* **2008**, *13*, 69–80. [CrossRef]
46. Kim, E.; Kim, M.; Woo, D.-H.; Shin, Y.; Shin, J.; Chang, N.; Oh, Y.T.; Kim, H.; Rheey, J.; Nakano, I.; et al. Phosphorylation of EZH2 activates STAT3 signaling via STAT3 methylation and promotes tumorigenicity of glioblastoma stem-like cells. *Cancer Cell* **2013**, *23*, 839–852. [CrossRef]
47. Fan, T.-Y.; Wang, H.; Xiang, P.; Liu, Y.-W.; Li, H.-Z.; Lei, B.-X.; Yu, M.; Qi, S.-T. Inhibition of EZH2 reverses chemotherapeutic drug TMZ chemosensitivity in glioblastoma. *Int. J. Clin. Exp. Pathol.* **2014**, *7*, 6662–6670.
48. Amara, S.; Tiriveedhi, V. Inflammatory role of high salt level in tumor microenvironment (review). *Int. J. Oncol.* **2017**, *50*, 1477–1481. [CrossRef]
49. Leslie, T.K.; James, A.D.; Zaccagna, F.; Grist, J.T.; Deen, S.; Kennerley, A.; Riemer, F.; Kaggie, J.D.; Gallagher, F.A.; Gilbert, F.J.; et al. Sodium homeostasis in the tumour microenvironment. *Biochim. Biophys. Acta (BBA)-Rev. Cancer* **2019**, *1872*, 188304. [CrossRef]
50. Kahle, K.T.; Rinehart, J.; Lifton, R.P. Phosphoregulation of the Na-K-2Cl and K-Cl cotransporters by the WNK kinases. *Biochim. Biophys. Acta (BBA)-Mol. Basis Dis.* **2010**, *1802*, 1150–1158. [CrossRef]
51. Gagnon, K.B.E.; England, R.; Delpire, E. Characterization of SPAK and OSR1, regulatory kinases of the Na-K-2Cl cotransporter. *Mol. Cell Biol.* **2006**, *26*, 689–698. [CrossRef] [PubMed]
52. Dowd, B.F.X.; Forbush, B. PASK (proline-alanine-rich STE20-related kinase), a regulatory kinase of the Na-K-Cl cotransporter (NKCC1). *J. Biol. Chem.* **2003**, *278*, 27347–27353. [CrossRef] [PubMed]
53. Nico, B.; Mangieri, D.; Tamma, R.; Longo, V.; Annese, T.; Crivellato, E.; Pollo, B.; Maderna, E.; Ribatti, D.; Salmaggi, A. Aquaporin-4 contributes to the resolution of peritumoural brain oedema in human glioblastoma multiforme after combined chemotherapy and radiotherapy. *Eur. J. Cancer* **2009**, *45*, 3315–3325. [CrossRef] [PubMed]
54. Bortner, C.D.; Hughes, F.M.; Cidlowski, J.A. A primary role for K^+ and Na^+ efflux in the activation of apoptosis. *J. Biol. Chem.* **1997**, *272*, 32436–32442. [CrossRef]
55. Lenart, B.; Kintner, D.B.; Shull, G.E.; Sun, D. Na-K-Cl cotransporter-mediated intracellular Na^+ accumulation affects Ca^{2+} signaling in astrocytes in an in vitro ischemic model. *J. Neurosci.* **2004**, *24*, 9585–9597. [CrossRef]
56. Chen, H.; Sun, D. The role of Na-K-Cl co-transporter in cerebral ischemia. *Neurol. Res.* **2005**, *27*, 280–286. [CrossRef]
57. Ben-Ari, Y. Excitatory actions of gaba during development: The nature of the nurture. *Nat. Rev. Neurosci.* **2002**, *3*, 728–739. [CrossRef]
58. Yamada, J.; Okabe, A.; Toyoda, H.; Kilb, W.; Luhmann, H.J.; Fukuda, A. Cl^- uptake promoting depolarizing GABA actions in immature rat neocortical neurones is mediated by NKCC1. *J. Physiol.* **2004**, *557*, 829–841. [CrossRef]
59. Owens, D.F.; Kriegstein, A.R. Is there more to GABA than synaptic inhibition? *Nat. Rev. Neurosci.* **2002**, *3*, 715–727. [CrossRef]
60. Di Cristo, G.; Awad, P.N.; Hamidi, S.; Avoli, M. KCC2, epileptiform synchronization, and epileptic disorders. *Prog. Neurobiol.* **2018**, *162*, 1–16. [CrossRef]
61. Yang, W.; Warrington, N.M.; Taylor, S.J.; Whitmire, P.; Carrasco, E.; Singleton, K.W.; Wu, N.; Lathia, J.D.; Berens, M.E.; Kim, A.H.; et al. Sex differences in GBM revealed by analysis of patient imaging, transcriptome, and survival data. *Sci. Transl. Med.* **2019**, *11*, eaao5253. [CrossRef] [PubMed]
62. SF8628 Human DIPG H3.3-K27M Cell Line SF8628 Pediatric Diffuse Intrinsic Pontine Glioma (DIPG) Cell Line Harbors the Histone H3.3 Lys 27-to-Methionine (K27M) Mutation and Can Support Research and Drug Development Efforts Targeting DIPG. | Sigma-Aldrich. Available online: http://www.sigmaaldrich.com/ (accessed on 8 December 2021).
63. Mueller, S.; Hashizume, R.; Yang, X.; Kolkowitz, I.; Olow, A.K.; Phillips, J.; Smirnov, I.; Tom, M.W.; Prados, M.D.; James, C.D.; et al. Targeting wee1 for the treatment of pediatric high-grade gliomas. *Neuro-Oncol.* **2014**, *16*, 352–360. [CrossRef] [PubMed]
64. Alonso, M.M.; Gomez-Manzano, C.; Bekele, B.N.; Yung, W.K.A.; Fueyo, J. Adenovirus-based strategies overcome temozolomide resistance by silencing the O6-methylguanine-DNA methyltransferase promoter. *Cancer Res.* **2007**, *67*, 11499–11504. [CrossRef] [PubMed]

65. Ortiz, R.; Perazzoli, G.; Cabeza, L.; Jiménez-Luna, C.; Luque, R.; Prados, J.; Melguizo, C. Temozolomide: An updated overview of resistance mechanisms, nanotechnology advances and clinical applications. *Curr. Neuropharmacol.* **2021**, *19*, 513–537. [CrossRef]
66. Livak, K.J.; Schmittgen, T.D. Analysis of relative gene expression data using real-time quantitative PCR and the 2(-delta delta C(T)) method. *Methods* **2001**, *25*, 402–408. [CrossRef]

International Journal of
Molecular Sciences

MDPI

Article

A Novel Role of BIRC3 in Stemness Reprogramming of Glioblastoma

Qiong Wu [1], Anders E. Berglund [2], Robert J. MacAulay [3] and Arnold B. Etame [1,*]

1 Department of Neuro-Oncology, H. Lee Moffitt Cancer Center and Research Institute, Tampa, FL 33612, USA; qiong.wu@moffitt.org
2 Department of Biostatistics and Bioinformatics, H. Lee Moffitt Cancer Center and Research Institute, Tampa, FL 33612, USA; anders.berglund@moffitt.org
3 Departments of Anatomic Pathology, H. Lee Moffitt Cancer Center and Research Institute, Tampa, FL 33612, USA; robert.macaulay@moffitt.org
* Correspondence: arnold.etame@moffitt.org; Tel.: +1-813-745-3871

Abstract: Stemness reprogramming remains a largely unaddressed principal cause of lethality in glioblastoma (GBM). It is therefore of utmost importance to identify and target mechanisms that are essential for GBM stemness and self-renewal. Previously, we implicated BIRC3 as an essential mediator of therapeutic resistance and survival adaptation in GBM. In this study, we present novel evidence that BIRC3 has an essential noncanonical role in GBM self-renewal and stemness reprogramming. We demonstrate that BIRC3 drives stemness reprogramming of human GBM cell lines, mouse GBM cell lines and patient-derived GBM stem cells (GSCs) through regulation of BMP4 signaling axis. Specifically, BIRC3 induces stemness reprogramming in GBM through downstream inactivation of BMP4 signaling. RNA-Seq interrogation of the stemness reprogramming hypoxic (pseudopalisading necrosis and perinecrosis) niche in GBM patient tissues further validated the high BIRC3/low BMP4 expression correlation. BIRC3 knockout upregulated BMP4 expression and prevented stemness reprogramming of GBM models. Furthermore, siRNA silencing of BMP4 restored stemness reprogramming of BIRC3 knockout in GBM models. In vivo silencing of BIRC3 suppressed tumor initiation and progression in GBM orthotopic intracranial xenografts. The stemness reprograming of both GSCs and non-GSCs populations highlights the impact of BIRC3 on intra-tumoral cellular heterogeneity GBM. Our study has identified a novel function of BIRC3 that can be targeted to reverse stemness programming of GBM.

Keywords: brain tumor; GBM; cancer stem cell; BIRC3; BMP4; stemness

Citation: Wu, Q.; Berglund, A.E.; MacAulay, R.J.; Etame, A.B. A Novel Role of BIRC3 in Stemness Reprogramming of Glioblastoma. *Int. J. Mol. Sci.* **2022**, *23*, 297. https://doi.org/10.3390/ijms23010297

Academic Editors: Nives Pećina-Šlaus and Ivana Jovčevska

Received: 11 November 2021
Accepted: 24 December 2021
Published: 28 December 2021

1. Introduction

Glioblastoma (GBM) is a highly resistant and lethal brain cancer with limited treatment options. The current multimodal therapy of maximal-safe surgical resection, radiation therapy (RT) and concurrent temozolomide (TMZ) leads to a median survival of only 14 months [1]. A classic hallmark of GBM is the rapid acquisition of therapeutic resistance leading to lethality. There is therefore a significant unmet need for effective anti-GBM therapies that prevent early acquisition of resistance.

Stemness adaptation is a leading hypothesis for therapeutic failures in GBM [2]. GBM cells with stem-like phenotype known as GBM stem-like cells or GBM stem cells (GSCs) drive resistance to RT and TMZ treatment [3]. Evidence from GBM mouse models implicates GSCs in the repopulation of tumors following TMZ and RT treatment [4]. Plasticity towards tumor repopulation is the basis for recurrence, cellular intra-tumoral heterogeneity and disease progression. GSCs have tumor-initiating capabilities and can induce tumors in vivo that recapitulate the molecular features of the parental GBM tumor [4–6]. Stemness phenotype in GBM is characterized by expressions of CD133 and the neural stem cell marker Nestin [7–11]. From a transcriptomics perspective, GSCs exploit normal neural stem cell developmental transcription mechanisms towards tumorigenesis adaptation [12–14]. Given the central role of stemness reprogramming in GBM resistance, it is therefore of utmost importance to identify and target essential stemness reprogramming mechanisms in GBM. Strategies that prevent or reverse stemness reprogramming would undoubtedly have a significant impact in GBM.

Previously, we implicated the anti-apoptotic protein, BIRC3 as an essential mediator of therapeutic resistance and survival adaptation in GBM [15,16]. BIRC3 is an inhibitor of apoptosis protein with established canonical anti-apoptosis function through inhibition of caspase activation [17,18]. Using GBM cell lines and GBM patient tissue samples, we have established that BIRC3 contributed toward TMZ and RT resistance in GBM through PI3K and STAT3 signaling activation [15]. We also showed that BIRC3 expression increased during GBM treatment, GBM recurrence and adversely impacted upon GBM patient survival [15]. In support of our findings, another group independently implicated BIRC3 as a facilitator of malignant progression in GBM [19]. In a subsequent study, we reported that BIRC3 was an important contributor to GBM hypoxia adaptation and mesenchymal phenotype [16]. Although the preponderance of evidence thus far supports a role for BIRC3 in GBM resistance adaptation, it remains unclear if BIRC3 has any role in GBM stemness adaptation.

Based on our previous discoveries of several novel noncanonical functions of BIRC3 in GBM survival adaptation, we hypothesized that BIRC3 was also critical for stemness reprogramming. In support of this hypothesis, we present novel evidence that BIRC3 regulates stemness reprogramming, tumor initiation and tumor progression in GBM. Importantly, we demonstrate that BIRC3 drives stemness reprogramming of human GBM cell lines, mouse GBM cell line and patient-derived GSCs through regulation of BMP4 signaling axis. RNA-Seq interrogation of the stemness reprogramming hypoxic (pseudopalisading necrosis and perinecrosis) niche in GBM patient tissues further validated the high BIRC3/low BMP4 expression correlation. Our findings represent the first implication of an anti-apoptotic protein in GBM cell-fate stemness reprogramming. We have therefore identified a novel noncanonical function of BIRC3 that can be targeted to reverse stemness programming of GBM.

2. Results

2.1. BIRC3 Expression Correlates with Stem Cell Markers Expression and Self-Renewal in Both Human and Mouse GBM Cells

We previously validated BIRC3 as a novel anti-GBM target for therapeutic resistance [15]. We were therefore interested in determining if BIRC3 played a role in GBM cell fate stemness reprogramming. We therefore established BIRC3 overexpressing and BIRC3 knockout lines in U251 and U87 human GBM cell lines. BIRC3 protein expression was validated by western blot (Figure 1A). In order to evaluate the impact of BIRC3 on stemness reprogramming, we evaluated the impact of BIRC3 gain-of-function and loss-of-function on the ability of GBM cells to form neurospheres. GSCs grow as neurospheres and therefore neurosphere formation serves as a surrogate for stemness [20]. Equal cell numbers were seeded into wells and allowed to grow into colonies under stem cell culture conditions (U251: 5000 cells/well; U87: 2000 cells/well). BIRC3 overexpression significantly enhanced neurosphere formation in both U251 and U87 GBM cells (Figure 1B,C, $p < 0.05$). Moreover, we observed a significant reduction in neurosphere formation in U87 BIRC3 knockout GBM cells compare to control cells (Figure 1C, $p < 0.05$). In U251 GBM cells, BIRC3 knockout significantly impacted neurosphere formation but to a lesser extent compared to U87 GBM cells (Figure 1B, $p < 0.05$). To further determine the effect of BIRC3 on maintenance of GBM stemness, we examined two different stem cell gene markers *CD133* and *ABCG2* expression by real-time PCR [21]. CD133 is a glycoprotein that is the most employed marker for isolation of cancer stem cell population from different tumors, especially various gliomas [7,8,10,22]. ABCG2 (ATP-binding cassette super-family G member 2) is a membrane-associated protein also a known cancer stem cell marker in gliomas [23–25].

Interestingly, BIRC3 overexpression significantly induced CD133 and ABCG2 expressions in both U251 and U87 cell lines (Figure 1D, $p < 0.05$). BIRC3 knockout cells was associated with a significant reduction in both CD133 and ABCG2 expressions compared to control wild type cells (Figure 1D, $p < 0.05$). Moreover, we also examined another stemness marker ALDH1A3. ALDH1A3 (Aldehyde dehydrogenase 1 family member A3) an isozyme metabolizes aldehydes to their respective carboxylic acid and is higher expressed in the cancer stem cell niche of GBM and other cancers [26,27]. In both U251 and U87 GBM cells, BIRC3 overexpression significantly increased ALDH1A3 expression compared to wild type control cells, while BIRC3 knockout only reduced ALDH1A3 expression in U87 cells (Figure S1, $p < 0.05$). To further characterize the role of BIRC3 in GBM stemness reprogramming, we evaluated Nestin expression via confocal immunocytochemistry analysis. Nestin expression is a prerequisite for the maintenance of stemness [10,11]. There was a direct and strong association between BIRC3 and Nestin expressions in both U251 and U87 GBM cells whereby BIRC3 overexpression induced higher Nestin expressions compared to BIRC3 knockout cells (Figure 1E). In order to validate our findings in a mouse GBM cell model, we repeated these experiments using CT-2A cell line, which is a murine glioma cell line [28]. BIRC3 expression enhanced neurosphere formation in CT-2A cells as we had observed in the human glioma cell lines. Mouse BIRC3 protein expression and knockout efficiency were validated by western blot (Figure 1F). BIRC3 overexpression significantly increased neurosphere formation capacity (Figure 1G, $p < 0.05$) while BIRC3 knockout significantly reduced neurosphere formation capacity (Figure 1G, $p < 0.05$). Importantly, CD133 and ABCG2 were significantly upregulated in BIRC3 overexpressed cells and downregulated in BIRC3 knockout cells (Figure 1H, $p < 0.05$). Collectively, these findings suggest that BIRC3 expression is critical for GBM cell self-renewal and stemness maintenance.

2.2. Human GBM Stem Cell Self-Renewal Is Regulated by BIRC3 Expression

In an effort to ascertain if BIRC3 had a similar impact on human GBM stem cells (GSCs), we established BIRC3 overexpression and BIRC3 knockout lines in three patient-derived GSCs (Figure 2A). First, we examined the effect of BIRC3 expression on GSC stemness maintenance by evaluating CD133 and ABCG2 expression levels in GSCs. Real-time PCR results revealed that BIRC3 overexpression significantly induced higher expressions of CD133 in GSC-2 (Figure 2B, $p < 0.05$); and higher expressions of ABCG2 in both GSC-1 and GSC-2 to maintain self-renewal and stemness (Figure 2B, $p < 0.05$). Interestingly, CD133 and ABCG2 expressions were significantly downregulated in all BIRC3 knockout GSCs including GSC-1, GSC-2 and GSC-3 (Figure 2B, $p < 0.05$). Moreover, ALDH1A3 expression was significantly downregulated in all BIRC3 knockout GSCs compared to wild type control, while its expression was induced only in BIRC3 overexpressed GSC-1 and GSC-3 (Figure S2). Next, in order to further understand and validate the stemness phenotype induced by BIRC3, we cultured all three GSCs under differentiating media conditions. BIRC3 overexpression facilitated formation of neurospheres and self-renewal in differentiated GSCs, whereas BIRC3 knockout prevented neurosphere formation and self-renewal (Figure 2C, $p < 0.05$) in differentiated GSCs. A similar trend was noted with real-time PCR analysis of CD133 and ABCG2 expressions. A significant fraction of BIRC3 overexpression GSCs demonstrated enhanced stemness marker expression; enhanced self-renewal capabilities; and enhanced stemness maintenance (Figure 2D, $p < 0.05$). BIRC3 knockout significantly inhibited stemness marker expression; self-renewal capabilities; and stemness maintenance compared to wild type cells in GSC-2 and GSC-3. Moreover, BIRC3 overexpression significantly increased ABCG2 expression, while BIRC3 knockout significantly suppressed ABCG2 expression (Figure 2D, $p < 0.05$). Furthermore, confocal immunocytochemistry analysis revealed that BIRC3 expression was sufficient in strongly inducing Nestin expression in all 3 GSCs (Figure 2E). Therefore, BIRC3 serves as a critical regulator in GSC self-renewal and stemness maintenance, even in differentiated GSCs.

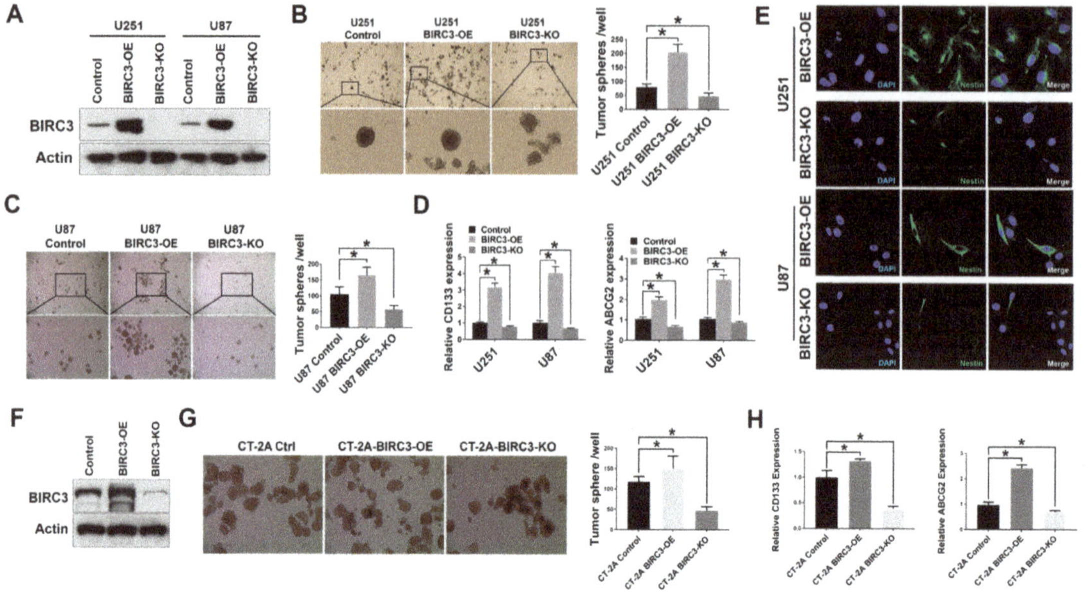

Figure 1. Expression of BIRC3 correlates with stem cell markers expression and self-renewal in both human and mouse GBM cells (**A**). Protein expression of BIRC3 in U251 and U87 GBM cells. Each cell line includes control, BIRC3 overexpression (BIRC3-OE) and BIRC3 knockout (BIRC3-KO) groups. Specific antibodies as indicated. β-actin acts as internal control. (**B**,**C**). Control, BIRC3-OE or BIRC3-KO of U251 and U87 cells were seeded in 6 well plates and cultured in neurosphere formation medium. The number of neurospheres were observed and calculated under microscope. (**B**): Representative images are under 4× magnification (top raw) and 20× magnification (bottom raw). (**C**): Representative images are under 4× magnification (top raw) and 10× magnification (bottom raw). n = 5, * $p < 0.05$. (**D**). CD133 and ABCG2 mRNA expression analyzed by real-time PCR in U251/U87 control, BIRC3-OE and BIRC3-KO cells. n = 3, * $p < 0.05$. (**E**). Immunofluorescence staining of Nestin in U251/U87 BIRC3-OE and BIRC3-KO cells. Blue: DAPI; Green: Nestin. (**F**). Protein expression of mBIRC3 in CT-2A mouse GBM cells including control, BIRC3-OE and BIRC3-KO groups. Specific antibodies as indicated. β-actin acts as internal control. (**G**). Control, BIRC3-OE or BIRC3-KO of CT-2A cells were seeded in 6 well plates and cultured in neurosphere formation medium. The number of neurospheres were observed and calculated under microscope. Representative images are under 10× magnification. $n = 5$, * $p < 0.05$. (**H**). Mouse CD133 and ABCG2 mRNA expression were analyzed by real-time PCR in CT-2A cells. $n = 3$, * $p < 0.05$.

2.3. BIRC3 Regulates BMP4 Signaling Inhibition in GBM

BMP4 is strongly associated with GBM stem cell differentiation [29–31], and has been reported as a potential anti-GBM target [32]. Since BIRC3 expression is associated with GBM cell self-renewal and stemness maintenance, we wanted to determine if there was any correlation between BIRC3 expression and BMP4 signaling activation. We sought to examine the relationship between *BIRC3* and *BMP4* using patient GBM tissue data. We initially estimated mRNA expression correlation between *BIRC3* and *BMP4* using TCGA (The Cancer Genome Atlas) GBM PanCan dataset. This dataset contains whole tumor data and therefore does not account for regional heterogeneity in GBM. *BIRC3* and *BMP4* had a low correlation in GBM PanCan dataset (Figure S3). However, considering the GBM intra-tumoral heterogeneity and BIRC3 regional expression [16], we then analyzed BIRC3 and BMP4 expression using the IVY Glioblastoma Atlas dataset that has RNA-Seq datasets from MRI-distinct GBM regions. We examined hypoxic (pseudopalisading necrosis and perinecrosis) and vascular (hyperplastic blood vessels and vascular proliferative) regions of GBM. The hypoxic region has been established as a critical niche for stemness reprogramming in GBM [33–36]. There was a negative correlation between *BIRC3* and *BMP4* expressions in both the vascular and hypoxic niches in GBM (Pearson r = −0.475; spearman r = −0449; Figure 3A). The stemness reprogramming hypoxic niche demonstrated high BIRC3/low BMP4 expression profile. Conversely, the vascular niche demonstrated low BIRC3/high BMP4 expression profile. To further confirm if the impact of BIRC3 on GBM cell stemness and self-renewal is correlated with BMP4 expression, we first performed real-time PCR analysis to evaluate BMP4 expression in both BIRC3 overexpressing and knockout GBM cell lines. We found that BIRC3 overexpression significantly inhibited BMP4 expression in GBM cell lines compare to wild type control. Interestingly, we found that knockout of BIRC3 significantly activated BMP4 expression (Figure 3B, $p < 0.05$). Moreover, similar observations were made in undifferentiated GSCs, differentiated GSCs and CT-2A mouse GBM cells (Figure 3C–E). BIRC3 significantly suppressed BMP4 expression in GSC-2 and GSC-3; and, furthermore, depletion of BIRC3 induced BMP4 expression in GSC-1 and GSC-2 (Figure 3C, $p < 0.05$). In differentiated GSCs, high levels of BIRC3 significantly inhibited BMP4 expression (Figure 3D, $p < 0.05$). However, depletion of BIRC3 increased BMP4 expression only in GSC-2 (Figure 3D, $p < 0.05$). We observed a similar gene expression pattern in CT-2A cells as well as human GBM cell lines (Figure 3E, $p < 0.05$). Furthermore, western blot results indicated that in BIRC3 knockout U251 and U87 cells, BMP4 signaling was strongly activated through SMAD1/5 phosphorylation and BIRC3 overexpression suppressed this activation effectively (Figure 3F). These results suggested that BIRC3 could directly suppress BMP4 signaling activation in GBM cell lines and stem cells.

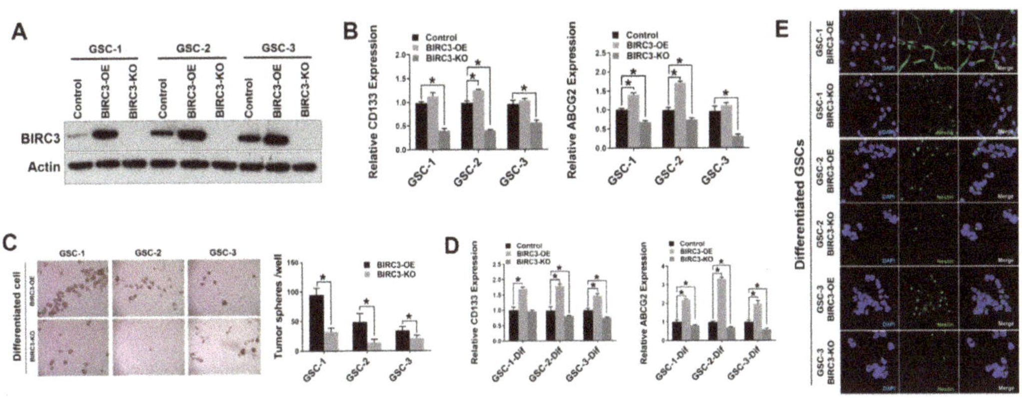

Figure 2. Human GBM stem cell self-renewal is regulated by BIRC3 expression. (**A**). Protein expression of BIRC3 in three different GSCs. Each GSC includes control, BIRC3-OE and BIRC3-KO groups. Specific antibodies as indicated. β-actin acts as internal control. (**B**). CD133 and ABCG2 mRNA expression analyzed by real time PCR in control, BIRC3-OE and BIRC3-KO GSCs. $n = 3$, * $p < 0.05$. (**C**). Control, BIRC3-OE or BIRC3-KO of differentiated GSCs were seeded in 6-well plate and cultured in neurosphere formation medium. The number of neurospheres were observed and calculated by microscope. Representative images are under 4× magnification. $n = 5$, * $p < 0.05$. (**D**). CD133 and ABCG2 mRNA expression analyzed by real time PCR in control, BIRC3-OE and BIRC3-KO of differentiated GSCs. $n = 3$, * $p < 0.05$. (**E**). Immunofluorescence staining of Nestin in differentiated BIRC3-OE and BIRC3-KO GSCs.

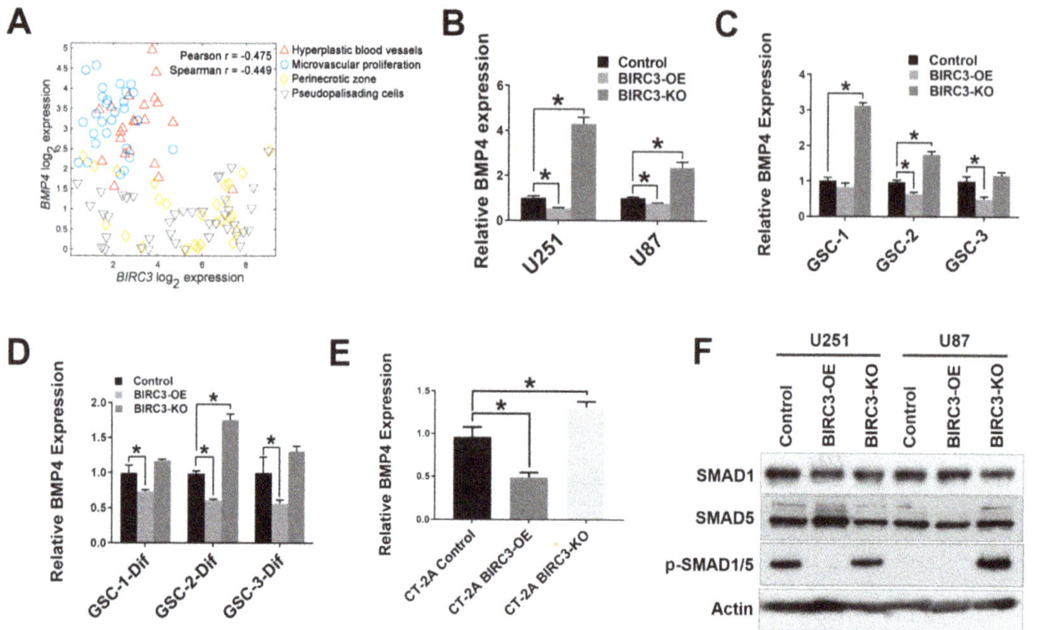

Figure 3. BIRC3 directs BMP4 signaling inhibition in GBM. (**A**). Analysis of correlation between *BIRC3* and *BMP4* expression in different regions of GBM using IVY dataset. Analyzed regions include hyperplastic blood vessels, microvascular proliferation region, perinecrotic zone and pseudopalisading cells region. (**B–D**). Human BMP4 mRNA expression analyzed by real time PCR in control, BIRC3-OE and BIRC3-KO cells including U251/U87 GBM cell lines, GSCs and differentiated GSCs. $n = 3$, * $p < 0.05$. (**E**). Mouse BMP4 mRNA expression analyzed by real time PCR in control, BIRC3-OE and BIRC3-CT-2A cells. $n = 3$, * $p < 0.05$. (**F**). Protein expression of SMAD1, SMAD5 and phosphorylated SMAD1/5 in U251/U87 control, BIRC3-OE and BIRC3-KO cells. Specific antibodies as indicated. β-actin acts as internal control.

2.4. BIRC3 Mediated Stemness Reprogramming in GBM Cells Is Dependent on BMP4 Suppression

Since BIRC3 knockout inhibits GBM cell self-renewal, we were interested in determining if this process was directly driven by BMP4 signaling activation. When we silenced BMP4 in BIRC3 knockout U251 and U87 cells using selective siRNAs, we observed a significant blockade and reversal of low-BIRC3 induced up-regulation of BMP4 expression (Figure 4A, $p < 0.05$). Moreover, we were also interested in determining if silencing BMP4 in BIRC3 knockout cell could restore GBM cell stemness reprograming. We therefore first examined the relative expressions of stemness markers CD133 and ABCG2 following BMP4 silencing. In BIRC3 knockout U251 cells, siRNA silencing of BMP4 in BIRC3 knockout cells significantly increased CD133 and ABCG2 expressions more than 2-fold (Figure 4B, $p < 0.05$). A similar trend in CD133 and ABCG2 expressions was noted in U87 cells (Figure 4C, $p < 0.05$). Interestingly, silencing of BMP4 in BIRC3 knockout U251 and U87 cells reduced SMAD1/5 phosphorylation, which had initially been induced by BIRC3 knockout, and further suppressed BMP4 signaling activation (Figure 4D). Next, we performed tumor sphere assay to determine if BMP4 signaling suppression directly impacted upon GBM cell self-renewal phenotype and neurosphere formation. The tumor sphere formation assay results revealed that silencing BMP4 in BIRC3 knockout U251 cells significantly induced tumor sphere formation and rescued the loss of stemness which was previously induced by BIRC3 knockout (Figure 4E, $p < 0.05$). These results suggested that low expression of BIRC3 suppresses GBM cell self-renewal through BMP4 signaling activation and that silencing of BMP4 could significantly restore stemness. Hence, BIRC3 drives stemness reprogramming in GBM through suppression of BMP4 signaling activation.

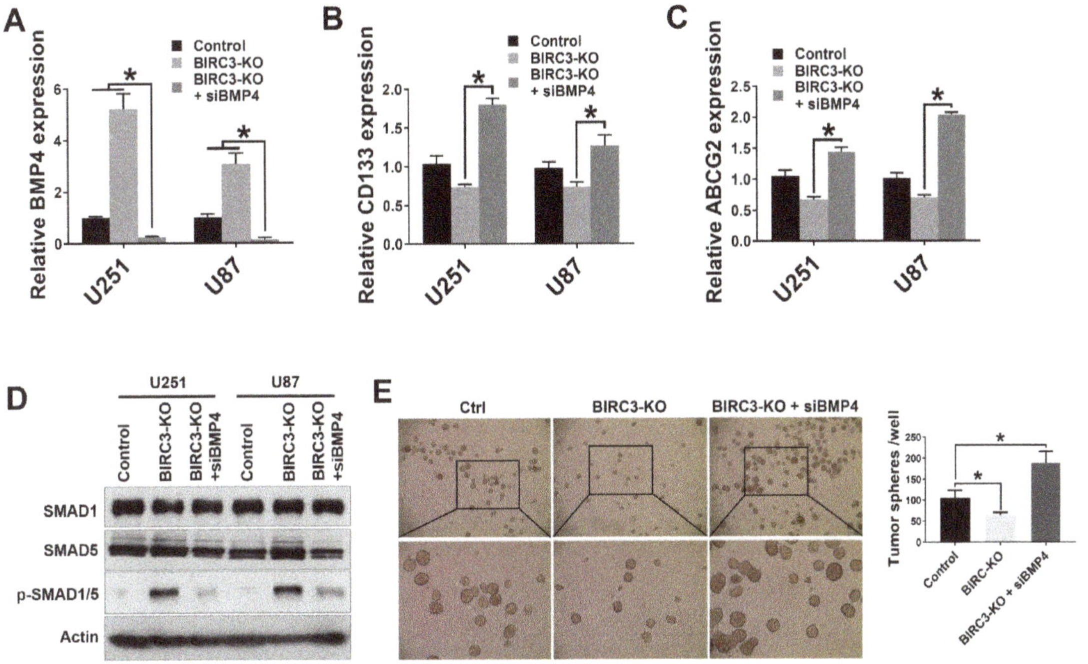

Figure 4. BIRC3 impacts GBM cell self-renewal and is dependent on BMP4 suppression. (**A**). Human BMP4 mRNA expression analyzed by real time PCR in control, BIRC3-KO, and BMP4-siRNA silenced BIRC3-KO U251/U87 GBM cell lines. $n = 3$, * $p < 0.05$. (**B**). Human CD133 mRNA expression analyzed by real time PCR in control, BIRC3-KO, and BMP4-siRNA silenced BIRC3-KO U251/U87 GBM cell lines. $n = 3$, * $p < 0.05$. (**C**). Human ABCG2 mRNA expression analyzed by real time PCR in control, BIRC3-KO, and BMP4-siRNA silenced BIRC3-KO U251/U87 GBM cell lines. $n = 3$, * $p < 0.05$. (**D**). Protein expression of SMAD1, SMAD5 and phosphorylated SMAD1/5 in control, BIRC3-KO, and BMP4-siRNA silenced BIRC3-KO U251/U87 GBM cell lines. Specific antibodies as indicated. β-actin acts as internal control. (**E**). Control or BIRC3-KO of U251 cells were seeded in 6 well plates and cultured in neurosphere formation medium. The BIRC3-KO cells had been treated with control siRNA and BMP4 siRNA separately 1 day before seeding. The number of neurospheres were observed and calculated by microscope. Representative images are under 4× magnification (top raw) and 10× magnification (bottom raw). $n = 5$, * $p < 0.05$.

2.5. BIRC3 Influences Tumor Initiation and Progression in GBM Orthotopic Xenograft Model

We wanted to determine if our in vitro data on BIRC3 stemness reprogramming phenotype had any in vivo relevance. We established orthotopic intracranial mouse xenografts consisting of including wild-type, BIRC3 overexpression and BIRC3 knockout U251 GBM cells. Tumor cells were stereotactically implanted into the brains of mice. Intracranial xenografts were monitored with MRI for tumor formation and progression. The MRI results indicated that BIRC3 expression significantly facilitated GBM tumor initiation and progression, whereas BIRC3 knockout significantly inhibited tumor initiation and progression (Figure 5A,B). Kaplan-Meier survival curve was recorded at desired time points (Figure 5C, $n = 5$). BIRC3 overexpression accelerated tumor progression and significantly decreased survival (Figure 5C, $p < 0.023$), while BIRC3 knockout significantly increased survival (Figure 5C, $p = 0.00008$). We further examined the xenograft tissues by H&E

staining and immunohistochemistry for BIRC3. We confirmed a marked high expression of BIRC3 in BIRC3 overexpressing U251 GBM xenograft compared to wild-type control U251 GBM xenograft (Figure 5D). To validate BIRC3 modulation of stemness genes in vitro, we further measured relative expression of CD133, ABCG2 and BMP4 using mRNA samples extracted from xenograft tumor tissues. BIRC3 overexpression significantly induced both CD133 and ABCG2 expressions, and reduced BMP4 expression (Figure 5E, $p < 0.05$), while BIRC3 knockout resulted in the opposite effect (Figure 5E, $p < 0.05$). Taken together, these data suggest that BIRC3 could impact on GBM tumor initiation, stemness and progression.

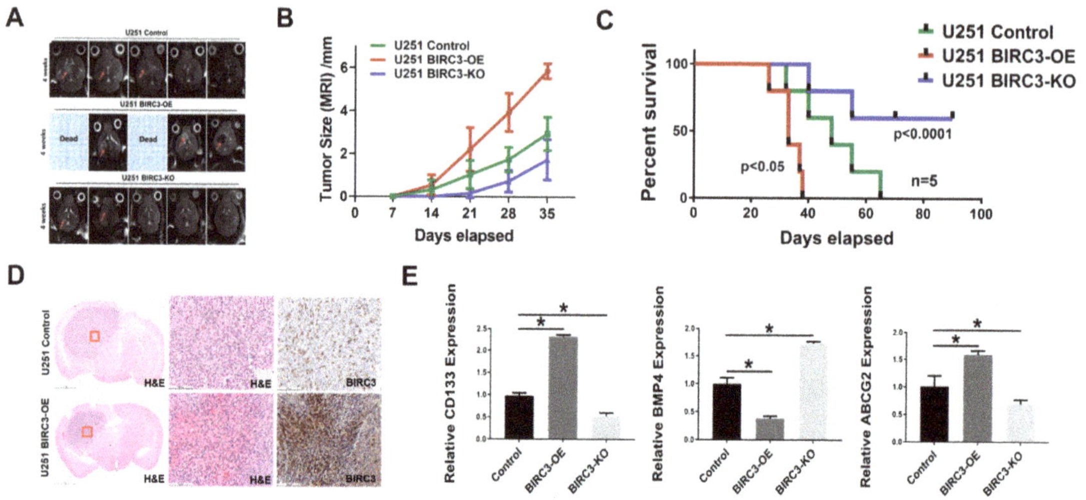

Figure 5. BIRC3 influences tumor initiation and progression in GBM orthotopic xenograft model. GBM Intracranial models with control, BIRC3-OE and BIRC3-KO U251 cells. (**A**). Horizontal axial MRI scan of mouse brain tumors 4 weeks after implantation. Two of BIRC3-OE mice were already dead at 4 weeks. (**B**). Tumor size calculation from MRI scan. $n = 5$. (**C**). Kaplan-Meier survival curve of U251 control BIRC3-OE and BIRC3-KO intracranial injection mice. n = 5 mice/group. (**D**). Mice were sacrificed at different timepoints and brain tissues of U251 control and BIRC3-OE groups were fixed in 10% neutral formalin. H&E staining and BIRC3 immunohistochemistry was performed as described in the Material and Methods Section 4. Five mice were included in this histological study and similar results were observed in each animal. (**E**). When mice were sacrificed, part of tumor tissues were isolated. mRNA from tumor tissues were extracted. BMP4, CD133 and ABCG2 mRNA expression analyzed by real-time PCR in extracted tumor tissues. $n = 3$, * $p < 0.05$.

3. Discussion

GBM is highly lethal cancer largely due to persistence and propagation of GSCs with enhanced stemness phenotype despite TMZ and RT. Hence identifying mechanisms of GBM stemness is very important in advancing our understanding and targeting of GBM resistance. In this study, we report the novel discovery that BIRC3 expression promotes GBM stemness and tumorigenicity of GSCs through inactivation of BMP4 signaling pathway. Using a combination of GBM cell lines, patient-derived GSCs and GBM patient tissue regional RNA-Seq data, we established the association between BIRC3 expression and GBM stemness maintenance. Specifically, we demonstrate that BIRC3 induced stemness and self-renewal through downstream inactivation of BMP4 signaling. Furthermore, the loss of stemness associated with BIRC3 knockout can be reversed or rescued through siRNA silencing of BMP4 signaling. Lastly, we demonstrate that depletion of BIRC3 significantly suppressed tumor initiation and progression in GBM intracranial xenografts. Our discovery reveals a novel function of BIRC3 that has never been described and that appears to be independent of the canonical anti-apoptotic functions of BIRC3 in GBM. Our findings therefore have several significant implications.

BMP4 signaling has been implicated in GSC differentiation and inhibition of GSC self-renewal and tumorigenicity [29–31]. BMP4 suppresses CD133 expression and CD133-positive GSC populations [30,31]. Furthermore, it has been reported BMP4 could inhibit GSC self-renewal and tumorigenicity through SMAD1/5 phosphorylation [29]. Hence BMP4 is an important driver of GSC differentiation and loss of stemness. Interestingly, in both our in vitro and in vivo studies, it appears that BIRC3 is a critical negative regulator of BMP4 signaling activation in GBM. In this role BIRC3 can therefore directly impact upon GBM cell self-renewal and differentiation. Hence BIRC3 emerges as a robust GBM stemness regulator. Our work for the first time also shows that high BIRC3 expression could significantly induce GBM cell self-renewal and stemness maintenance. We further present new evidence that depletion of BIRC3 significantly enhances activation of BMP4-SMAD1/5 signaling in GBM.

The identification of elevated BIRC3 expression as a GBM stemness maker is novel. Interestingly, our analysis indicates that BIRC3 is an independent biomarker for stemness not only in human/mouse GBM cell lines but also patient-derived GSCs. Our data indicated that BIRC3 could significantly increase neurosphere formation ability in both human and mouse GBM cell lines and patient-derived GSCs. Even upon differentiation of patient-derived GSCs, subsequent upregulation of BIRC3 restored GSCs self-renewal and stem-like phenotype. Hence BIRC3 contributed to stemness even in differentiated GBM cells. A major implication of this finding is that through propagation of stemness in both GSCs and non-GSCs populations, BIRC3 appears to be a major driver of intra-tumoral cellular heterogeneity in GBM. We previously demonstrated that BIRC3 was upregulated in GBM recurrence, TMZ-resistance, RT treatment and GBM hypoxia [15,16]. Our current study would imply that upregulation of BIRC3 in the above context is a central mechanism for treatment and microenvironment induced stemness reprogramming.

Moreover, our results suggest that BIRC3 could maintain GBM cell self-renewal and stemness through inhibiting BMP4 expression and further inactivating downstream SMAD1/5 phosphorylation. Suppression of BMP4 signaling will result in cell differentiation inhibition and expression of GBM stem cell marker CD133 and ABCG2. Hence, within the context of BIRC3/BMP4 axis, BMP4 antagonizes BIRC3-induced stemness in GBM. This is further supported by the observation that depletion of BMP4 in BIRC3 knockout cells could significantly restore stemness. Interestingly, high *BIRC3* expression correlated with low *BMP4* expression within the hypoxic niche. Given the critical role of the hypoxic niche in GBM stemness reprogramming [33–36], our findings provide further support for the hypothesis that BIRC3/BMP4 axis regulates stemness reprograming in GBM. Further studies are necessary to fully understand the molecular underpinnings of BMP4 pathway signaling with respect to BIRC3.

Lastly, we evaluated the impact of BIRC3 on tumor initiation. We demonstrated in orthotopic intracranial xenografts that high BIRC3 expression could significantly promote tumor initiation and propagation and moreover BIRC3 depletion could enhance survival though suppressing of tumor initiation and growth. We believe this is in line with the impact of BIRC3 on GBM stemness. The strategy of preventing tumor initiation through depletion of BIRC3 is of clinical importance and addresses a major reason for treatment failures in GBM.

In summary, our studies have shed some lights with respect to the regulation of GBM stem cell self-renewal and stemness maintenance. In particular, a novel translational function of BIRC3 in GBM and GSCs has been uncovered. Our data supports targeting BIRC3/BMP4 axis as a relevant therapeutic approach in addressing GBM stemness reprogramming. Further mechanistic elucidation of BIRC3/BMP4 signaling will undoubtedly provide new therapeutic avenues for GBM patients.

4. Methods and Materials

4.1. Cell Culture and Reagents

U251 and U87 human glioblastoma cell lines (ATCC) were cultured in DMEM (Life Technologies) supplemented with 10% fetal bovine serum (Sigma-Aldrich, St. Louis, MO, USA), 100 units/mL penicillin and -100 ug/mL streptomycin (Life Technologies, NY, USA). The cultures were maintained at 37 °C in a humidified atmosphere containing 5% CO_2. The patient-derived GSCs used in this study were isolated from GBM patients and were well characterized. The patient-derived GSCs were culture in NS-A medium (90% NeuroCult NS-A Basal Medium Human plus 10% Human NeuroCult NS-A proliferation Supplements, StemCell Technologies). Complete medium was supplied with recombinant human epidermal growth factor (R&D system, Minneapolis, MN, USA), and 100 units/mL penicillin plus 100 ug/mL streptomycin (Life Technologies, NY, USA). For differentiation, GSCs were cultured in NS-A medium supplied with 10% fetal bovine serum. Anti-BIRC3 antibody was obtained from R&D system; anti-b-actin IgG-HRP was obtained from Santa Cruz Biotech; and anti-SMAD1, anti-SMAD5, anti-p-SMAD1/5, Goat anti-Rabbit IgG-HRP and Goat anti-mouse IgG-HRP were obtained from CellSignal.

4.2. Gene Overexpression in GBM Cell

Human BIRC3 expression and empty vector constructs were obtained from Genecopoeia (Rockville, MD, USA). A single bacteria clone was picked from a freshly streaked LB plate containing100 μg/mL ampicillin and inoculated to a culture of 5 mL LB medium containing 100 μg/mL ampicillin, which was then incubated for 16 h at 37 °C with vigorous shaking. Plasmid was purified using QuickLyse Miniprep Kit (Qiagen, Germantown, MD, USA). U251, U87 cells and GSCs (2×10^5) was seeded in 6-well plate 24 h before transfection. BIRC3 expression plasmid was transfected by Lipofectmine 3000 kit (for U251 and U87, ThermoFisher, Waltham, MA, USA) and Lipofectmine Stem reagent (for GSCs, ThermoFisher, Waltham, MA, USA) following manufacture's protocol. The cells were then incubated at 37 °C in a CO_2 incubator for 48 h. G418 sulfate (500 μg/mL) was used for selection 48 h after transfection and the G418 sulfate concentration was then reduced to 200 μg/mL 7 days later for maintenance. The overexpression of BIRC3 was verified by western blot.

4.3. Gene Silencing by CRISPR/Cas9 System

CRISPR/Cas9 vectors lentiCRISPR-v2-puro was obtained from Addgene. sgRNA targeting human/mouse BIRC3 and sgRNA control were cloned into lentiCRISPR-v2-puro. Human BIRC3 targeting forward primer: CACCGTATTTCAGTTCAAACGTGT, reverse primer: AAACACACGTTTGAACTGAAATAC; Mouse BIRC3 targeting forward primer: CACCGTTCCGGCGCGCCGAGTCCTT, reverse primer: AAACAAGGACTCGGCGCGC-CGGAAC; control sgRNA cloning forward primer: CACCGCACTCACATCGCTACATCA, reverse primer: AAACTGATGTAGCGATGTGAGTGC. Lentivirus was packed by 293T cells through 2nd generation lentivirus packaging system. U251, U87, CT-2A cells and GSCs were next infected with Lenti-sgBIRC3-puro or Lenti-sgControl-puro followed by extensive selection with 1 μg/mL puromycin (InvivoGen, San Diego, CA, USA). To confirm CRISPR silencing efficiency, we harvested protein from cell lysis and tested them with western blot.

4.4. Real-Time PCR

Total RNA was extracted using RNeasy mini-prep kit (Qiagen, Germantown, MD, USA). RNA was quantified with Nanodrop 2000 (Thermo Scientific, Waltham, MA, USA). cDNA was synthesized using 1ug total RNA with the iScript cDNA Synthesis kit (Bio-Rad, Hercules, CA, USA). Real-time PCR was performed using iQ SYBR green Supermix buffer system (Bio-Rad, Hercules, CA, USA) and the Bio-Rad CFX96 Touch Real-Time PCR Detection system. Human CD133 forward primer: ACTCCCATAAAGCTG-GACCC, reverse primer: TCAATTTTGGATTCATATGCCTT; human ABCG2 forward primer: AGCAGCAGGTCAGAGTGTGG, reverse primer: GATCGATGCCCTGCTTTACC; human ALDH1A3 forward primer: TGGATCAACTGCTACAACGC, reverse primer: CACT TCTGTGTATTCGGCCA; human BMP4 forward primer: GCCGGAGGGCCAAGCGTAGC-CCTAAG, reverse primer: CTGCCTGATCTCAGCGGCACCCACATC; human GAPDH was used as the internal control, GAPDH forward primer: ACCACAGTCCATGCCATCAC, reverse primer: TCCACCACCCTGTTGCTGT. Mouse CD133 forward primer: TTGGTG-CAAATGTGGAAAAG, reverse primer: ATTGCCATTGTTCCTTGAGC; mouse ABCG2 forward primer: CAGTTCTCAGCAGCTCTTCGAC, reverse primer: TCCTCCAGAGAT-GCCACGGATA; mouse BMP4 forward primer: GCCGAGCCAACACTGTGAGGA, reverse primer: GATGCTGCTGAGGTTGAAGAGG; mouse GAPDH was used as the internal control, GAPDH forward primer: ATGGTGAAGGTCGGTGTGA, reverse primer: AATCTC-CACTTTGCCACTGC. The PCR program was as follow: 95 °C 10 min, 1 cycle; 95 °C 15 s, → 60 °C 30 s → 72 °C 30 s, 40 cycles; 72 °C 10 min, 1 cycle.

4.5. Western Blot Analysis

50–100 µg of heat-denatured proteins were loaded on 4–15% precast polyacrylamide gel (Bio-Rad, Hercules, CA, USA). The proteins were then transferred to PVDF membranes (Bio-Rad, Hercules, CA, USA), which were blocked with 5% non-fat milk solutions for 1 hour at room temperature. The target proteins were then detected by the primary antibody at 4 °C overnight, washed with 0.1% Tween-TBS and incubated with appropriate secondary antibody for 1 hour at room temperature. The membranes were then washed, and the target proteins were detected with luminol reagent and X-ray film (Santa Cruz Biotechnology, Dallas, TX, USA).

4.6. Tumor Sphere Formation Assay

U251 and U87 Cells were collected, counted, and seeded in DMEM/F12 medium with B27 supplement, 20 ng/mL human recombinant epidermal growth factor (EGF, ThermoFisher, Waltham, MA, USA), and 10 ng/mL human recombinant basic fibroblast growth factor (FGF-2, Sigma-Aldrich, St. Louis, MO, USA), 100 units/mL-penicillin-100 ug/mL streptomycin (Life Technologies, NY, USA). Differentiated GSCs were collected, counted, and seeded in NS-A medium, 20 ng/mL human recombinant epidermal growth factor (EGF, ThermoFisher, Waltham, MA, USA), and 10 ng/mL human recombinant basic fibroblast growth factor (FGF-2, Sigma-Aldrich, St. Louis, MO, USA), 100 units/mL-penicillin-100 ug/mL streptomycin (Life Technologies, NY, USA). The cells were subsequently cultured in ultra-low attachment 6-well plates (Corning, NY, USA) at a density of 2000–5000 cells/well. Half of the culture medium was replaced or supplemented with additional growth factors twice a week. To propagate spheres in vitro, the cells were collected by gentle centrifugation, dissociated by Accutase (StemCell Technologies, Vancouver, Canada) into single-cell suspensions and cultured to allow the regeneration of spheres. Third-generation spheres were used for all subsequent experiments except siRNA knockdown sphere formation. For siRNA knockdown cells, first-generation spheres were used. The total number of tumor spheres was counted following 10 days of culture. Images are taken under 4× magnification or 10× magnification observation.

4.7. Immunocytochemistry Analysis

Cells were seeded onto poly-L-lysine–coated slides and cultured for 24 h. Then, cells were fixed using Cytofix/Cytoperm (BD Biosciences, San Jose, CA, USA) according to the manufacturer's protocol. Cells were washed in phosphate-buffered saline (PBS), blocked with PBS buffer containing 2% BSA and 0.1% TritonX-100 for 1 h, and incubated with Alexa Fluor 488 anti-Nestin (BioLegend, San Diego, CA, USA) overnight. Next, cells were washed with PBS and slides were mounted onto coverslips over a drop of Vectashield mounting medium with DAPI (Vector Laboratories, Burlingame, CA, USA). Cells were examined with an automated Zeiss Observer Z.1 inverted microscope through a 63X/1.4NA objective and DAPI and FITC filters. Multi-channel images were captured using the AxioCam

MRm3 CCD camera and Axiovision version 4.7 software suite (Carl Zeiss Inc., Oberkochen, Germany).

4.8. siRNA Knockdown

U251 and U87 cells were transfected with predesigned BMP4 small interfering RNA (siRNA; 30 nM, Millipore-Sigma, St. Louis, MO, USA) or control siRNA (30 nM, Millipore-Sigma, St. Louis, MO, USA) using Lipofectamine RNAiMAX Transfection Reagent (ThermoFisher, Waltham, MA, USA). Briefly, one day prior to transfection, the cells were seeded in 6-well plate (2×10^5) with 10% FBS DMEM without antibiotics. siRNAs were prepared according to the manufacturer's instructions and added to the cells. For tumor sphere formation the cells were then exposed to tumor sphere formation culture medium after 24 h.

4.9. Mice and GBM Orthotopic Xenograft Model

Female NCRNU athymic mice of 6–8 weeks were ordered from Taconic Biosciences. All animals were housed in the American Association for Laboratory Animal Care-accredited Animal Resource Center at Moffitt Cancer Center. All animal procedures and Experiments were carried out under protocols approved by the Institutional Animal Care and Use Committee of the University of South Florida and Moffitt Cancer Center. All animal studies were performed in accordance with relevant guidelines and regulations of University of South Florida and Moffitt Cancer Center. Tumors were established by injecting 2×10^5 U251 control, BIRC3 overexpression and BIRC3 knockout cells in a 4 μL volume of PBS in the right striatum of mice (n = 5/group) on a Stoelting Digital Stereotaxic Instrument (Stoelting, IL, USA). The tumor progression was monitored by MRI (Bruker Biospec 7T, Billerica, MA, USA) every week. For survival studies, animals were followed until they lost 20% of body weight or had trouble ambulating, feeding, or grooming.

4.10. Immunohistochemistry

Tumor samples were fixed with 10% neutral-formalin buffer for 72 h. The samples were then dehydrated, paraffin-embedded and sectioned. Sections were dewaxed, treated with 3% H_2O_2 for 10 min and incubated with anti-BIRC3 antibody (1:100 dilutions) overnight at 4 °C. Biotinylated secondary antibody (1:200 dilutions) was added at room temperature for 1 h, followed by the incubation with ABC-peroxidase for additional 1 h. After washing with Tris-buffer, the sections were incubated with DAB (3, 30 diaminobenzidine, 30 mg dissolved in 100 mL Tris-buffer containing 0.03% H_2O_2) for 5 min, rinsed in water and counterstained with hematoxylin.

4.11. Bioinformatics and Statistics

The IVY data was downloaded and log2 transformed. The TCGA GBM samples was extracted from the normalized and debatched PanCan RNA-Seq data and log2 transformed. Student's *t*-test (for 2 condition experiments) and ANOVA (for multiple condition experiments) was employed. Survival was assessed using Kaplan-Meier analysis with statistical comparisons made by log rank (Mantel-Cox) test. All statistical tests were considered significant at $p < 0.05$. * means $p < 0.05$.

Supplementary Materials: The following supporting information can be downloaded at: https://www.mdpi.com/article/10.3390/ijms23010297/s1.

Author Contributions: A.B.E. designed the study; A.E.B. performed bioinformatics analysis; Q.W., A.E.B. and A.B.E. performed statistical analysis; Q.W. performed the experiments; and Q.W., A.E.B., R.J.M. and A.B.E. analyzed the results and wrote the manuscript. All authors have read and agreed to the published version of the manuscript.

Funding: The study was supported by the Moffitt Cancer Center Foundation (A.B.E); by National Institute of Health—National Institute of Neurological Disorders and Stroke Support Grant 1R21NS120800 (A.B.E.).

Institutional Review Board Statement: The animal study protocol was approved by the Institutional Animal Care and Use Committee (IACUC) of University of South Florida (#IS00006039, 12 January 2020).

Informed Consent Statement: Not applicable.

Data Availability Statement: The data that support the findings of this study are available from the corresponding authors upon reasonable request.

Acknowledgments: This work was supported in part by the Analytic Microscopy and Biostatistics and Bioinformatics Core facilities of the Moffitt Cancer Center.

Conflicts of Interest: The authors declare that they have no competing interests, or any other interests that might be perceived to influence the results and/or discussion reported in this paper.

References

1. Stupp, R.; Mason, W.P.; Bent, M.V.D.; Weller, M.; Fisher, B.; Taphoorn, M.J.; Belanger, K.; Brandes, A.; Marosi, C.; Bogdahn, U.; et al. Radiotherapy plus Concomitant and Adjuvant Temozolomide for Glioblastoma. *N. Engl. J. Med.* **2005**, *352*, 987–996. [CrossRef] [PubMed]
2. Bao, S.; Wu, Q.; McLendon, R.E.; Hao, Y.; Shi, Q.; Hjelmeland, A.B.; Dewhirst, M.W.; Bigner, D.D.; Rich, J.N. Glioma stem cells promote radioresistance by preferential activation of the DNA damage response. *Nat. Cell Biol.* **2006**, *444*, 756–760. [CrossRef] [PubMed]
3. Alves, A.L.V.; Gomes, I.N.F.; Carloni, A.C.; Rosa, M.N.; da Silva, L.S.; Evangelista, A.F.; Reis, R.M.; Silva, V.A.O. Role of glioblastoma stem cells in cancer therapeutic resistance: A perspective on antineoplastic agents from natural sources and chemical derivatives. *Stem Cell Res. Ther.* **2021**, *12*, 206. [CrossRef]
4. Singh, S.K.; Hawkins, C.; Clarke, I.D.; Squire, J.; Bayani, J.; Hide, T.; Henkelman, R.M.; Cusimano, M.; Dirks, P.B. Identification of human brain tumour initiating cells. *Nat. Cell Biol.* **2004**, *432*, 396–401. [CrossRef] [PubMed]
5. Galli, R.; Binda, E.; Orfanelli, U.; Cipelletti, B.; Gritti, A.; De Vitis, S.; Fiocco, R.; Foroni, C.; DiMeco, F.; Vescovi, A. Isolation and Characterization of Tumorigenic, Stem-like Neural Precursors from Human Glioblastoma. *Cancer Res.* **2004**, *64*, 7011–7021. [CrossRef]
6. Hemmati, H.D.; Nakano, I.; Lazareff, J.A.; Masterman-Smith, M.; Geschwind, D.H.; Bronner, M.; Kornblum, H.I. Cancerous stem cells can arise from pediatric brain tumors. *Proc. Natl. Acad. Sci. USA* **2003**, *100*, 15178–15183. [CrossRef]
7. Ahmed, S.I.; Javed, G.; Laghari, A.A.; Bareeqa, S.B.; Farrukh, S.; Zahid, S.; Samar, S.S.; Aziz, K. CD133 Expression in Glioblastoma Multiforme: A Literature Review. *Cureus* **2018**, *10*, 3439. [CrossRef]
8. Beier, C.P.; Beier, D. CD133 negative cancer stem cells in glioblastoma. *Front. Biosci.* **2011**, *3*, 701–710. [CrossRef]
9. Cho, J.-H.; Kim, A.-R.; Kim, S.-H.; Lee, S.-J.; Chung, H.; Yoon, M.-Y. Development of a novel imaging agent using peptide-coated gold nanoparticles toward brain glioma stem cell marker CD133. *Acta Biomater.* **2017**, *47*, 182–192. [CrossRef]
10. Liu, G.; Yuan, X.; Zeng, Z.; Tunici, P.; Ng, H.; Abdulkadir, I.R.; Lu, L.; Irvin, D.; Black, K.L.; Yu, J.S. Analysis of gene expression and chemoresistance of CD133+ cancer stem cells in glioblastoma. *Mol. Cancer* **2006**, *5*, 67. [CrossRef]
11. Wang, Q.; Wu, H.; Hu, J.; Fu, H.; Qu, Y.; Yang, Y.; Cai, Q.; Efimov, A.; Wu, M.; Yen, T.; et al. Nestin is required for spindle assembly and cell cycle progression in glioblastoma cells. *Mol. Cancer Res.* **2021**, *19*, 1651–1665. [CrossRef] [PubMed]
12. Couturier, C.P.; Ayyadhury, S.; Le, P.U.; Nadaf, J.; Monlong, J.; Riva, G.; Allache, R.; Baig, S.; Yan, X.; Bourgey, M.; et al. Single-cell RNA-seq reveals that glioblastoma recapitulates a normal neurodevelopmental hierarchy. *Nat. Commun.* **2020**, *11*, 3406. [CrossRef]
13. Swartling, F.; Bolin, S.; Phillips, J.J.; Persson, A.I. Signals that regulate the oncogenic fate of neural stem cells and progenitors. *Exp. Neurol.* **2014**, *260*, 56–68. [CrossRef]
14. Wang, X.; Zhou, R.; Xiong, Y.; Zhou, L.; Yan, X.; Wang, M.; Li, F.; Xie, C.; Zhang, Y.; Huang, Z.; et al. Sequential fate-switches in stem-like cells drive the tumorigenic trajectory from human neural stem cells to malignant glioma. *Cell Res.* **2021**, *31*, 684–702. [CrossRef] [PubMed]
15. Wang, D.; Berglund, A.; Kenchappa, R.S.; Forsyth, P.A.; Mulé, J.J.; Etame, A.B. BIRC3 is a novel driver of therapeutic resistance in Glioblastoma. *Sci. Rep.* **2016**, *6*, 21710. [CrossRef] [PubMed]
16. Wang, D.; Berglund, A.; Kenchappa, R.S.; Macaulay, R.; Mulé, J.J.; Etame, A.B. BIRC3 is a biomarker of mesenchymal habitat of glioblastoma, and a mediator of survival adaptation in hypoxia-driven glioblastoma habitats. *Sci. Rep.* **2017**, *7*, 1–11. [CrossRef]
17. Diessenbacher, P.; Hupe, M.; Sprick, M.R.; Kerstan, A.; Geserick, P.; Haas, T.L.; Wachter, T.; Neumann, M.; Walczak, H.; Silke, J.; et al. NF-kappaB inhibition reveals differential mechanisms of TNF versus TRAIL-induced apoptosis upstream or at the level of caspase-8 activation independent of cIAP2. *J. Investig. Dermatol.* **2008**, *128*, 1134–1147. [CrossRef]
18. Wang, C.Y.; Mayo, M.W.; Korneluk, R.G.; Goeddel, D.V.; Baldwin, A.S. NF-kappaB antiapoptosis: Induction of TRAF1 and TRAF2 and c-IAP1 and c-IAP2 to suppress caspase-8 activation. *Science* **1998**, *281*, 1680–1683. [CrossRef]
19. Gressot, L.V.; Doucette, T.; Yang, Y.; Fuller, G.N.; Manyam, G.; Rao, A.; Latha, K.; Rao, G. Analysis of the inhibitors of apoptosis identifies BIRC3 as a facilitator of malignant progression in glioma. *Oncotarget* **2017**, *8*, 12695–12704. [CrossRef]
20. Zhang, S.-C.; Lundberg, C.; Lipsitz, D.; O'Connor, L.T.; Duncan, I.D. Generation of oligodendroglial progenitors from neural stem cells. *J. Neurocytol.* **1998**, *27*, 475–489. [CrossRef]

21. Mesrati, M.H.; Behrooz, A.B.; Abuhamad, A.Y.; Syahir, A. Understanding Glioblastoma Biomarkers: Knocking a Mountain with a Hammer. *Cells* **2020**, *9*, 1236. [CrossRef]
22. Lottaz, C.; Beier, D.; Meyer, K.; Kumar, P.; Hermann, A.; Schwarz, J.; Junker, M.; Oefner, P.J.; Bogdahn, U.; Wischhusen, J.; et al. Transcriptional Profiles of CD133+ and CD133− Glioblastoma-Derived Cancer Stem Cell Lines Suggest Different Cells of Origin. *Cancer Res.* **2010**, *70*, 2030–2040. [CrossRef] [PubMed]
23. Chen, L.; Shi, L.; Wang, W.; Zhou, Y. ABCG2 downregulation in glioma stem cells enhances the therapeutic efficacy of demethoxycurcumin. *Oncotarget* **2017**, *8*, 43237–43247. [CrossRef]
24. Chu, L.; Huang, Q.; Zhai, D.-Z.; Zhu, Q.; Huo, H.-M.; Dong, J.; Qian, Z.-Y.; Wang, A.-D.; Lan, Q.; Gao, Y.-L. Expression and significance of ABCG2 in human malignant glioma. *Ai Zheng Aizheng Chin. J. Cancer* **2007**, *26*, 1090–1094.
25. Patrawala, L.; Calhoun, T.; Schneider-Broussard, R.; Zhou, J.; Claypool, K.; Tang, D. Side Population Is Enriched in Tumorigenic, Stem-Like Cancer Cells, whereas ABCG2+ and ABCG2− Cancer Cells Are Similarly Tumorigenic. *Cancer Res.* **2005**, *65*, 6207–6219. [CrossRef] [PubMed]
26. Cheng, P.; Wang, J.; Waghmare, I.; Sartini, S.; Coviello, V.; Zhang, Z.; Kim, S.-H.; Mohyeldin, A.; Pavlyukov, M.S.; Minata, M.; et al. FOXD1–ALDH1A3 Signaling Is a Determinant for the Self-Renewal and Tumorigenicity of Mesenchymal Glioma Stem Cells. *Cancer Res.* **2016**, *76*, 7219–7230. [CrossRef] [PubMed]
27. Sullivan, K.E.; Rojas, K.; Cerione, R.A.; Nakano, I.; Wilson, K.F. The stem cell/cancer stem cell marker ALDH1A3 regulates the expression of the survival factor tissue transglutaminase, in mesenchymal glioma stem cells. *Oncotarget* **2017**, *8*, 22325–22343. [CrossRef] [PubMed]
28. Ecsedy, J.A.; Manfredi, M.G.; Yohe, H.C.; Seyfried, T. Ganglioside biosynthetic gene expression in experimental mouse brain tumors. *Cancer Res.* **1997**, *57*, 1580–1583.
29. Ciechomska, I.A.; Gielniewski, B.; Wojtas, B.; Kaminska, B.; Mieczkowski, J. EGFR/FOXO3a/BIM signaling pathway determines chemosensitivity of BMP4-differentiated glioma stem cells to temozolomide. *Exp. Mol. Med.* **2020**, *52*, 1326–1340. [CrossRef]
30. Koguchi, M.; Nakahara, Y.; Ito, H.; Wakamiya, T.; Yoshioka, F.; Ogata, A.; Inoue, K.; Masuoka, J.; Izumi, H.; Abe, T. BMP4 induces asymmetric cell division in human glioma stem-like cells. *Oncol. Lett.* **2019**, *19*, 1247–1254. [CrossRef] [PubMed]
31. Piccirillo, S.G.M.; Vescovi, A.L. Bone Morphogenetic Proteins Regulate Tumorigenicity in Human Glioblastoma Stem Cells. *Cancer Stem Cells* **2007**, 59–81.
32. Nayak, S.; Mahenthiran, A.; Yang, Y.; McClendon, M.; Mania-Farnell, B.; James, C.D.; Kessler, J.A.; Tomita, T.; Cheng, S.-Y.; Stupp, S.I.; et al. Bone Morphogenetic Protein 4 Targeting Glioma Stem-Like Cells for Malignant Glioma Treatment: Latest Advances and Implications for Clinical Application. *Cancers* **2020**, *12*, 516. [CrossRef] [PubMed]
33. Bar, E.E.; Lin, A.; Mahairaki, V.; Matsui, W.; Eberhart, C.G. Hypoxia Increases the Expression of Stem-Cell Markers and Promotes Clonogenicity in Glioblastoma Neurospheres. *Am. J. Pathol.* **2010**, *177*, 1491–1502. [CrossRef]
34. Seidel, S.; Garvalov, B.K.; Wirta, V.; Von Stechow, L.; Schänzer, A.; Meletis, K.; Wolter, M.; Sommerlad, D.; Henze, A.-T.; Nistér, M.; et al. A hypoxic niche regulates glioblastoma stem cells through hypoxia inducible factor 2α. *Brain* **2010**, *133 Pt 4*, 983–995. [CrossRef]
35. Semenza, G.L. Defining the role of hypoxia-inducible factor 1 in cancer biology and therapeutics. *Oncogene* **2009**, *29*, 625–634. [CrossRef]
36. Soeda, A.; Park, M.; Lee, D.; Mintz, A.; Androutsellis-Theotokis, A.; McKay, R.D.; Engh, J.; Iwama, T.; Kunisada, T.; Kassam, A.B.; et al. Hypoxia promotes expansion of the CD133-positive glioma stem cells through activation of HIF-1alpha. *Oncogene* **2009**, *28*, 3949–3959. [CrossRef] [PubMed]

International Journal of *Molecular Sciences*

MDPI

Article

Profiling Glioblastoma Cases with an Expression of DCX, OLIG2 and NES

Adrian Odrzywolski [1,2], Bożena Jarosz [3], Michał Kiełbus [1], Ilona Telejko [1], Dominik Ziemianek [3], Sebastian Knaga [4] and Radosław Rola [3,*]

1 Department of Biochemistry and Molecular Biology, Medical University of Lublin, 20-093 Lublin, Poland; adrian.odrzywolski@umlub.pl (A.O.); michal.kielbus@umlub.pl (M.K.); ilona.telejko@umlub.pl (I.T.)
2 Laboratory for Cytogenetics and Genome Research, Department of Human Genetics, KU Leuven, B-3000 Leuven, Belgium
3 Department of Neurosurgery and Pediatric Neurosurgery, Medical University of Lublin, 20-090 Lublin, Poland; bozena.jarosz@umlub.pl (B.J.); ziemianek.dominic@gmail.com (D.Z.)
4 Institute of Biological Bases of Animal Production, University of Life Sciences, 20-950 Lublin, Poland; sebastian.knaga@up.lublin.pl
* Correspondence: rola.radoslaw@gmail.com

Abstract: Glioblastoma (GBM) remains the leading cause of cancer-related deaths with the lowest five-year survival rates among all of the human cancers. Multiple factors contribute to its poor outcome, including intratumor heterogeneity, along with migratory and invasive capacities of tumour cells. Over the last several years Doublecortin (DCX) has been one of the debatable factors influencing GBM cells' migration. To resolve DCX's ambiguous role in GBM cells' migration, we set to analyse the expression patterns of DCX along with Nestin (NES) and Oligodendrocyte lineage transcription factor 2 (OLIG2) in 17 cases of GBM, using immunohistochemistry, followed by an analysis of single-cell RNA-seq data. Our results showed that only a small subset of DCX positive (DCX^+) cells was present in the tumour. Moreover, no particular pattern emerged when analysing DCX^+ cells relative position to the tumour margin. By looking into single-cell RNA-seq data, the majority of DCX^+ cells were classified as non-cancerous, with a small subset of cells that could be regarded as glioma stem cells. In conclusion, our findings support the notion that glioma cells express DCX; however, there is no clear evidence to prove that DCX participates in GBM cell migration.

Keywords: Glioblastoma; Doublecortin; DCX; OLIG2; NES; single cell RNA-seq; immunohistochemistry; immunofluorescence; brain tumours

Citation: Odrzywolski, A.; Jarosz, B.; Kiełbus, M.; Telejko, I.; Ziemianek, D.; Knaga, S.; Rola, R. Profiling Glioblastoma Cases with an Expression of DCX, OLIG2 and NES. *Int. J. Mol. Sci.* **2021**, *22*, 13217. https://doi.org/10.3390/ijms222413217

Academic Editors: Nives Pećina-Šlaus and Ivana Jovčevska

Received: 17 October 2021
Accepted: 1 December 2021
Published: 8 December 2021

1. Introduction

Even though malignant brain tumours account only for a small percentage of all adult cancers, they lead to an extensive amount of cancer-related deaths [1]. Moreover, the five-year survival rates are among the lowest for all human cancers [2], regardless of treatment modality [3]. This remarkable resistance results mostly from tumour heterogeneity and its high propensity for malignant progression. One of the vital pathophysiologic features contributing to this dismal prognosis is their strong migrational capacity [4] for significant dispersal beyond the macroscopic tumour borders [5]. Interestingly, a similar migratory ability is one of the principal features of neuronal progenitor cells (NPC) during CNS development [6,7]. Given that, it is worth noticing that data generated by the Cancer Genome Atlas Research Network proved that one of the main gene profiles of GBM, a proneural profile, involves DCX expression [8].

DCX itself, when it is mutated, is responsible for an X-linked form of lissencephaly, affecting the organisation of neocortical layering in the cerebral cortex [9]. Subsequent studies have shown that DCX directly binds to microtubules, thereby regulating their polymerisation and stabilisation [10]. This process is crucial for a multipolar mode of neuroblast migration [11], a transient stage in neuronal progenitor migration where migrating cells search for environmental signals that will determine their mode of migration [12]. DCX expression has been restricted to migrating neuroblasts in developing and adult animals [12]. However, Daou et al. [13] proved that DCX expression might be found in various neuroepithelial origin tumours.

Interestingly, DCX was highly expressed in both high-grade and low-grade invasive tumours. Moreover, invasive tumours have been shown to express higher levels of DCX when compared to circumscribed tumours; no expression in normal brain tissue surrounding the tumour was found. A more recent follow-up study evaluated the sensitivity and specificity of DCX immunostaining to detect infiltrating glioma cells [14]. It confirmed that DCX is explicitly expressed in infiltrating gliomas but not in reactive astrocytes. Santra et al. presented a different hypothesis [15]. Their data indicate that DCX mRNA transcripts were not detected in primary glioma cells, while DCX expressing cells were revealed in tumour penumbra. DCX positive cells within glioma tumours, in their opinion, were either infiltrating neuroblasts or pre-existing neuronal cells.

According to Verhaak et al. [8], one of the proneural subtype's signature genes is Oligodendrocyte lineage transcription factor 2 (OLIG2), a family member of basic helix–loop–helix transcription factors. It plays a crucial role in the early stages of brain development in oligodendrocyte precursor cells and neural progenitor cells by enhancing mitosis and limiting cell differentiation [16]. In gliomas, OLIG2 was expressed by glioma stem cells [17]. On top of that, cells expressing Rai (ShcC/N-Shc), a member of the family of Shc-like adaptor proteins, are involved in non-neoplastic cell migration, co-expressed DCX, and OLIG2 [18]. Accordingly, whenever DCX and OLIG2 is observed in migrating cells, it would imply that these cells have stem-cell capabilities.

On the other hand, Bott et al. [19] recently found that DCX function in complex with nestin (NES) is a marker of neural progenitors. Although initially detected in neuronal stem cells, its presence in multiple other tissues (including gliomas) has recently been described. NES is a type VI intermediate filament. It plays a role in several key aspects of primary cell functioning: self-renewal, proliferation, survival, differentiation, and migration. With regard to the later, Bott et al. established that cdk5/p35 selectively phosphorises DCX due to the DCX-NES complex's presence. That could directly influence the growth cone during migration.

Based on the observations mentioned above, we contend that the role of DCX expression in glioma cells' migration is still a matter of debate. Therefore, in order to further clarify this role, we decided to conduct an experiment focusing on the distribution of glioma cells, both inside and in the tumour's margin. Simultaneously, in order to elucidate the possible role of DCX in the context of the proneuronal subtype pathogenesis with its potential interactions with microtubules and intermediate filaments, we have marked OLIG2 and NES.

2. Results

Our heuristic approach to quantify the number of cells with each of the antigens allowed us to obtain data that can be compared with the results from single-cell RNA-seq. The fundamental component of this method is the determination of cell nuclei in the examined ROI. Cellular nuclei (Figure 1A) act as seeds to assess adjacent fluorescent signals: OLIG2 was expected to overlap with nuclei (Figure 1B,F,H), whereas NES (Figure 1C,E,H) and DCX (Figure 1D,G,H) signal should tightly adhere to it.

As expected, cell counts were significantly different in various parts of the tumour. More cells were tumour-adjacent to the margin or in non-specific tumour sites than in the margin ($p = 1 \times 10^{-4}$ and $p = 1.9 \times 10^{-7}$, respectively)—Figure 2H. No significant cell count change was detected while comparing tumours adjacent to margin and non-specific tumour sites ($p = 0.91$).

The DCX$^+$ cells were only a small subpopulation of cells in total, although they were present in all types of tissues—Figure 2A. The median frequency of DCX$^+$ cells was 3.1% in margin tissue and 1.1% in adjacent tumour tissue, without significant difference ($p = 0.82$).

The majority of ROIs did not have a context of margin tissue. Thus, it was impossible to assess the exact placement within the tumour. They were grouped and called other/non-specific tumour tissue. A batch comparison of DCX^+ cells in other tumour tissue versus tumour-adjacent to margin did not show statistical importance ($p = 0.24$).

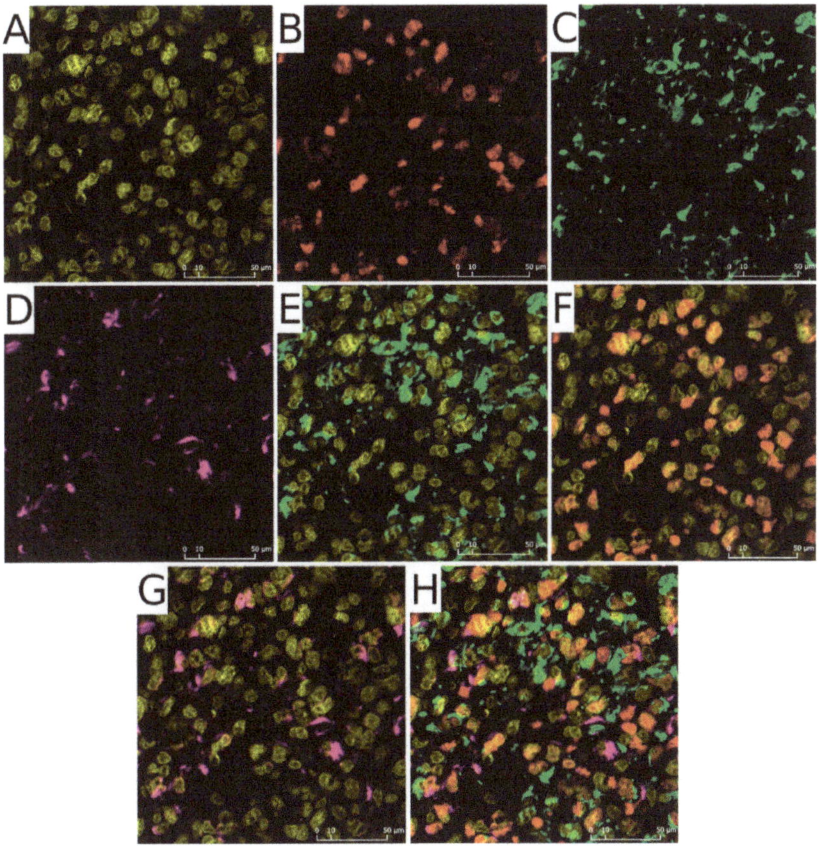

Figure 1. Staining patterns. (**A**)—Nuclei (yellow), (**B**)—OLIG2 signal (red), (**C**)—NES signal (green), (**D**)—DCX signal (magenta), (**E**)—NES signal with nuclei, (**F**)—OLIG2 signal with nuclei, (**G**)—DCX signal with nuclei (**H**)—composition of (**A**–**D**).

Contrary to the frequency of DCX^+ cells, NES^+ cells were much more abundant in studied cases (Figure 2B). The median frequency was 17.7% in margin tissue, 20.6% in adjacent tumour tissue, and 25.4% in the rest of ROIs tumours, yet without significant difference (the lowest $p = 0.78$ in the batch comparison, and $p = 0.30$ when accounting pairs of margins vs. adjacent tumour).

$OLIG2^+$ cells were more frequent than DCX^+ cells (Figure 2C). Non-specific tumour sites and tumours close to the margin had a similar median (8.2% and 5.1%, respectively). The median of the tumour margin, on the other hand, was 19.8%. Pairwise comparison of the margin and the tumour-adjacent to the margin shows a significant change in $OLIG2^+$ cells frequency ($p = 4.9 \times 10^{-4}$).

No significant changes in frequency in cells expressing more than one of the analysed markers were detected. The majority of measured frequencies were next to 0% (Figure 2D–G).

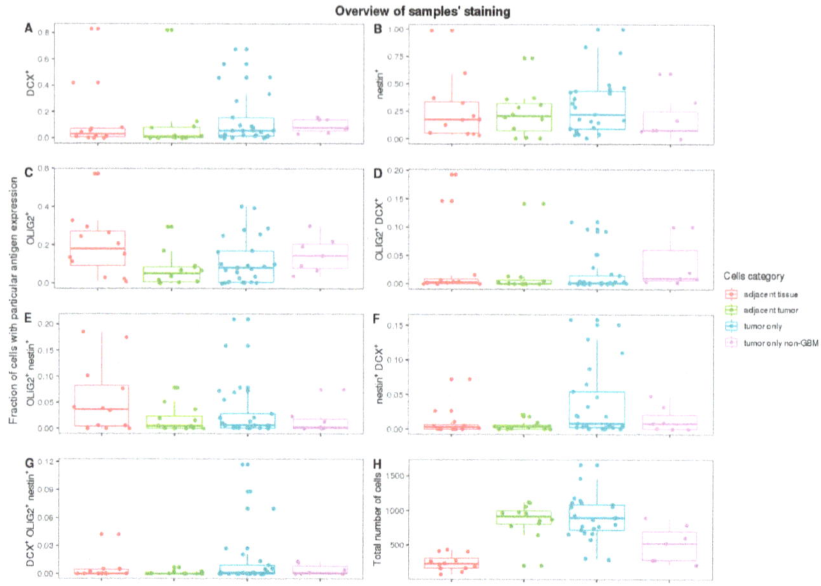

Figure 2. Overview of samples' staining and staining overlap. Frequency of cells with different stainings with respect to tissues: (**A**)—DCX$^+$ cells, (**B**)—NES$^+$ cells, (**C**)—OLIG2$^+$ cells. Frequency of cells with mixtures of all stainings with respect to tissues: (**D**)—OLIG2$^+$DCX$^+$ cells, (**E**)—OLIG2$^+$NES$^+$ cells, (**F**)—NES$^+$DCX$^+$ cells, (**G**)—DCX$^+$OLIG2$^+$NES$^+$ cells. (**H**)—Total number of cells irrespective of antigen staining.

Furthermore, we tested if subsequent tumour resections impacted cell count with specific staining (Figure 3A–G). We have focused on the first three subsequent resections, as they have a representative number of cases. There were no significant trends in all stainings (DCX$^+$: p-value = 0.90, OLIG2$^+$: 0.72, NES$^+$: p-value = 0.89; df = 2).

Then, we tested if there are any correlations between the frequency of cells with different antigens and sites (Supplementary Figure S1). The frequency of cells expressing either NES or OLIG2 had a strong positive correlation between margins and tumour sites adjacent to margins. Although cells expressing DCX lacked this correlation, there was a strong negative correlation between the frequency of DCX$^+$ cells in the margin and the frequency of NES$^+$ cells both in tumour and margin sites.

Finally, we did not observe any change in overall survival regarding tested antigens (Supplementary Figure S2).

To better understand the function of DCX, we opted for the single-cell approach, which allowed annotating the cells as "cancer" and "normal" according to the number of GBM specific copy number variations (CNV) prediction (Supplementary Figure S3) based on transcriptomics of single cells (Figure 4A). To avoid false-positive identification of malignant cells, we considered only the canonical aberrations for GBM, which are a duplication of chromosome 7 or chromosome 10 loss. The cells defined as "cancer" and "normal" were 39.77% and 60.23%. The cells identified as 'normal' were clustered together, mainly in the two big clusters—one of them was specific for the cells derived from the foetal origin, whereas the second big cluster contained the non-cancer brain cells (Figure 4B–D). The 'cancer' cells, which were defined as GBM specific CNVs containing, had been clustered into the number of overlapping and sample-specific clusters, which shows their high heterogeneity (Figure 4A–C).

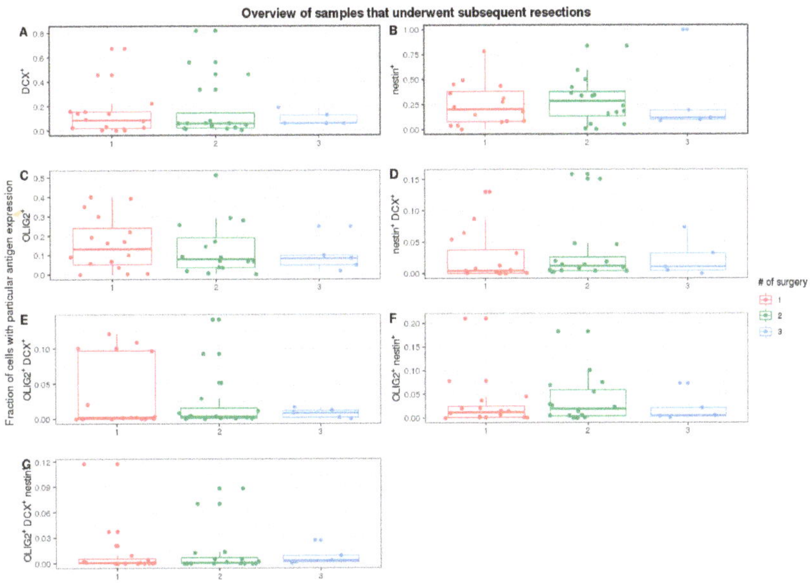

Figure 3. Overview of samples underwent subsequent resections. Frequency of cells with different stainings for tissues: (**A**)—DCX$^+$ cells, (**B**)—NES$^+$ cells, (**C**)—OLIG2$^+$ cells. Frequency of cells with mixtures of all stainings for tissues: (**D**)—OLIG2$^+$DCX$^+$ cells, (**E**)—OLIG2$^+$NES$^+$ cells, (**F**)—NES$^+$DCX$^+$ cells, (**G**)—DCX$^+$OLIG2$^+$NES$^+$ cells.

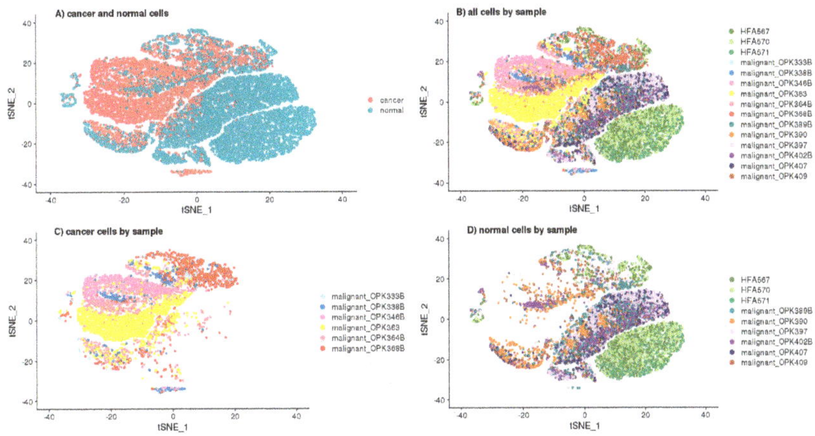

Figure 4. tSNE projection of single cells. (**A**)—Cancer and normal cells, (**B**)—All the cells coloured by sample name, (**C**)—"Cancer" cells only coloured by sample name, (**D**)—"Normal" cells only coloured by sample name.

Based on the single-cell transcriptomics, we also annotated the cluster cells to define the analysed cells' phenotype (Figure 5A–C). The complete list of cluster-specific genes are attached to the Supplementary Data chapter as a CSV file, whereas the distribution of the cells among the clusters by their "cancer" and "normal" status are shown in Supplementary Figure S4. These annotations were used to predict the role of DCX, OLIG2, and NES in tumour biology. We visualised the expressions of these genes that were shown among defined cell clusters in Figure 6. The cells defined as "normal" showed the expression of specific markers for neurons, astrocytes, dendritic cells, macrophages, plasmacytoid dendritic cells, circulating foetal cells, undifferentiated cells, and oligodendrocytes. These cells that had been labelled as "cancer" expressed the markers specific for circulating foetal cells, basophils, astrocytes, dendritic cells, neurons, proliferative cells, undifferentiated cells, and oligodendrocytes. The cells' distribution by their "cancer" or "normal" status is shown in Supplementary Figure S4.

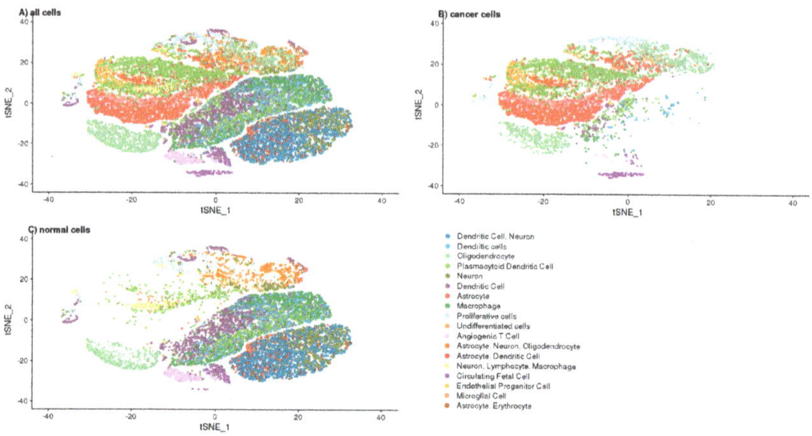

Figure 5. tSNE projection of annotated single cells by expression of markers specific for various cell types. (**A**)—Cancer and normal cells, (**B**)—Cancer cells, (**C**)—Normal cells.

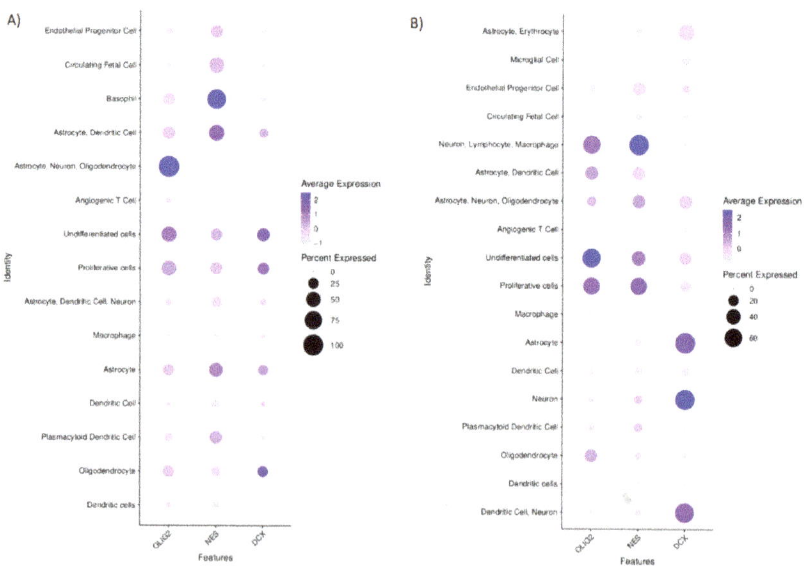

Figure 6. The mean expression of OLIG2, NES, and DCX transcripts among the annotated clusters of single cells that express markers for each cell type. (**A**) the cells annotated as "cancer", (**B**) the cells annotated as "normal".

OLIG2 level in the cells annotated as "cancer" was found only in this cluster of cells which also expressed the markers specific for astrocytes, neurons, and oligodendrocytes; however, it also was found in the cells with undifferentiated and proliferative properties.

NES was the most specific for the cells annotated as 'cancer', for which basophil markers are characteristic. DCX in the "cancer" identified cells was expressed in these clusters, identified according to the transcriptomics signatures specific for oligodendrocytes, undifferentiated, and proliferative cells (Figure 6A). In normal cells, OLIG2 was found mainly in the undifferentiated cells and in lower levels in clusters of proliferative cells and this group of cells with the gene signature that is specific for neurons, lymphocytes, and macrophages. NES expression was mostly seen in the clusters of cells that expressed the markers of neurons, lymphocytes, and macrophages and those that harboured the transcripts specific for proliferative and undifferentiated cells. DCX in normal cells was specific for neurons and astrocytes (Figure 6B).

We also evaluated the correlations between the expression of DCX, OLIG2, and NES genes at a single-cell level in the population of cells derived from the tumour tissue to show the potential relation of these genes. We found significant ($p < 0.01$) but rather weak positive correlations between transcript levels of these genes conducting Spearman's rank test: DCX vs. OLIG2 (r = 0.2179733), NES vs. OLIG2 (r = 0.3019327), NES vs. DCX (r = 0.1535854). When we look closer at the populations of these cells that had been annotated as 'normal', simultaneous expression of OLIG2 and DCX was observed in the clusters of these cells that were expressing markers specific for highly proliferative or undifferentiated cells as well as in the cluster of cells expressing markers that are common for astrocytes, oligodendrocytes and neurons. On the other hand, in the population of cancerous cells, NES and DCX were simultaneously expressed by the cells that were positive mostly for markers of: proliferative, undifferentiated or astrocytic cells.

3. Discussion

DCX status in GBM has been a subject of debate since Rich [20] and Daou [13] reported the expression of DCX on mRNA and protein levels in 2005. While Rich et al. correlated the mRNA level of DCX with poor diagnosis, Daou et al. proved that DCX had more intense staining towards the margin of the tumour using immunostaining. Not all samples of GBM had the DCX expression, however. Interestingly, our results did not confirm most of those observations. Although DCX was detected in most samples, DCX$^+$ cells were only a small subset of the GBM's cell population, both in tumours and margins. Only in a few cases were there many DCX$^+$ cells. Moreover, there was no shift in DCX$^+$ cell frequency towards one of the sites. Although inconclusive, this may imply that DCX is not directly connected to GBM cell migration.

On the other hand, Santra et al. [15,21] reported DCX as a marker for a favourable patient outcome. They found that cells with DCX overexpression had lower invasion abilities, thus, the authors concluded that DCX positive cells in glioma either infiltrate neuroblasts or pre-existing neuronal cells. Importantly, our data partially supports this idea. Although most of the cells expressing DCX within the tumour were marked as non-cancerous, four subpopulations were marked as GBM origin. They followed expression patterns similar to oligodendrocytes, astrocytes, as well as proliferative and undifferentiated cells. This aligns with the idea that DCX is expressed by GSC.

In general, NES$^+$ and OLIG2$^+$ cells were more frequent than DCX$^+$ ones, which is consistent with the bulk values reported in The Human Protein Atlas [22]. Interestingly, cells expressing NES or OLIG2 were present both in the margin and in tumours, while being highly correlated. The lack of a significant drop in NES$^+$ cells is in contradiction with a study by Smith et al. [23]; this might be explained by a different definition of margin adopted in our study. Smith et al. defined margin as a region with 5-aminolevulinic acid (5ALA) fluorescence during surgery that is beyond the T1 enhancing region on magnetic resonance imaging (MRI) [23]. However, when considering regions by their relative position to tumour sites, both studies are consistent. We also found that OLIG2$^+$ cells frequency differed among sites: more cells with OLIG2 expression were in margin than in tumour.

On the other hand, the frequency of NES^+ cells on the margin and within the tumour were reversely correlated with DCX^+ cells in the margin. A negative correlation seems not to support the existence of the DCX-NES complex in the tumour. It is worth keeping in mind that there is a gross difference between the frequency of DCX- and NES-positive cells, and sample size does not compensate for that. When we look closer at the populations of these cells that had been annotated as 'normal', simultaneous expression of OLIG2 and DCX was observed in the clusters of these cells that were expressing markers specific for highly proliferative or undifferentiated cells as well as in the cluster of cells expressing markers that are common for astrocytes, oligodendrocytes and neurons. On the other hand, in the population of cancerous cells, NES and DCX were simultaneously expressed by the cells that were positive mostly for markers of: proliferative, undifferentiated or astrocytic cells.

Another interesting finding was that subsequent resections did not significantly influence assessed markers' frequency. In fact, all three of them seemed to be associated with some form of undifferentiated cells. It was reported previously that glioma stem cells have higher invasion capabilities [24], suggesting that stemness might be a phenotypic response to changing the tumour's microenvironment [25]. Our data also support the notion that there is a set proportion of cells expressing DCX/NES/OLIG2, regardless of treatment, guided by tumour plasticity.

The GBM cells revealed high heterogeneity, which we saw as clustering into several different groups of cells derived by transcriptomics. Others also reported a similar relationship [26], which confirms the value of our analytical approach. However, annotating the cells to "cancer" or "normal" subpopulations using our quite simple approach may not be entirely accurate as we might omit other possible genetic aberrations than those used canonically for GMB. Nonetheless, our approach seems to be valuable, for it limits the false positive detection of malignant cells. Moreover, the algorithms used in our investigation allowed us to properly annotate the neurons and the immune cells (i.e., macrophages, dendritic and T cells) as non-cancer cells. However, we also identified a distinct cluster of cells that we recognised as "cancer" because of genetic aberrations parallel to CD63 basophil marker expression, while not being positive for other canonical basophil markers (e.g., CD123). That suggests that these cells should not be classified as immune cells [27].

On the technical side, our primary consideration was to distinguish individual cells. Fluorescent signals from cell nuclei and OLIG2 were easy to partition between cells, as in the vast majority of cases, it was single point luminescence. On the other hand, DCX fluorescence was more challenging to evaluate because of the branched structure of a cell's cytoskeleton. Finally, the hardest to assess was NES: not only were the signal figures branched, but NES+ cells were also more prevalent, with a tendency to be clumped.

In conclusion, our findings support the notion that DCX is indeed expressed by glioma cells, but there is no clear evidence to prove that it may participate in GBM cell migration. Other GSC markers: NES and OLIG2 are in much higher abundance and are present both in tumours and their margin.

4. Materials and Methods

4.1. Samples Collection and Selection of Regions of Interest (ROIs)

Paraffin-embedded samples were obtained from the Department of Neurosurgery and Paediatric Neurosurgery archive, Medical University of Lublin (Poland). We selected only those samples subjected to at least two resections, and at least one sample from each patient was diagnosed with GBM. Moreover, we have focused on case studies since 2011. For each of the samples, a trained neuropathologist performed second-hand diagnosis, along with marking ROIs. The goal was to mark regions either with representative tumour tissue or with the border between tumour and margin tissue. In total, we collected and assessed 17 cases over 46 paraffin-embedded tissues, marking 60 ROIs.

Additionally, all ROIs were classified using two types of categories: tissue origin and surgery sequence number. Tissue origin contains four subcategories: GBM cases (1) when a tumour was adjacent to margin, (2) tissue adjacent to margin, (3) tumour-only when tumour ROI was far from margin/margin was not detected, and (4) tumour-only non-GBM, with ROIs containing a lower-grade astrocytoma. The second category showed surgery sequence number corresponding to samples taken during the first, second or third surgery.

4.2. Multiplex Immunofluorescence Staining

Paraffin-embedded tissues were cut into three μm sections and placed on glass slides (Thermo Fisher Scientific, Waltham, MA, USA: 10149870). Slides were baked overnight at 60 °C. The next day, slides were deparaffinised and hydrated in a series of xylene and ethyl alcohol. Antigens were retrieved by microwave-HIER in a citrate buffer (pH 6.0) over 20 min, followed by another 20 min in RT to cool down. Endogenous peroxide activity was quenched by a mixture of 1% H2O2 and 1% sodium azide for 20 min. All non-specific binding sites were blocked by incubation with blocking solution: PBS (Merck, Darmstadt, Germany: P4417-50TAB) with 1% BSA (Jackson ImmunoResearch, Cambridgeshire, UK: 001-000-161), 5% NDS (Cambridgeshire, UK: 017-000-121) and 1% Triton X-100, for 1 h. OLIG2, NES, and DCX were marked sequentially. Firstly, sections were incubated with appropriate primary antibodies (anti-NES: 1:100, Novus Biologicals, Centennial, CO, USA, MAB1259; anti-OLIG2: 1:100, AF2418; anti-DCX: 1:100, Cambridge, United Kingdom, ab18723; diluted in PBS with 1% BSA, 1% Triton X-100). After washing, corresponding secondary antibodies conjugated with HRP were used (anti-Mouse–Rabbit–Goat: 1:500, Cambridgeshire, UK, 715-036-150, 711-036-152, 705-036-147; diluted in PBS with 1% BSA, 1% Triton X-100). The visualisation was carried out using the Tyramide Signal Amplification (Thermo Fisher Scientific, Waltham, MA, USA: B40953, B40958, B40955) setup. In between fluorescent stainings, HRP activity was quenched by a mixture of 1% H2O2 and 1% sodium azide for 20 min to not interfere with subsequent staining. Finally, nuclei were detected with Hoechst-33342 (Thermo Fisher Scientific, Waltham, MA, USA: H3570), and slides were mounted (Thermo Fisher Scientific, Waltham, MA, USA: P36982). During incubations, slides were covered with a hand-cut parafilm piece, and all washing between steps was conducted with PBS 3 × 5 min. Incubations with primary and secondary antibodies lasted 1 h each.

4.3. Image Acquisition

All images were captured using a Nikon Ti Confocal microscope, using four lasers for fluorescence: 405, 488, 563, and 647 nm. The NISelements (ver 3.22.08, Melville, NY, USA) software was used to set up analyses. The corresponding primary antibody's negative control was used to control non-specific staining for each batch of analysed slides. In this regard, specific laser intensity and gain parameters were chosen to remove any signal corresponding to 488, 563 and 647 nm. In addition, epifluorescence site conformation was done before confocal imaging to ensure the highest quality of the images.

4.4. Image Analysis

If possible, three different random square spots with a side of 600 pixels from each tissue image were selected to perform further analysis. In total, 152 random square spots were generated across 60 ROIs.

First, stack images were divided by channels corresponding to the wavelength used to detect each antigen: NES, OLIG2, DCX, and nuclei. Next, all features were marked, applying the following three heuristic assumptions:

- Nucleus—any oval and coherent figure with a signal pattern corresponding to Hoechst-33342;
- DCX/NES—nucleus with an adjacent signal corresponding to DCX or NES; in case of a signal adjacent to more than one nucleus, all were counted as positive;
- OLIG2—nucleus overlapping with a signal corresponding to OLIG2.

A nuclei drove Voronoi's diagram was created to find all positive cells for each of the fluorescent signals. Finally, DCX/NES/OLIG2 layers with marks were used to find and count all overlaps with cells in Voronoi's diagram. The analysis was carried out in ImageJ (Fiji) [28].

4.5. Image Deposition and Sharing

We set up a local data share service based on Digital Slide Archive (DSA) [29]. All images in TIFF format were converted to pyramidal TIFF format using the ImageMagick tool [30] and uploaded to DSA along with all necessary metadata.

4.6. Single-Cell RNA-Seq and Data Processing

The data used here were generated and published previously by Couturier et al. [26]. It was further processed using the CellRanger pipeline. We included data for 12 CRC patients (originally named as OPK333B, OPK338B, OPK346B, OPK363, OPK364B, OPK368B, OPK389B, OPK390, OPK397, OPK402B, OPK407, OPK409) and three samples derived from the foetal brain (HFA567, HFA570, HFA571). First, the raw gene expression matrix was filtered and normalised using the Seurat R package. Then, the dataset was filtered according to the following criteria: cells with >1000 unique molecular identifier (UMI) counts; >500 genes and <5000 genes; and <10% of mitochondrial gene expression in UMI counts. The gene expression matrices from 15,000 randomly down-sampled cells were log-normalised to the total UMI counts per cell, scaled and finally clustered and visualised using t-SNE projection. The major cell types were characterised by comparing the canonical marker genes found in tissue-specific cell taxonomy reference database CellMatch and the differentially expressed genes for each cluster using the scCATCH automatic annotation algorithm [31].

To identify evidence for somatic large-scale chromosomal copy number alterations, we used inferCNV of the Trinity CTAT Project algorithm [32] in the reference with non-cancer, foetal brain cells. The approach used to annotate the cells as malignant ('cancer'), was based on the presence of the most common canonical aberration for GBM (duplication of chromosome 7 or chromosome 10 loss), and it is line with the analytical strategy used previously by Couturier et al. [26].

4.7. Statistics

Data files from image analysis were imported and analysed in Rstudio, along with ggplot, corrplot, and dplyr packages. Pairwise comparisons using the Wilcoxon rank-sum test was used to check the cell count in different regions of samples. The Wilcoxon signed-rank test was used to find any consequential difference between margin and adjacent tumour tissue in terms of any staining. Mann–Whitney-Wilcoxon Test compared any other tumour tissue to the ones that were adjacent to the margin. The Kruskal–Wallis rank-sum test allowed to check any change in staining after subsequent tumour resections. Spearman's rank correlation coefficient measured correlations. To check overall survival, Cox proportional-hazards models were used. In all tests, $p < 0.05$ was considered significant. All subsamples of the same image were averaged.

Supplementary Materials: The following are available online at https://www.mdpi.com/article/10.3390/ijms222413217/s1.

Author Contributions: Conceptualisation, R.R., B.J. and A.O.; Methodology, A.O., M.K. and I.T.; Software, A.O.; Validation, R.R., B.J. and A.O.; Formal Analysis, A.O. and M.K.; Investigation, A.O., B.J., M.K., I.T., and S.K.; Resources, B.J., and D.Z.; Data Curation, R.R. and A.O.; Writing—Original Draft Preparation, A.O., and M.K.; Writing—Review and Editing, R.R., and B.J.; Visualization, A.O., M.K.; Supervision, R.R.; Project Administration, R.R.; Funding Acquisition, A.O. and R.R. All authors have read and agreed to the published version of the manuscript.

Funding: This research was funded by the Polish National Science Centre, grant number 2015/17/N/NZ4/02839.

Institutional Review Board Statement: The study material was collected for histopathological diagnosis and came from the archives of Independent Public Clinical Hospital No. 4 in Lublin, Poland. In accordance with article 26(4) of the Act on Patient's Rights and Patient's Ombudsman of 2008, November the 6th (i.e., Journal of Laws of 2016, item 186 as amended) medical records can also be shared with universities or research institutes to be used for scientific purposes without revealing surnames or other data enabling identification of the person to which it relates. According to the judgment of the Supreme Court number VCSK 256/10, tissue sections from human organism and histopathological specimen constitute medical records. Thereby, tissue sections and specimen can be shared with medical colleges according to the law under conditions other than patient's consent. Therefore, no Institutional Review Board Statement is required under Polish Law.

Informed Consent Statement: The study material was collected for histopathological diagnosis and came from the archives of Independent Public Clinical Hospital No. 4 in Lublin, Poland. In accordance with article 26(4) of the Act on Patient's Rights and Patient's Ombudsman of 2008, November the 6th (i.e., Journal of Laws of 2016, item 186 as amended) medical records can also be shared with universities or research institutes to be used for scientific purposes without revealing surnames or other data enabling identification of the person to which it relates. According to the judgment of the Supreme Court number VCSK 256/10, tissue sections from human organism and histopathological specimen constitute medical records. Thereby, tissue sections and specimen can be shared with medical colleges according to the law under conditions other than patient's consent. Therefore, no Informed Consent from the patients is required under Polish Law.

Data Availability Statement: Single-cell RNA-seq data (EGAD00001006206) can be accessed via European Genome-Phenome Archive. A.O. acquired written permission to use mentioned data.

Acknowledgments: We would like to thank Ewa Hordyjewska-Kowalczyk for allowing us to use the workstation funded by the Polish National Science Centre [2017/27/N/NZ5/02940 to EHK].

Conflicts of Interest: The authors declare no conflict of interest.

References

1. Louis, D.N.; Perry, A.; Reifenberger, G.; von Deimling, A.; Figarella-Branger, D.; Cavenee, W.K.; Ohgaki, H.; Wiestler, O.D.; Kleihues, P.; Ellison, D.W. The 2016 World Health Organization Classification of Tumors of the Central Nervous System: A summary. *Acta Neuropathol.* **2016**, *131*, 803–820. [CrossRef] [PubMed]
2. Park, J.; de Lomana, A.; Marzese, D.; Juarez, T.; Feroze, A.; Hothi, P.; Cobbs, C.; Patel, A.; Kesari, S.; Huang, S.; et al. A Systems Approach to Brain Tumor Treatment. *Cancers* **2021**, *13*, 3152. [CrossRef] [PubMed]
3. Zhang, H.; Wang, R.; Yu, Y.; Liu, J.; Luo, T.; Fan, F. Glioblastoma Treatment Modalities besides Surgery. *J. Cancer* **2019**, *10*, 4793–4806. [CrossRef]
4. Vollmann-Zwerenz, A.; Leidgens, V.; Feliciello, G.; Klein, C.A.; Hau, P. Tumor Cell Invasion in Glioblastoma. *Int. J. Mol. Sci.* **2020**, *21*, 1932. [CrossRef] [PubMed]
5. Petrecca, K.; Guiot, M.-C.; Panet-Raymond, V.; Souhami, L. Failure pattern following complete resection plus radiotherapy and temozolomide is at the resection margin in patients with Glioblastoma. *J. Neuro-Oncol.* **2013**, *111*, 19–23. [CrossRef]
6. Alvarez-Buylla, A.; Temple, S. Stem cells in the developing and adult nervous system. *J. Neurobiol.* **1998**, *36*, 105–110. [CrossRef]
7. Gage, F.H. Mammalian Neural Stem Cells. *Science* **2000**, *287*, 1433–1438. [CrossRef] [PubMed]
8. Verhaak, R.G.W.; Hoadley, K.A.; Purdom, E.; Wang, V.; Qi, Y.; Wilkerson, M.D.; Miller, C.R.; Ding, L.; Golub, T.; Mesirov, J.P.; et al. Integrated Genomic Analysis Identifies Clinically Relevant Subtypes of Glioblastoma Characterized by Abnormalities in PDGFRA, IDH1, EGFR, and NF1. *Cancer Cell* **2010**, *17*, 98–110. [CrossRef] [PubMed]
9. Gleeson, J.G.; Allen, K.M.; Fox, J.W.; Lamperti, E.D.; Berkovic, S.; Scheffer, I.; Cooper, E.C.; Dobyns, W.; Minnerath, S.R.; Ross, M.; et al. Doublecortin, A Brain-Specific Gene Mutated in Human X-Linked Lissencephaly and Double Cortex Syndrome, Encodes a Putative Signaling Protein. *Cell* **1998**, *92*, 63–72. [CrossRef]
10. Gleeson, J.G.; Lin, P.T.; Flanagan, L.A.; Walsh, C.A. Doublecortin Is a Microtubule-Associated Protein and Is Expressed Widely by Migrating Neurons. *Neuron* **1999**, *23*, 257–271. [CrossRef]
11. Tabata, H.; Nakajima, K. Multipolar migration: The third mode of radial neuronal migration in the developing cerebral cortex. *J. Neurosci.* **2003**, *23*, 9996–10001. [CrossRef] [PubMed]
12. LoTurco, J. Doublecortin and a Tale of Two Serines. *Neuron* **2004**, *41*, 175–177. [CrossRef]
13. Daou, M.-C.; Smith, T.W.; Litofsky, N.S.; Hsieh, C.C.; Ross, A.H. Doublecortin is preferentially expressed in invasive human brain tumors. *Acta Neuropathol.* **2005**, *110*, 472–480. [CrossRef] [PubMed]
14. Masui, K.; Mawatari, S.Y.; Suzuki, S.O.; Iwaki, T. Evaluation of sensitivity and specificity of Doublecortin immunostatining for the detection of infiltrating glioma cells. *Brain Tumor Pathol.* **2008**, *25*, 1–7. [CrossRef] [PubMed]
15. Santra, M.; Santra, S.; Roberts, C.; Zhang, R.L.; Chopp, M. Doublecortin induces mitotic microtubule catastrophe and inhibits glioma cell invasion. *J. Neurochem.* **2009**, *108*, 231–245. [CrossRef]

16. Mitew, S.; Hay, C.; Peckham, H.; Xiao, J.; Koenning, M.; Emery, B. Mechanisms regulating the development of oligodendrocytes and central nervous system myelin. *Neuroscience* **2014**, *276*, 29–47. [CrossRef] [PubMed]
17. Trépant, A.-L.; Bouchart, C.; Rorive, S.; Sauvage, S.; Decaestecker, C.; Demetter, P.; Salmon, I. Identification of OLIG2 as the most specific Glioblastoma stem cell marker starting from comparative analysis of data from similar DNA chip microarray platforms. *Tumor Biol.* **2014**, *36*, 1943–1953. [CrossRef] [PubMed]
18. Ortensi, B.; Setti, M.; Osti, D.; Pelicci, G. Cancer stem cell contribution to Glioblastoma invasiveness. *Stem Cell Res. Ther.* **2013**, *4*, 18. [CrossRef] [PubMed]
19. Bott, C.J.; McMahon, L.P.; Keil, J.M.; Yap, C.C.; Kwan, K.Y.; Winckler, B. Nestin selectively facilitates the phosphorylation of the Lissencephaly-linked protein Doublecortin (DCX) by cdk5/p35 to regulate growth cone morphology and Sema3a sensitivity in developing neurons. *J. Neurosci.* **2020**, *40*, 3720–3740. [CrossRef]
20. Rich, J.N.; Hans, C.; Jones, B.; Iversen, E.S.; McLendon, R.E.; Rasheed, B.K.A.; Dobra, A.; Dressman, H.K.; Bigner, D.D.; Nevins, J.R.; et al. Gene Expression Profiling and Genetic Markers in Glioblastoma Survival. *Cancer Res.* **2005**, *65*, 4051–4058. [CrossRef] [PubMed]
21. Santra, M.; Zheng, X.; Roberts, C.; Santra, S.; Lu, M.; Panda, S.; Jiang, F.; Chopp, M. Single Doublecortin gene therapy significantly reduces glioma tumor volume. *J. Neurosci. Res.* **2010**, *88*, 304–314. [CrossRef] [PubMed]
22. Uhlén, M.; Zhang, C.; Lee, S.; Sjöstedt, E.; Fagerberg, L.; Bidkhori, G.; Benfeitas, R.; Arif, M.; Liu, Z.; Edfors, F.; et al. A pathology atlas of the human cancer transcriptome. *Science* **2017**, *357*, eaan2507. [CrossRef] [PubMed]
23. Smith, S.J.; Diksin, M.; Chhaya, S.; Sairam, S.; Estevez-Cebrero, M.A.; Rahman, R. The Invasive Region of Glioblastoma Defined by 5ALA Guided Surgery Has an Altered Cancer Stem Cell Marker Profile Compared to Central Tumour. *Int. J. Mol. Sci.* **2017**, *18*, 2452. [CrossRef]
24. Cheng, L.; Wu, Q.; Guryanova, O.A.; Huang, Z.; Huang, Q.; Rich, J.N.; Bao, S. Elevated invasive potential of Glioblastoma stem cells. *Biochem. Biophys. Res. Commun.* **2011**, *406*, 643–648. [CrossRef] [PubMed]
25. Dirkse, A.; Golebiewska, A.; Buder, T.; Nazarov, P.V.; Muller, A.; Poovathingal, S.; Brons, N.H.C.; Leite, S.; Sauvageot, N.; Sarkisjan, D.; et al. Stem cell-associated heterogeneity in Glioblastoma results from intrinsic tumor plasticity shaped by the microenvironment. *Nat. Commun.* **2019**, *10*, 1787. [CrossRef]
26. Couturier, C.P.; Ayyadhury, S.; Le, P.U.; Nadaf, J.; Monlong, J.; Riva, G.; Allache, R.; Baig, S.; Yan, X.; Bourgey, M.; et al. Single-cell RNA-seq reveals that Glioblastoma recapitulates a normal neurodevelopmental hierarchy. *Nat. Commun.* **2020**, *11*, 3406. [CrossRef]
27. Kim, Z.; Choi, B.S.; Kim, J.K.; Won, D.I. Basophil Markers for Identification and Activation in the Indirect Basophil Activation Test by Flow Cytometry for Diagnosis of Autoimmune Urticaria. *Ann. Lab. Med.* **2016**, *36*, 28–35. [CrossRef]
28. Schindelin, J.; Arganda-Carreras, I.; Frise, E.; Kaynig, V.; Longair, M.; Pietzsch, T.; Preibisch, S.; Rueden, C.; Saalfeld, S.; Schmid, B.; et al. Fiji: An open-source platform for biological-image analysis. *Nat. Methods* **2012**, *9*, 676–682. [CrossRef]
29. Gutman, D.A.; Khalilia, M.; Lee, S.; Nalisnik, M.; Mullen, Z.; Beezley, J.; Chittajallu, D.R.; Manthey, D.; Cooper, L.A.D. The Digital Slide Archive: A Software Platform for Management, Integration, and Analysis of Histology for Cancer Research. *Cancer Res.* **2017**, *77*, e75–e78. [CrossRef]
30. The ImageMagick Development Team, ImageMagick. 2021. Available online: https://imagemagick.org (accessed on 15 October 2021).
31. Shao, X.; Liao, J.; Lu, X.; Xue, R.; Ai, N.; Fan, X. scCATCH: Automatic Annotation on Cell Types of Clusters from Single-Cell RNA Sequencing Data. *iScience* **2020**, *23*, 100882. [CrossRef]
32. Tickle, T.; Tirosh, I.; Georgescu, C.; Brown, M.; Haas, B. *inferCNV of the Trinity CTAT Project*; Klarman Cell Observatory, Broad Institute of MIT and Harvard: Cambridge, MA, USA, 2019. Available online: https://github.com/broadinstitute/inferCNV (accessed on 15 October 2021).

International Journal of *Molecular Sciences*

Article

Microglia-Derived Olfactomedin-like 3 Promotes Pro-Tumorigenic Microglial Function and Malignant Features of Glioma Cells

Ryan G. Toedebusch, Christopher A. Lucchesi, Eshetu T. Debebe, Luke A. Wittenburg, Xinbin Chen and Christine M. Toedebusch *

Department of Surgical and Radiological Sciences, School of Veterinary Medicine, University of California, Davis, CA 95616, USA; rgtoed@ucdavis.edu (R.G.T.); calucchesi@ucdavis.edu (C.A.L.); etdebebe@ucdavis.edu (E.T.D.); lwittenburg@ucdavis.edu (L.A.W.); xbchen@ucdavis.edu (X.C.)

* Correspondence: cmtoedebusch@ucdavis.edu; Tel.: +1-530-754-0798

Abstract: Under the influence of transforming growth factor-beta (TGFβ), glioma-associated microglia produce molecules that promote glioma growth and invasion. Olfactomedin-like 3 (*Olfml3*), a novel, secreted glycoprotein, is known to promote several non-CNS cancers. While it is a direct TGFβ1 target gene in microglia, the role of microglia-derived OLFML3 in glioma progression is unknown. Here, we tested the hypotheses that microglial *Olfml3* is integral to the pro-tumorigenic glioma-associated microglia phenotype and promotes glioma cell malignancy. Using an *Olfml3* knockout microglial cell line (N9), we demonstrated that *Olfml3* is a direct target gene of all TGFβ isoforms in murine microglia. Moreover, loss of *Olfml3* attenuated TGFβ-induced restraint on microglial immune function and production of cytokines that are critical in promoting glioma cell malignancy. Importantly, microglia-derived OLFML3 directly contributes to glioma cell malignancy through increased migration and invasion. While exposure to conditioned medium (CM) from isogenic control microglia pre-treated with TGFβ increased mouse glioma cell (GL261) migration and invasion, this effect was abolished with exposure to CM from TGFβ-treated *Olfml3*$^{-/-}$ microglia. Taken together, our data suggest that *Olfml3* may serve as a gatekeeper for TGFβ-induced microglial gene expression, thereby promoting the pro-tumorigenic microglia phenotype and glioma cell malignancy.

Keywords: microglia; glioblastoma; olfactomedin-like 3; TGFβ

Citation: Toedebusch, R.G.; Lucchesi, C.A.; Debebe, E.T.; Wittenburg, L.A.; Chen, X.; Toedebusch, C.M. Microglia-Derived Olfactomedin-like 3 Promotes Pro-Tumorigenic Microglial Function and Malignant Features of Glioma Cells. *Int. J. Mol. Sci.* **2021**, *22*, 13052. https://doi.org/10.3390/ijms222313052

Academic Editors: Nives Pećina-Šlaus and Ivana Jovčevska

Received: 31 October 2021
Accepted: 29 November 2021
Published: 2 December 2021

Publisher's Note: MDPI stays neutral with regard to jurisdictional claims in published maps and institutional affiliations.

1. Introduction

Glioblastoma (GBM) is the most common and aggressive primary brain tumor of adults, with a 5-year survival rate of approximately 5% [1]. Although immunotherapy has advanced the treatment of non-central nervous system (CNS) tumors, it has failed to overcome the substantial barrier of immune resistance in the glioma microenvironment. While immunoresistance and tumor progression are conferred by a confluence of factors, glioma-associated microglia/macrophages (GAM) play a critical role. As the most abundant infiltrating cells [2], GAM infiltration has been positively correlated with glioma grade [3], invasiveness [4], and resistance to therapy [5,6].

While the glioma-GAM signaling axis is complex, transforming growth factor-beta (TGFβ) isoforms have been recognized to substantially influence the pro-tumorigenic effects of GAM. β1 stimulates GAM to produce cytokines and growth factors promoting glioma growth [7] and invasion [8,9], whereas β2 suppresses GAM immune responses [10]. β3 promotes tumor invasion and augments β1 and β2 signaling [11]. A variety of approaches to inhibit TGFβ signaling [12] and glioma recruitment of GAM [13] have failed to show therapeutic efficacy in GBM [12], underscoring the need for refined therapeutic targets.

Intriguingly, β1 was recently found to induce transcription of a novel gene, *Olfml3*, in mouse microglia. *Olfml3*, encoding the secreted glycoprotein olfactomedin-like 3 (OLFML3), is expressed in microglia but not macrophages [14]. Following exposure to β1, primary mouse microglia increased *Olfml3* mRNA expression 20-fold. Although the role of *Olfml3* in microglia function is unknown, olfactomedin-family proteins modulate the Wnt pathway [15,16], which plays an important role in microglial phenotype determination [17] and gliomagenesis [18]. Importantly, OLFML3 has broad relevance to cancer progression. OLFML3 is a disease biomarker in colon cancer [19] and has been shown to promote neoangiogenesis [20,21], epithelial-to-mesenchymal transition [22], and metastasis [22] in several cancers.

While the role of OLFML3 in GBM has just begun to be explored, depletion of *OLFML3* in human glioma cells reduced GAM infiltration and extended survival in a glioma xenograft mouse model [23]. However, the function of microglia-derived *Olfml3*, and its contribution to the TGFβ-induced pro-tumorigenic GAM phenotype, is unknown. Therefore, this study aimed to (1) define the function of *Olfml3* in microglia phenotype determination and (2) determine the effect of microglia-derived OLFML3 on the malignant phenotype of murine glioma cells.

2. Results

2.1. OLFML3 Is Up-Regulated in GBM and Is a TGFβ Target Gene in Microglia

Examination of The Cancer Genome Atlas (TCGA) transcriptomic datasets revealed that *OLFML3* mRNA expression increased with increasing glioma tumor malignancy. While low-grade gliomas (LGG, $n = 592$) had increased *OLFML3* mRNA levels relative to normal brain ((Normal, $n = 1141$; $p < 0.001$), glioblastomas (GBM, $n = 166$) had increased *OLFML3* mRNA expression relative to both LGG ($p < 0.001$) and normal brain ($p < 0.001$) (Figure 1A).

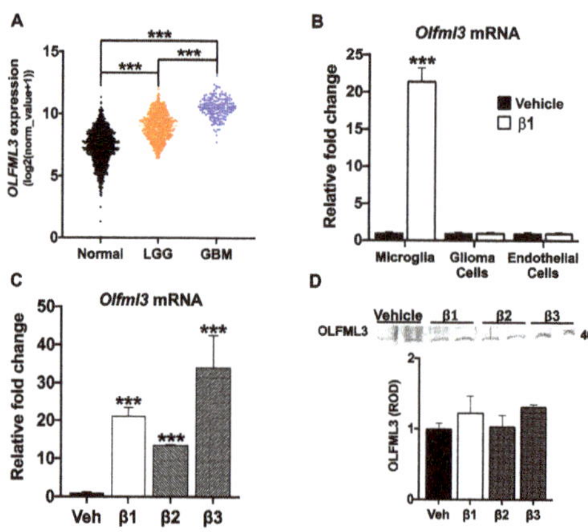

Figure 1. *OLFML3* is increased in GBM and is regulated by TGFβ in microglia. (**A**) *OLFML3* mRNA is increased in low-grade glioma (LGG; $n = 529$) and glioblastoma (GBM; $n = 166$) relative to normal brain (Normal; $n = 1141$) in TCGA patient datasets. (**B**) Exposure to β1 (5 ng/mL; 48 h) increased *Olfml3* mRNA 20-fold in a microglial cell line (N9) but did not affect mRNA expression in a mouse glioma cell line (GL261) or primary mouse brain endothelial cells. Fold was calculated via ΔΔCt and normalized to GAPDH; *** $p < 0.001$. (**C**) Exposure to each TGFβ isoform increased *Olfml3* mRNA (5 ng/mL; 48 h); *** $p < 0.001$. (**D**) Representative immunoblot for OLFML3 protein in N9 cell lysate following exposure to vehicle (Veh) and TGFβ isoforms (5 ng/mL; 48 h). The optical density of OLFML3 protein in cell lysates was measured and normalized to the Ponceau stain. Relative optical densities (ROD) were expressed relative to vehicle-treated cells. No differences were measured between groups ($p = 0.17$). Comparisons based on one-way ANOVA with Tukey's Multiple Comparison Test. Bars represent group mean with standard error of the mean (SEM); data represent one of three independent experiments.

To begin to explore putative sources for increased *OLFML3* in GBM, we confirmed *Olfml3* expression in a mouse microglia cell line (N9) [24], a mouse glioma cell line (GL261) [25], and primary mouse brain endothelial cells. As previously demonstrated [26], exposure to β1 increased *Olfml3* mRNA 22-fold in N9 cells relative to vehicle-treated cells ($p < 0.001$). However, neither GL261 nor endothelial cell *Olfml3* mRNA levels were affected by β1 treatment (Figure 1B). Importantly, exposure to all three TGFβ isoforms increased N9 *Olfml3* mRNA (β1: 20-fold, $p < 0.001$; β2: 13-fold, $p < 0.001$; β3: 33-fold, $p < 0.001$) (Figure 1C). Exposure to TGFβ isoforms did not alter OLFML3 protein in N9 cell lysate ($p = 0.17$; Figure 1D). Given these findings, it is possible that increased *OLFML3* mRNA expression in GBM is derived from microglia.

2.2. CRISPR/Cas9-Mediated Knockout of Olfml3 in Microglia

To determine the function of *Olfml3* and its contribution to the TGFβ-induced pro-tumorigenic phenotype in mouse microglial cells, we performed CRISPR-Cas9-mediated *Olfml3* gene editing in N9 cells. Due to alternative splicing within the *Olfml3* gene (Figure 2A), exon 1 was targeted using the guide RNA, as outlined in Table 1. Forty base-pairs were deleted in Exon 1 (Figure 2A) and verified via Sanger sequencing. This deletion resulted in an immediate stop codon. Knockout of *Olfml3* was validated via qRT-PCR and Western blot (Figure 2B,C).

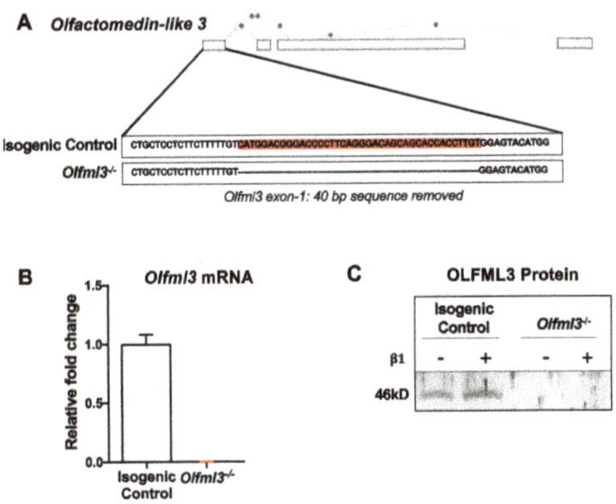

Figure 2. Development and validation of CRISPR-Cas9-mediated *Olfml3* gene editing in microglia. (**A**) Depiction of mouse *Olfml3* with predicted splice variants, demonstrating targeted deletion of 40 bases within Exon 1. (**B**) *Olfml3* mRNA was detected in isogenic control, but not *Olfml3*$^{-/-}$, microglia. (**C**) Representative immunoblot for OLFML3, demonstrating immunoreactivity at the predicted molecular weight (46 kD) in isogenic control, but not *Olfml3*$^{-/-}$, N9 cells. Bars represent group mean with standard error of the mean (SEM); data represent one of three independent experiments.

Table 1. Primer and guide-RNA sequences used for PCR and CRISPR-Cas9, respectively.

Gene.	Sequence-F (5′ to 3′)	Sequence-R (5′ to 3′)	Use
Olfml3 s1	GCTAACGGGCTGGAGGGAAA	AGTGGTACCATCCCATCCGA	PCR
Olfml3 s2	AGCTGCCTTAGAGGAACGG	CCTCCCTTTCAAGACGGTCC	qPCR
H2-Ab1	AGCCCCATCACTGTGGAGT	GATGCCGCTCAACATCTTGC	qPCR
Nos2	TTCTCAGCCACCTTGGTGAAG	AAGTGAAATCCGATGTGGCC	qPCR
Pdgfa	GAGGAAGCCGAGATACCCC	TGCTGTGGATCTGACTTCGAG	qPCR
Rpl22	AGCAGGTTTTGAAGTTCACCC	CAGCTTTCCCATTCACCTTGA	qPCR
Olfml3-gRNA	TCATGGACGGGACCCCTTCA		CRISPR

2.3. Loss of *Olfml3* Impaired Microglial Phagocytosis and Chemotaxis

Microglia are actively recruited to and proliferate within the glioma microenvironment [27]. Therefore, we evaluated several key microglial functions following *Olfml3* deletion. First, we established that loss of *Olfml3* did not alter microglial morphology (Figure 3A). Moreover, loss of *Olfml3* did not alter cellular viability, as assessed via Cell Titer-Glo® assay (100.0 ± 17.2 vs. 95.4 ± 13.5; $p = 0.53$) and MTS assay (100.0 ± 3.2 vs. 103.9 ± 0.5; $p = 0.27$) (Figure 3B). However, phagocytosis of pHrodo™ *Escherichia coli* bioparticles was reduced in *Olfml3*$^{-/-}$ microglia (1.0 ± 0.04 vs. 0.7 ± 0.06; $p < 0.001$) (Figure 3C). Similarly, loss of *Olfml3* altered microglial response to chemotactic cues. Using transwell migration assays, *Olfml3*$^{-/-}$ microglia had reduced migration toward fetal bovine serum (FBS; 10%) compared to isogenic control cells (9.4 ± 6 vs. 35.6 ± 7; $p < 0.001$) (Figure 3D,E) but not the potent chemoattractant human recombinant c-c motif chemokine ligand 2 [28] (rhCCL2) (43.7 ± 11 vs. 45.4 ± 10; $p = 0.54$) (Figure 3D,F). Moreover, while isogenic control microglia exhibited increased chemotaxis toward adenosine triphosphate (ATP)-supplemented medium (Veh: 24.4 ± 1 vs. 50 μM: 36.5 ± 1, $p < 0.0001$; 100 μM: 32.3 ± 1, $p < 0.001$) (Figure 3G), this effect was abolished in *Olfml3*$^{-/-}$ microglia (50 μM: 4.8 ± 0.2, $p < 0.001$; 100 μM: 4.9 ± 0.3, $p < 0.001$) (Figure 3G). Taken together, these findings suggest that loss of *Olfml3* induces specific perturbations in microglial response to environmental stimuli.

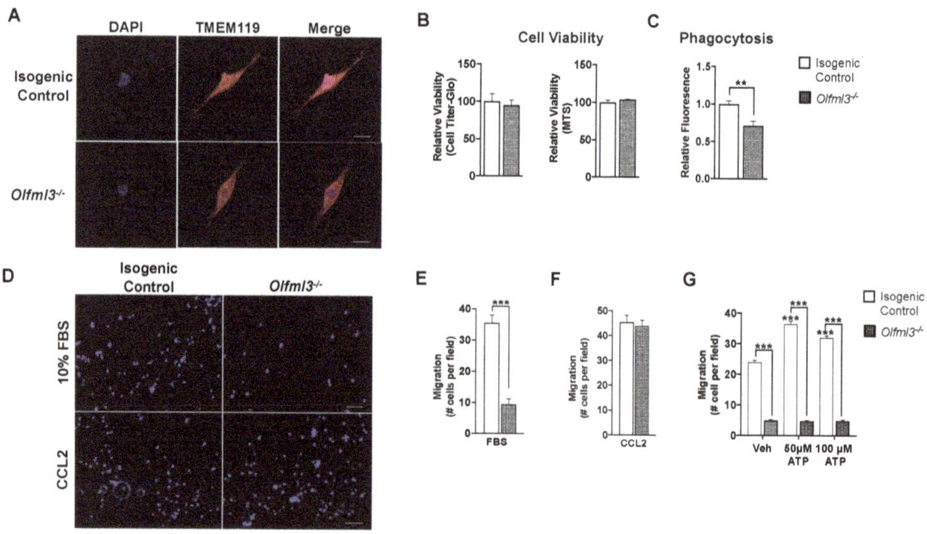

Figure 3. Loss of *Olfml3* impairs microglial phagocytosis and chemotaxis. (**A**) There were no morphological differences detected between isogenic control and *Olfml3*$^{-/-}$ microglia; scale bar 5 μm. (**B**) Cellular viability, as assayed by Cell Titer-Glo® ($p = 0.53$) and MTS ($p = 0.27$) assays, was not altered following deletion of *Olfml3* in N9 microglia. (**C**) Microglial phagocytosis of pHrodo™ *Escherichia coli* bioparticles was reduced by 30% in *Olfml3*$^{-/-}$ microglia relative to isogenic control cells. Comparisons based on students t-test; ** $p < 0.01$. (**D**) Representative images of isogenic control and *Olfml3*$^{-/-}$ microglial migration toward fetal bovine serum (FBS; 10%) or human recombinant C-C motif chemokine ligand 2 (rhCCL2; 10 ng/mL); scale bar 20 μm. (**E**) Loss of *Olfml3* markedly attenuated microglial migration toward FBS relative to isogenic control cells. Comparisons based on students *t*-test; *** $p < 0.001$. (**F**) The rhCCL2 elicited equivalent chemotaxis between isogenic control and *Olfml3*$^{-/-}$ N9 cells. Comparisons based on students *t*-test; $p = 0.54$. (**G**) Chemotaxis toward ATP (50 μM, 100 μM) increased relative to vehicle in isogenic control, but not *Olfml3*$^{-/-}$, microglia. Comparisons based on one-way ANOVA with Tukey's Multiple Comparison Test; *** $p < 0.001$. Bars represent group mean with standard error of the mean (SEM); data represent one of three independent experiments.

2.4. Microglial Olfml3 Is Necessary for the Pro-Tumorigenic GAM Phenotype

Once recruited to the glioma microenvironment, GAMs provide a major source of cytokines to support glioma growth [29]. In the absence of *Olfml3*, microglial secretion of key cytokines promoting microglial invasion and GBM growth were reduced. Secreted levels of colony stimulating factor-1 (CSF-1), a cytokine critical for microglial recruitment and glioma growth [30], were reduced in *Olfml3*$^{-/-}$ microglia media compared to isogenic control media following exposure to vehicle (5.5 ± 0.3 vs. 33.8 ± 7, $p < 0.01$) and TGFβ treatment (β1: 12.6 ± 10 vs. 2.1 ± 0.3, $p < 0.05$; β2: 17.0 ± 4 vs. 3.2 ± 0.4, $p < 0.01$; β3: 11.4 ± 1 vs. 2.1 ± 0.4, $p = 0.06$) (Figure 4A). Similar to CSF-1, granulocyte–macrophage colony stimulating factor (GM-CSF) is a key molecule promoting microglial proliferation [31] and glioma progression [32]. Under the influence of β1, loss of *Olfml3* attenuated secretion of GM-CSF relative to β1-stimulated isogenic control cells (2.2 ± 0.3 vs. 11.9 ± 3; $p < 0.05$) (Figure 4B).

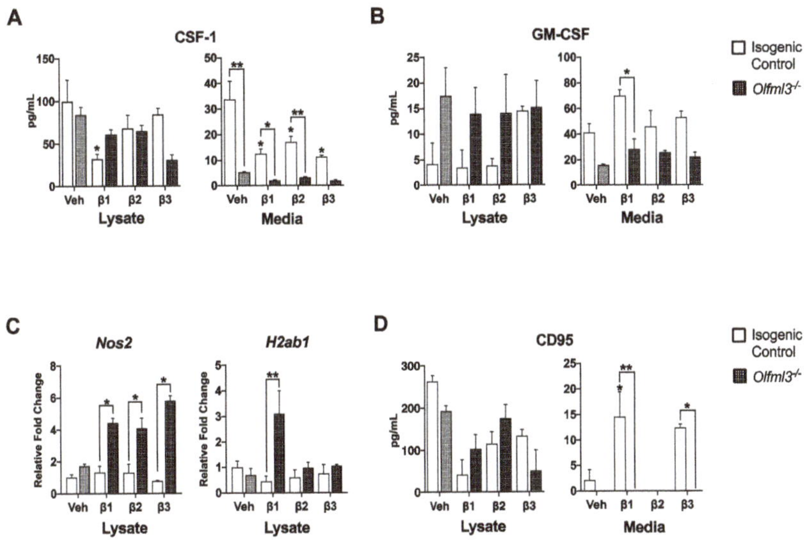

Figure 4. *Olfml3* deletion attenuated TGFβ-induced microglial immunosuppression. (**A**) Secretion of colony stimulating factor-1 (CSF-1) was attenuated in *Olfml3*$^{-/-}$ microglia relative to isogenic control microglia in all conditions; * $p < 0.05$, ** $p < 0.01$. (**B**) While exposure to β1 increased secretion of granulocyte–macrophage colony stimulating factor (GM-CSF) in isogenic control microglia, secretion was reduced in *Olfml3*$^{-/-}$ microglia; * $p < 0.05$. (**C**) The mRNA levels of the pro-inflammatory genes *Nos2* and *H2ab1* were increased in *Olfml3* microglia relative to isogenic control cells following treatment with TGFβ; * $p < 0.05$, ** $p < 0.01$. (**D**) Secretion of CD95, a potent inducer of cytotoxic T cell apoptosis, increased in isogenic control microglia following exposure to β1 and β3, but was undetectable in the media of *Olfml3*$^{-/-}$ microglia across all conditions; * $p < 0.05$, ** $p < 0.01$. Comparisons based on one-way ANOVA with Tukey's Multiple Comparison Test. Bars represent group mean with standard error of the mean (SEM); data represent one of three independent experiments.

Coinciding with production of tumor supportive cytokines, GAM's anti-tumor immunity is suppressed in GBM. Remarkably, *Nos2* mRNA, which encodes inducible nitric oxide synthase to generate cytotoxic nitric oxide [33], was increased in *Olfml3*$^{-/-}$ microglia relative to isogenic control microglia following exposure to β1 (4-fold; $p < 0.05$), β2 (4-fold; $p < 0.05$), and β3 (6-fold; $p < 0.01$) (Figure 4C). Moreover, *H2-Ab1* mRNA, encoding major histocompatibility class II [34], increased 3-fold in *Olfml3*$^{-/-}$ microglia relative to isogenic control microglia following exposure to β2 ($p < 0.05$) (Figure 4C). Microglial secretion of CD95, a Fas ligand, has been implicated in immune evasion and induction of cytotoxic T cell apoptosis [35]. While CD95 was increased in the media of isogenic control cells following exposure to β1 (14.5 ± 5 vs. 2.1 ± 2; $p < 0.05$) and β3 (12.4 ± 1; $p < 0.05$), it remained undetectable in *Olfml3*$^{-/-}$ microglial media under all conditions (Figure 4D).

2.5. *OLFML3 Promotes Glioma Cell Migration and Invasion*

To determine the effect of OLFML3 on the glioma cell malignancy, we exposed GL261 mouse glioma cells to recombinant human OLFML3 (rhOLFML3) for 48 h. Using transwell assays, we observed a dose-dependent increase in GL261 migration following exposure to rhOLFML3 (Figure 5A,B). Exposure to 1 ng/mL rhOLFML3 increased GL261 migration compared to vehicle-treated cells (134.1 ± 19 vs. 265.1 ± 0.6; $p < 0.05$) (Figure 5B). Migration was further increased following exposure to 10 ng/mL rhOLFML3 (384.3 ± 14; $p < 0.001$) (Figure 5B). Similarly, GL261 invasion was increased following exposure to 1 ng/mL rhOLFML3 compared to vehicle-treated cells (22.6 ± 7 vs. 101.4 ± 6; $p < 0.001$) (Figure 5A,C). However, this effect was lost following exposure to 10 ng/mL rhOLFML3 (40.6 ± 6 vs. 22.6 ± 7; $p = 0.25$) (Figure 5C). Interestingly, rhOLFML3 did not act as a chemoattract for GL261 cells, as neither GL261 migration nor invasion was altered by rhOLFML3-supplemented medium in the bottom chamber of a transwell assay (Supplemental Figure S1). Moreover, GL261 viability was not affected by exposure to rhOLFML3 (100.0 ± 6 vs. 100.0 ± 7 vs. 100.0 ± 3; $p = 0.14$) (Figure 5D).

To determine the contribution of TGFβ-induced, microglia-derived OLFML3 on GL261 migration and invasion, GL261 cells were exposed to conditioned medium (CM) from isogenic control and *Olfml3*$^{-/-}$ microglia following vehicle or β1 pre-treatment (5 ng/mL; 48 h). Migration was similar between GL261 cells exposed to CM from vehicle-treated isogenic control and *Olfml3*$^{-/-}$ microglia (26.4 ± 0.8 vs. 23.6 ± 2; $p = 0.84$) (Figure 5E). As expected, GL261 migration increased following exposure to CM from isogenic control microglia pre-treated with β1 vs. vehicle (42.0 ± 3; $p < 0.0001$) (Figure 5E). However, loss of *Olfml3* abolished this effect, as migration rates were similar between GL261 cells exposed to CM from β1 pre-treated *Olfml3*$^{-/-}$ microglia (31.6 ± 1) and vehicle-treated isogenic control ($p = 0.42$) and *Olfml3*$^{-/-}$ microglia ($p = 0.09$) (Figure 5E). Similarly, GL261 invasion was increased following exposure to CM from isogenic control microglia pre-treated with β1 vs. vehicle (4.9 ± 0.7 vs. 8.4 ± 2; $p < 0.001$) (Figure 5F). Again, this effect was abolished in the absence of microglial *Olfml3*, with similar invasion rates between GL261s treated with CM from β1-treated *Olfml3*$^{-/-}$ microglia and vehicle-treated isogenic control microglia (6.9 ± 0.5 vs. 4.9 ± 0.7; $p = 0.06$) (Figure 5F). Cellular viability was not affected by exposure to microglial CM under any condition (81.3 ± 1 vs. 91.0 ± 5; $p > 0.999$) (Figure 5G).

In addition to the loss of OLFML3 in microglial CM on GL261 malignancy, *Olfml3* deletion significantly reduced microglial secretion of key cytokines that promote GBM invasion. While there were no differences in cell lysate concentrations, secretion of interleukin-6 (IL-6) was markedly attenuated in *Olfml3*$^{-/-}$ microglia compared to isogenic control cells following exposure to vehicle (244.6 ± 49 vs. 721.5 ± 61; $p < 0.001$) and β2 (193.8 ± 7 vs. 458.4 ± 21; $p < 0.05$) (Figure 5H). Moreover, secretion of platelet factor 4 (PF4), a growth factor critical in GBM invasion [36], was reduced following vehicle treatment in *Olfml3*$^{-/-}$ microglia compared to isogenic control cells (426.0 ± 437 vs. 721.2 ± 206; $p < 0.05$) (Figure 5I). Similarly, loss of microglial *Olfml3* abolished the TGFβ-induced increase in *Pdgfa* mRNA, a key negative prognostic indicator in GBM [37] (β1: 2.3 ± 0.6 vs. 1.6 ± 0.6, $p < 0.05$); β2: 2.7 ± 0.5 vs. 1.4 ± 0.1; β3: 3.3 ± 0.6 vs. 1.2 ± 0.2, $p < 0.01$) (Figure 5J).

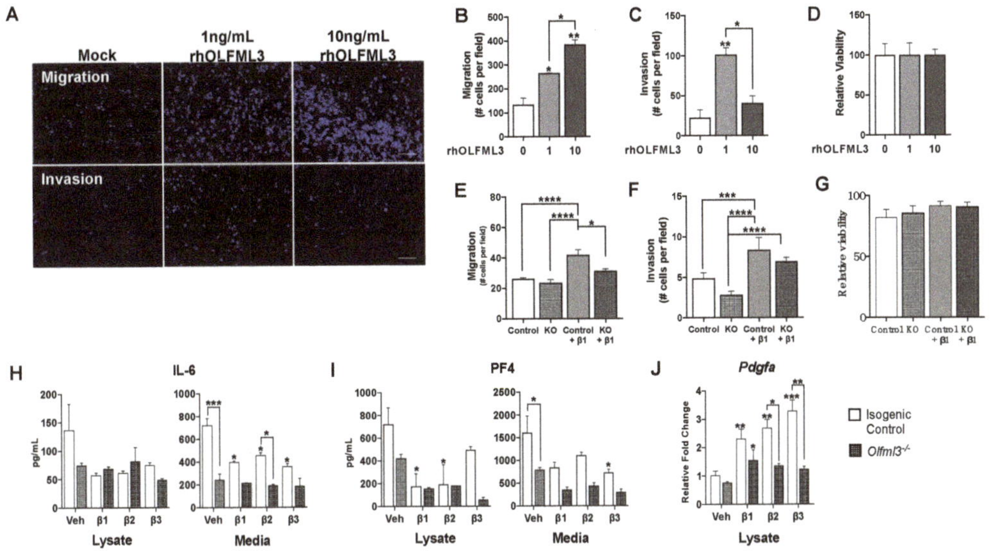

Figure 5. OLFML3 promotes glioma cell migration and invasion. (**A**) Representative images of glioma cell (GL261) migration and invasion following treatment with human recombinant OLFML3 (rhOLFML3; 1 ng/mL, 10 ng/mL); scale bar 20 μm. (**B**) Exposure to rhOLFML3 induced a concentration-dependent increase in GL261 migration relative to vehicle-treated (0) cells (1: 1 ng/mL, 10: 10 ng/mL; 48 h); * $p < 0.05$, ** $p < 0.01$. (**C**) GL261 invasion was increased following exposure to 1 ng/mL, but not 10 ng/mL, rhOLFML3 relative to vehicle-treated cells (48 h); * $p < 0.05$, ** $p < 0.01$. (**D**) Cell viability was not affected by rhOLFML3 (1 ng/mL, 10 ng/mL; 48 h). (**E**) Migration was similar between GL261 cells exposed to conditioned media (CM) from vehicle-treated isogenic control and *Olfml3*$^{-/-}$ microglia (48 h). However, exposure to CM from isogenic control, but not *Olfml3*$^{-/-}$, microglia pre-treated with β1 increased GL261 migration (48 h; 5 ng/mL); * $p < 0.05$, **** $p < 0.0001$. (**F**) GL261 invasion was increased following exposure to CM from isogenic control microglia pre-treated with β1 (48 h; 5 ng/mL) relative to CM from vehicle-treated isogenic control and *Olfml3*$^{-/-}$ microglia (48 h). Exposure to CM from *Olfml3*$^{-/-}$ microglia pre-treated with β1 increased GL261 invasion relative to CM from vehicle-treated *Olfml3*$^{-/-}$ microglia; *** $p < 0.001$, **** $p < 0.0001$. (**G**) GL261 viability was not affected by exposure to microglia CM under any condition ($p = 0.4925$). (**H**) Interleukin-6 (IL-6) was reduced in the media of *Olfml3*$^{-/-}$ microglia relative to isogenic control microglia following exposure to vehicle and β1; * $p < 0.05$, *** $p < 0.001$. (**I**) Secretion of platelet factor 4 (PF4) was attenuated in *Olfml3*$^{-/-}$ microglia following exposure to vehicle and TGFβ isoforms; * $p < 0.05$. (**J**) Loss of microglial *Olfml3* abolished the TGFβ-induced increase in *Pdgfa* mRNA (48 h; 5 ng/mL); * $p < 0.05$, ** $p < 0.01$, *** $p < 0.001$. Comparisons based on one-way ANOVA with Tukey's Multiple Comparison Test. Bars represent group mean with standard error of the mean (SEM); data represent one of three independent experiments.

3. Discussion

In this study, we began to uncover the role of *Olfml3* in microglial function and glioma cell malignancy. Our data showed that microglial *Olfml3* is a direct target gene of all TGFβ isoforms and plays a key role in TGFβ-induced, pro-tumorigenic microglia phenotype determination. Importantly, our data suggest that OLFML3 may directly contribute to glioma cell malignancy through increasing migration and invasion capacity. The myriad pro-tumorigenic effects of microglia-derived *Olfml3* illuminates the potential for therapeutic development targeting the TGFβ-GAM-*Olfml3* signaling axis in GBM.

OLFML3 is a secreted glycoprotein that belongs to the family of the olfactomedin domain-containing proteins [15]. It has been identified as an extracellular matrix protein [20], suggesting that the majority of OLFML3 is secreted. This aligns well with our observation that TGFβ exposure dramatically increases *Olfml3* mRNA but not protein expression in the cell lysate.

While the biological function of olfactomedin domain-containing proteins remains incompletely characterized, growing evidence indicates that they are important for intercellular signaling and protein–protein interaction during development and disease. In particular, olfactomedin 4 (OLFM4), a member of a closely related subfamily of OLFML3, negatively regulates pro-inflammatory responses. OLFM4 knockout mice have enhanced bacterial clearance of *Staphylococcus aureus* and *Escherichia coli* through modulation of neutrophil killing [38], as well as *Helicobacter pylori* through disinhibition of NF-kB [39]. Moreover, *Olfm4* deletion exacerbated inflammation and mucosal damage in a mouse model of colitis [40], further supporting its role in immune restraint. Similarly, our study suggests that *Olfml3* may restrict microglial immune responses, thereby contributing to the markedly immunosuppressed tumor microenvironment of GBM.

Anti-tumor immune responses in GBM are limited through the combination of GAM and T cell dysfunction. Within the glioblastoma microenvironment, GAMs exert immunosuppressive functions through direct cell–cell interactions and release of soluble factors. Importantly, microglia function as antigen-presenting cells in the CNS, requiring up-regulation of MHC II for T cell activation [41]. However, this activity is suppressed in GBM [34]. In fact, MHC I and MHC II molecules were absent in 50% of GBM samples [42], with specific suppression of GAM MHC II occurring through TGFβ signaling. In line with these findings, we demonstrated that *Olfml3* deletion abolished β1-mediated transcriptional suppression of MHC II, which may improve microglial antigen presentation function. Additionally, loss of *Olfml3* may mitigate T cell turnover. In the glioma microenvironment, GAM perpetuate $CD4^+/CD8^+$ T cell apoptosis through secretion of CD95 [35], the ligand for the T cell death receptor Fas, and IL-6, a potent inducer of Fas [43]. Strikingly, *Olfml3* deletion abolished microglial secretion of CD95. While exposure to TGFβ increased secretion in isogenic control cells, CD95 was undetectable in the media of *Olfml3*$^{-/-}$ in all conditions. Moreover, loss of *Olfml3* attenuated secretion of IL-6. These findings, coupled with the dependency of microglial *Olfml3* expression upon TGFβ1-SMAD2-mediated de novo protein synthesis [26], suggest that *Olfml3* functions as a gatekeeper for TGFβ-induced effects on microglia-mediated immunity.

Importantly, targeting the immunomodulatory effects of *Olfml3* may enhance efficacy of currently available immunotherapies. Expression of the immune checkpoint molecule programmed cell death ligand-1 (PD-L1) is inversely correlated with overall patient survival in GBM [44]. While there are many ongoing Phase I and II clinical trials targeting PD-1/PD-L1, preliminary results in patients with recurrent GBM demonstrate unpredictable efficacy, with meager to no survival benefit compared to standard therapies [45–47]. As IL-6 is necessary and sufficient for PD-L1 induction [48], we speculate that therapeutic targeting of *Olfml3* may enhance current immunotherapeutic approaches for GBM patients. In support of this hypothesis, recent work has demonstrated that anti-OLFML3 therapy in conjunction with anti-PD1 immunotherapy increased overall survival in a mouse model of colorectal cancer [21]. Thus, inhibition of microglial *Olfml3*, in tandem with immune checkpoint blockade, may yield improved patient survival in GBM.

Treatment resistance is also governed by the diffuse infiltrative capacity of glioblastoma. Our results support the hypothesis that microglia-derived OLFML3 acts as a paracrine factor facilitating glioma cell invasion. Glioma cell migration and invasion were only affected following 48 h exposure to rhOLFML3, suggesting that OLFML3 may regulate key signaling pathways in glioma cells. This is consistent with general properties of the olfactomedin protein family, which are known to interact with multiple protein binding partners and regulate several cell signaling pathways [16]. This effect is in contrast to recent work that demonstrated that glioma-derived OLFML3 is a GAM chemoattractant [23]. Thus, OLFML3 may have cell type-specific functions within the glioma microenvironment that collectively support tumor growth. Moreover, OLFML3 expression is likely regulated by multiple molecules. The circadian regulator CLOCK and its partner BMAL1 have been identified to promote transcriptional upregulation of OFLML3 in GBM cells [23]. Remarkably, TGFβ signaling is necessary for normal circadian clock function [49]. In fact,

TGFβ induces expression of the core clock gene *Per1* [50]. The interaction between *CLOCK, BMAL1,* and molecules of the canonical TGFβ signaling pathway in GBM is unknown. However, it is interesting to consider the interconnectedness of these systems and their possible synergistic promotion of OLFML3 expression in microglia and glioma cells alike.

Herein, our data demonstrated that microglia-derived *Olfml3* may contribute to glioma cell malignancy through intrinsic and extrinsic mechanisms. Silencing of *Olfml3* attenuated the pro-tumorigenic microglial secretome, as well as mitigating glioma cell malignancy in vitro. Together, these results provide a rationale for further exploration of anti-OLFML3 therapeutic strategies in GBM.

4. Materials and Methods

4.1. Cell Culture and Reagents

The N9 microglial cell line [24] was generously donated from Jyoti Watters at The University of Wisconsin School of Veterinary Medicine. N9 cells were submitted to ATCC for authentication and confirmed to be of murine origin. N9 cells were maintained in DMEM (Gibco^{TM}, ThermoFisher Scientific, Waltham, MA, USA) supplemented with 10% fetal bovine serum (FBS) (Gibco^{TM}, ThermoFisher Scientific, Waltham, MA, USA) and 1% Penicillin/Streptomycin (Gibco^{TM}, ThermoFisher Scientific, Waltham, MA, USA). The GL261 mouse glioma cell line was obtained from the Developmental Therapeutics Program Repository at the National Cancer Institute. GL261 cells were maintained in RPMI 1640 (Gibco^{TM}, ThermoFisher Scientific, Waltham, MA, USA) supplemented with 10% fetal bovine serum (FBS) (Gibco^{TM}, ThermoFisher Scientific, Waltham, MA, USA). All cells were confirmed to be Mycoplasma-free and maintained at 37 °C in a humidified incubator with 5% CO_2. All cells were used below passage 15 and within 1 month after thawing.

4.2. CRISPR/Cas9-Mediated Olfml3 Knockout

Generation of the *Olfml3*-knockout (*Olfml3*$^{-/-}$) microglial cell line was achieved using the CRISPR-Cas9 gene editing system. All reagents were purchased from Integrated DNA Technologies (IDT; Coralville, IA, USA) and used according to the manufacturer's recommendations. Briefly, a guide-RNA (Table 1), targeted to exon 1 of *Olfml3* and the tracrRNA-ATTO-550, was duplexed and mixed with recombinant Cas9 enzyme (IDT, Coralville IA, USA) to form the ribonucleoprotein (RNP) complex. The RNP complex was transfected into cells using Lipofectamine CRISPRMAX (ThermoFisher Scientific, Waltham, MA, USA) transfection reagent. Then, 24 h following transfection, cells were subjected to fluorescence-activated cell sorting and individual ATTO-550-positive cells were sorted into a single well of a 96-well plate. Each single cell created a clonal population, whereby Sanger sequencing confirmed *Olfml3* editing within the defined region of exon 1. Western blot analysis confirmed successful *Olfml3* knockout. An isogenic control line was generated using the same parameters described above without the addition of the gRNA for *Olfml3*.

4.3. Human Recombinant OLFML3 Protein Generation

The protein sequence for OFLML3, consisting of 406 amino acids, is 94.3% identical between human and mouse as determined by a protein BLAST through the National Center for Biotechnology Information. The OLFML3 sequence was cloned into pTXB1 Vector (NEB, N6707S) using *Olfml3* cDNA (Addgene, Wattertown, MA, USA) as template with the following primers: forward, 5′-GGTGGTCATATGGGGCCCAGCACCCCT-3′, and reverse, 5′-GGTGGTTGCTCTTCCGCAAACCTCCTCCTCTTTCTTCCTCAT-3′. The pTXB1-OLFML3 vector was electroporated into ClearColi® BL21 (DE3) Electrocompetent cells (Lucigen, Middleton, WI, USA). These cells have a genetically modified Lipopolysaccharide (LPS) that does not trigger endotoxic response in subsequent assays. Briefly, pTXB1-OLFML3-expressing ClearColi cells were induced (500 μM Isopropyl β-d-1-thiogalactopyranoside (IPTG)) at OD600 = 0.72 and incubated at 16 °C for 18 h. Cells were pelleted, lysed, and incubated with Chitin resin. After washing the beads, rhOLFML3 was cleaved using 50 mM Dithiothreitol (DTT) at 4 °C for 72 h. The rhOLFML3 was eluted and concentrated using Pierce™ Protein Concentrator PES column, 10,000 Da molecular weight cutoff (ThermoFisher Scientific, Waltham, MA, USA). The rhOLFML3 protein was subjected to Fast Protein Liquid Chromatography (FPLC) using HIPREP 16/60 SEPHACRYL S-200 column to remove residual DTT before BCA quantification and subsequent use in all experiments.

4.4. Generation of Anti-OLFML3 Antibody

We generated an anti-OLFML3 polyclonal antibody using the commercially available service from Cocalico Biologicals (Reamstown, PA, USA). Briefly, recombinant OLFML3 protein was generated as described above, purified, and electrophoresed on a 12% SDS-PAGE gel. The OLFML3 band was excised and sent to Cocalico for inoculation of rabbit host. Serum antibody titer was tested until endogenous OLFML3 was detectable using wild-type N9 microglia and rhOLFML3 as a positive control. Final exsanguination was carried out and antibody was purified from the final serum volume.

4.5. Quantitative Real-Time PCR

Cells were grown to 80% confluency and treated with human recombinant TGFβ isoforms (5 ng/mL; β1: 100–21, PeproTech, Cranbury, NJ, USA; β2: PHG9114, Life Technologies, ThermoFisher Scientific, Waltham, MA, USA; β3: SRP3171, Sigma-Aldrich, St. Louis, MO, USA) or vehicle (PBS; ThermoFisher Scientific, Waltham, MA, USA) once every 24 h for a total of two treatments (48-h total incubation). We evaluated *Olfml3* mRNA expression in murine microglia cells following 24-, 48-, and 72-h exposure to TGFβ. We observed the greatest increase in *Olfml3* mRNA at 48 h; thus, all subsequent experiments were performed at this timepoint. Cells were pelleted and RNA was isolated with the Direct-zol MiniPrep kit (Zymo Research, Irvine, CA, USA) according to manufacturer's specifications. Using one microgram-purified DNase-treated RNA, cDNA was reverse transcribed using the High-Capacity cDNA Reverse Transcription Kit (Thermo Fisher—Applied Biosystems, ThermoFisher Scientific, Waltham, MA, USA). Primer sets were designed using NCBI primer design (https://www.ncbi.nlm.nih.gov/tools/primer-blast/index.cgi) (accessed on 4 August 2019) and purchased through IDT (Table 1). Primer validation was performed using a 4× cDNA serial dilution series from isogenic control microglia as template. The efficiency and fit of the generated curves were evaluated; primer sets that did not produce efficiency of at least 0.9 and R^2 value of 0.95 from the cDNA dilution series were rejected. Only experimental quantification cycle (Cq) values that fell within the boundaries of the validated curves were used for analysis.

The qPCR reactions consisted of primer pairs at a final concentration of 200 nM, 50 ng cDNA template, and 2× SSoAdvanced Universal SYBR Green Superix (Bio-Rad, Hercules, CA, USA) per manufacturer's protocol on a CFXConnect (Bio-Rad, Hercules, CA, USA) machine as previously described [51]. All reactions were run as 20-μL triplicates, and the average Cq was used as the data point for a given sample. The mRNA expression values were quantified by the $2^{-\Delta\Delta Ct}$ method, whereby ΔCT = 18S Ct−gene of interest Ct.

4.6. Immunofluorescence and Confocal Microscopy

Isogenic control and *Olfml3*$^{-/-}$ microglia were cultured on sterile glass coverslips treated with fibronectin. Cells were fixed using 4% (*w/v*) paraformaldehyde (Millipore Sigma, Burlington, MA, USA), washed three times for 5 min at RT, and permeabilized with 0.1% Triton X-100-Tris-buffered saline (TBST) for 15 min at RT and blocked for 2 hours at RT with normal goat serum (5% *w/v*) and bovine serum albumin (1% *w/v*) in TBST. Cells were incubated in primary antibody solution (mouse monoclonal anti-TMEM119 (BioLegend, San Diego, CA, USA #853302; 1:1000) in fresh blocking buffer) overnight at 4 °C. Cells were washed three times for 5 min at RT and incubated in secondary antibody solution for 1 hour at RT (IgG (heavy and light) anti-mouse Alexa Fluor 555 (Molecular Probes, Invitrogen, Carlsbad, CA, USA; 1:1000) in fresh blocking buffer). Cells were washed three times for 5 min at RT and mounted with Vectashield with 4′5-diamidino-2phenylindole

(DAPI) (Vector Labs, Burlingame, CA, USA). Images were captured via Leica TCS Sp8 STED 3× confocal microscope.

4.7. Western Blot Analysis

Whole cell protein samples were lysed using RIPA buffer (50 mM Tris–HCl (pH 8.0), 150 mM NaCl, 1% NP-40, 0.5% sodium deoxycholate, and 1% SDS, 1× protease/phosphatase inhibitor cocktail (Thermo Fisher Scientific, Waltham, MA, USA)). The cellular homogenate was rotated for 30 min at 4 °C and centrifuged at 12,000× *g* for 10 min. Protein concentrations of the resultant supernatants were determined using the BCA assay (Pierce Biotechnology, Rockford, IL, USA). Forty micrograms of protein were loaded, electrophoresed on 15% SDS-PAGE gels, and transferred to nitrocellulose membranes overnight. All blots were incubated with Ponceau S (Sigma, St. Louis, MO, USA) to validate equal loading and transfer across all lanes. Membranes were blocked overnight at 4 °C in 5% fat-free milk. Anti-OLFML3 primary antibody was diluted (1:1000) in Tris-buffered saline + Tween-20 (TBST) with 1% fat-free milk and applied to the membrane overnight at 4 °C with gentle rocking. The membranes were washed three times in TBST and incubated in horseradish peroxidase (HRP)-conjugated secondary antibody (1:20,000; Cell Signaling Technology, Danvers, MA, USA) for 1 h at RT with gentle rocking. The HRP substrate for enhanced chemiluminescence (ThermoFisher Scientific, Waltham, MA, USA) was applied immediately prior to exposure. Band densitometry was performed using Image Lab (Bio-Rad, Hercules, CA, USA) and normalized to the Ponceau as a protein loading and transfer control. Optical densities were normalized to vehicle-treated conditions and expressed as relative optical densities (ROD). All experiments were independently repeated in triplicate.

4.8. Murine Protein Arrays

Isogenic control and *Olfml3*$^{-/-}$ microglia were grown to 80% confluency and treated with human recombinant TGFβ isoforms as described above (5 ng/mL; 48 h), followed by serum starvation for 12 h (0.1% FBS). The cell media were aspirated, centrifuged at 350 g for 5 min, and concentrated using Pierce PES protein concentrator columns (ThermoFisher Scientific, Waltham, MA, USA). Whole cell protein samples were treated as described for Western blot. Cell lysate and media samples were sent for analysis by RayBiotech Life (Peachtree Corners, GA, USA) with standard quality control. In brief, Quantibody®Mouse Full Testing Service (QAH-INF-1) utilized two non-overlapping arrays of antibody pairs to quantify selected molecules. RayBiotech confirmed no cross reactivity between antibody pairs and standard controls.

4.9. Cell Viability

Cell viability was performed using the Cell Titer Glo®2.0 Assay (Promega, Madison, WI, USA) according to the manufacturer's protocol. Cells (2.5×10^4) were seeded in 96-well, black-sided plates. Titer Glo®reagent was added to each well and the plate was incubated for 10 min at RT on a plate shaker, followed by luminescence recording via plate reader (BioTek800TS). Optical densities were recorded for six replicates per condition and the average optical density of media alone (blank) was subtracted from all experimental conditions. Three independent experiments were performed.

Similar to Cell Titer Glo®, cells (2.5×10^4) were seeded in 96-well, black-sided plates and cultured for 48 h. MTS reagent was added to each well and the plate was incubated for 10 min at RT on a plate shaker, followed by absorbance reading via plate reader (BioTek800TS) at 590 mm. Optical densities were recorded for six replicates per condition. Three independent experiments were performed.

4.10. Transwell Migration and Invasion Assays

The modified Boyden chamber assay was used for analysis of cell migration and invasion. Migration assays were performed using cells (microglia: 2×10^5; GL261: 5×10^4) suspended in serum-free culture medium and seeded into 24-well Transwell inserts with an 8-μm pore polycarbonate filter insert. Invasion assays were conducted similarly, with the addition of 50-μL Matrigel coating onto the 8-μm pore polycarbonate filter insert. FBS, serum-free medium with indicated factors (rhCCL2: 479-JE-010, R&D Systems; ATP: A6419-1G, Sigma, St. Louis, MO, USA) or CM, was added to the receiver wells. After 90 min (ATP) or 24 h (rhCCL2, FBS), inserts were removed and the top of each insert was swabbed to remove non-migrated cells. Remaining cells attached to the bottom of the insert were fixed using 4% paraformaldehyde. Membranes were excised from the inserts and mounted onto microscope slides using mounting medium containing 4′,6- diamidino-2-phenylindole (DAPI). Nine photographs were taken per membrane, with three technical replicates per experiment, using a brightfield microscope (Leica, DM5000 B). Cells were identified by positive DAPI immunoreactivity and quantified via an ImageJ custom macro. Three independent experiments were performed.

4.11. Phagocytosis Assay

Isogenic control and *Olfml3*$^{-/-}$ microglia were seeded in 96-well plates (2×10^5) and incubated overnight. The following day, media were removed and cells were incubated with pHrodo™Green E.Coli BioParticles (ThermoFisher Scientific, Waltham, MA, USA) for 1 h at 37 °C following the manufacturer's protocol. Luminescence was determined via microplate reader (Molecular Devices SpectraMax Gemini EM Microplate Reader 19745) at 509/533 nm. Percent phagocytosis was calculated as follows:

$$\% \text{ phagocytosis } = \frac{\text{net experimental phagocytosis } \times 100\%}{\text{net positive control phagocytosis}} \tag{1}$$

4.12. Statistics

Statistical analysis was performed with Prism GraphPad V9.0.2 software. Data are presented as the mean ± SEM. Cell culture experiments were performed in technical replicates, with three biological replicates. Data were tested for normality via Shapiro–Wilks test. Statistical significance was assessed via unpaired two-tailed Student's *t*-test or ANOVA with Tukey's multiple comparisons test. Results were regarded as statistically significant for $p < 0.05$.

Supplementary Materials: The following are available online at https://www.mdpi.com/article/10.3390/ijms222313052/s1.

Author Contributions: Conceptualization, R.G.T., C.A.L., L.A.W., X.C. and C.M.T.; methodology, R.G.T., C.A.L. and C.M.T.; investigation, R.G.T., E.T.D. and C.M.T.; writing—original draft preparation, R.G.T. and C.M.T.; writing—review and editing, R.G.T., C.A.L., E.T.D., L.A.W., X.C. and C.M.T.; funding acquisition, L.A.W. and C.M.T. All authors have read and agreed to the published version of the manuscript. This work has been partially presented in abstract form at the 2020 Experimental Biology Virtual Conference.

Funding: This research was supported in part by the UC Davis Paul Calabresi Career Development Award for Clinical Oncology as funded by the National Cancer Institute/National Institutes of Health through grant #2K12CA138464-11 (C..MT), the UC Davis Comprehensive Cancer Center through grant #CANCERCOP2-WT (L.A.W, C.M.T), and the Paul C. and Borghild T. Petersen Brain Tumor Foundation (R.G.T, E.T.D, C.M.T).

Institutional Review Board Statement: Not applicable.

Informed Consent Statement: Not applicable.

Data Availability Statement: Not applicable.

Acknowledgments: The authors thank Jyoti Watters from the University of Wisconsin for generously supplying N9 microglial cells with permission from Paola Ricciardi-Castagnoli. We also thank I. Brust-Mascher and the health Sciences District Advanced Imaging Facility for confocal microscopy assistance.

Conflicts of Interest: The authors declare no conflict of interest.

References

1. Stupp, R.; Taillibert, S.; Kanner, A.A.; Read, W.; Steinberg, D.M.; Lhermitte, B.; Toms, S.; Idbaih, A.; Ahluwalia, M.S.; Fink, K.; et al. Effect of Tumor-Treating Fields Plus Maintenance Temozolomide vs Maintenance Temozolomide Alone on Survival in Patients With Glioblastoma. *JAMA* **2017**, *318*, 2306–2316. [CrossRef]
2. Gutmann, D.H.; McLellan, M.D.; Hussain, I.; Wallis, J.W.; Fulton, L.L.; Fulton, R.S.; Magrini, V.; Demeter, R.; Wylie, T.; Kandoth, C.; et al. Somatic neurofibromatosis type 1 (NF1) inactivation characterizes NF1-associated pilocytic astrocytoma. *Genome Res.* **2013**, *23*, 431–439. [CrossRef] [PubMed]
3. Sørensen, M.D.; Dahlrot, R.H.; Boldt, H.B.; Hansen, S.; Kristensen, B.W. Tumour-associated microglia/macrophages predict poor prognosis in high-grade gliomas and correlate with an aggressive tumour subtype. *Neuropathol. Appl. Neurobiol.* **2018**, *44*, 185–206. [CrossRef]
4. Markovic, D.S.; Glass, R.; Synowitz, M.; van Rooijen, N.; Kettenmann, H. Microglia Stimulate the Invasiveness of Glioma Cells by Increasing the Activity of Metalloprotease-2. *J. Neuropathol. Exp. Neurol.* **2005**, *64*, 754–762. [CrossRef]
5. Yan, D.; Kowal, J.; Akkari, L.; Schuhmacher, A.J.; Huse, J.T.; West, B.; A J, S. Inhibition of colony stimulating factor-1 receptor abrogates microenvironment-mediated therapeutic resistance in gliomas. *Oncogene* **2017**, *36*, 6049–6058. [CrossRef]
6. Wang, W.; Cho, H.-Y.; Rosenstein-Sisson, R.; Ramos, N.I.M.; Price, R.; Hurth, K.; Schönthal, A.H.; Hofman, F.M.; Chen, T.C. Intratumoral delivery of bortezomib: Impact on survival in an intracranial glioma tumor model. *J. Neurosurg.* **2018**, *128*, 695–700. [CrossRef]
7. Brandenburg, S.; Müller, A.; Turkowski, K.; Radev, Y.T.; Rot, S.; Schmidt, C.; Bungert, A.D.; Acker, G.; Schorr, A.; Hippe, A.; et al. Resident microglia rather than peripheral macrophages promote vascularization in brain tumors and are source of alternative pro-angiogenic factors. *Acta Neuropathol.* **2015**, *131*, 365–378. [CrossRef] [PubMed]
8. Wesolowska, A.; Kwiatkowska, A.J.; Slomnicki, L.; Dembinski, M.; Master, A.M.; Sliwa, M.; Franciszkiewicz, K.; Chouaib, S.; Kaminska, B. Microglia-derived TGF-β as an important regulator of glioblastoma invasion—An inhibition of TGF-β-dependent effects by shRNA against human TGF-β type II receptor. *Oncogene* **2007**, *27*, 918–930. [CrossRef]
9. Coniglio, S.J.; Eugenin, E.; Dobrenis, K.; Stanley, E.R.; West, B.L.; Symons, M.H.; Segall, J.E. Microglial Stimulation of Glioblastoma Invasion Involves Epidermal Growth Factor Receptor (EGFR) and Colony Stimulating Factor 1 Receptor (CSF-1R) Signaling. *Mol. Med.* **2012**, *18*, 519–527. [CrossRef] [PubMed]
10. Li, W.; Graeber, M.B. The molecular profile of microglia under the influence of glioma. *Neuro-Oncology* **2012**, *14*, 958–978. [CrossRef]
11. Seystahl, K.; Papachristodoulou, A.; Burghardt, I.; Schneider, H.; Hasenbach, K.; Janicot, M.; Roth, P.; Weller, M. Biological Role and Therapeutic Targeting of TGF-β3 in Glioblastoma. *Mol. Cancer Ther.* **2017**, *16*, 1177–1186. [CrossRef] [PubMed]
12. Han, J.; Alvarez-Breckenridge, C.A.; Wang, Q.E.; Yu, J. TGF-beta signaling and its targeting for glioma treatment. *Am J Cancer Res.* **2015**, *5*, 945–955.
13. Butowski, N.; Colman, H.; De Groot, J.F.; Omuro, A.M.; Nayak, L.; Wen, P.Y.; Cloughesy, T.F.; Marimuthu, A.; Haidar, S.; Perry, A.; et al. Orally administered colony stimulating factor 1 receptor inhibitor PLX3397 in recurrent glioblastoma: An Ivy Foundation Early Phase Clinical Trials Consortium phase II study. *Neuro-Oncology* **2015**, *18*, 557–564. [CrossRef]
14. Haage, V.; Semtner, M.; Vidal, R.O.; Hernandez, D.P.; Pong, W.W.; Chen, Z.; Hambardzumyan, D.; Magrini, V.; Ly, A.; Walker, J.; et al. Comprehensive gene expression meta-analysis identifies signature genes that distinguish microglia from peripheral monocytes/macrophages in health and glioma. *Acta Neuropathol. Commun.* **2019**, *7*, 1–18. [CrossRef] [PubMed]
15. Tomarev, S.I.; Nakaya, N. Olfactomedin Domain-Containing Proteins: Possible Mechanisms of Action and Functions in Normal Development and Pathology. *Mol. Neurobiol.* **2009**, *40*, 122–138. [CrossRef]
16. Liu, W.; Rodgers, G.P. Olfactomedin 4 expression and functions in innate immunity, inflammation, and cancer. *Cancer Metastasis Rev.* **2016**, *35*, 201–212. [CrossRef]
17. Halleskog, C.; Dijksterhuis, J.P.; Kilander, M.B.C.; Becerril-Ortega, J.; Villaescusa, J.C.; Lindgren, E.; Arenas, E.; Schulte, G. Heterotrimeric G protein-dependent WNT-5A signaling to ERK1/2 mediates distinct aspects of microglia proinflammatory transformation. *J. Neuroinflammation* **2012**, *9*, 111. [CrossRef]
18. Zuccarini, M.; Giuliani, P.; Ziberi, S.; Carluccio, M.; Di Iorio, P.; Caciagli, F.; Ciccarelli, R. The Role of Wnt Signal in Glioblastoma Development and Progression: A Possible New Pharmacological Target for the Therapy of This Tumor. *Genes* **2018**, *9*, 105. [CrossRef]
19. Torres, S.; Bartolome, R.A.; Mendes, M.; Barderas, R.; Fernández-Aceñerp, M.J.; Peláez-García, A.; Peña, C.; Lopez-Lucendo, M.; Villar-Vázquez, R.; De Herreros, A.G.; et al. Proteome Profiling of Cancer-Associated Fibroblasts Identifies Novel Proinflammatory Signatures and Prognostic Markers for Colorectal Cancer. *Clin. Cancer Res.* **2013**, *19*, 6006–6019. [CrossRef]
20. Miljkovic-Licina, M.; Hammel, P.; Garrido-Urbani, S.; Lee, B.P.-L.; Meguenani, M.; Chaabane, C.; Bochaton-Piallat, M.-L.; Imhof, B.A. Targeting Olfactomedin-like 3 Inhibits Tumor Growth by Impairing Angiogenesis and Pericyte Coverage. *Mol. Cancer Ther.* **2012**, *11*, 2588–2599. [CrossRef]
21. Stalin, J.; Imhof, B.A.; Coquoz, O.; Jeitziner, R.; Hammel, P.; McKee, T.A.; Jemelin, S.; Poittevin, M.; Pocard, M.; Matthes, T.; et al. Targeting OLFML3 in Colorectal Cancer Suppresses Tumor Growth and Angiogenesis, and Increases the Efficacy of Anti-PD1 Based Immunotherapy. *Cancers* **2021**, *13*, 4625. [CrossRef]
22. Qiu, R.; Shi, H.; Wang, S.; Leng, S.; Liu, R.; Zheng, Y.; Huang, W.; Zeng, Y.; Gao, J.; Zhang, K.; et al. BRMS1 coordinates with LSD1 and suppresses breast cancer cell metastasis. *Am. J. Cancer Res.* **2018**, *8*, 2030–2045.

23. Chen, P.; Hsu, W.-H.; Chang, A.; Tan, Z.; Lan, Z.; Zhou, A.; Spring, D.J.; Lang, F.F.; Wang, Y.A.; Depinho, R.A. Circadian Regulator CLOCK Recruits Immune-Suppressive Microglia into the GBM Tumor Microenvironment. *Cancer Discov.* **2020**, *10*, 371–381. [CrossRef] [PubMed]
24. Righi, M.; Mori, L.; De Libero, G.; Sironi, M.; Biondi, A.; Mantovani, A.; Donini, S.D.; Ricciardi-Castagnoli, P. Monokine production by microglial cell clones. *Eur. J. Immunol.* **1989**, *19*, 1443–1448. [CrossRef]
25. Ausman, J.I.; Shapiro, W.R.; Rall, D.P. Studies on the chemotherapy of experimental brain tumors: Development of an experimental model. *Cancer Res.* **1970**, *30*, 2394–2400. [PubMed]
26. Neidert, N.; Von Ehr, A.; Zöller, T.; Spittau, B. Microglia-Specific Expression of Olfml3 Is Directly Regulated by Transforming Growth Factor β1-Induced Smad2 Signaling. *Front. Immunol.* **2018**, *9*, 1728. [CrossRef]
27. Platten, M.; Kretz, A.; Naumann, U.; Aulwurm, S.; Egashira, K.; Isenmann, S.; Weller, M. Monocyte chemoattractant protein-1 increases microglial infiltration and aggressiveness of gliomas. *Ann. Neurol.* **2003**, *54*, 388–392. [CrossRef] [PubMed]
28. Flores-Toro, J.A.; Luo, D.; Gopinath, A.; Sarkisian, M.R.; Campbell, J.J.; Charo, I.F.; Singh, R.; Schall, T.J.; Datta, M.; Jain, R.K.; et al. CCR2 inhibition reduces tumor myeloid cells and unmasks a checkpoint inhibitor effect to slow progression of resistant murine gliomas. *Proc. Natl. Acad. Sci. USA* **2020**, *117*, 1129–1138. [CrossRef]
29. Roesch, S.; Rapp, C.; Dettling, S.; Herold-Mende, C. When Immune Cells Turn Bad—Tumor-Associated Microglia/Macrophages in Glioma. *Int. J. Mol. Sci.* **2018**, *19*, 436. [CrossRef]
30. De, I.; Nikodemova, M.; Steffen, M.D.; Sokn, E.; Maklakova, V.I.; Watters, J.J.; Collier, L.S. CSF1 overexpression has pleiotropic effects on microgliain vivo. *Glia* **2014**, *62*, 1955–1967. [CrossRef]
31. Dikmen, H.O.; Hemmerich, M.; Lewen, A.; Hollnagel, J.-O.; Chausse, B.; Kann, O. GM-CSF induces noninflammatory proliferation of microglia and disturbs electrical neuronal network rhythms in situ. *J. Neuroinflammation* **2020**, *17*, 1–13. [CrossRef]
32. Revoltella, R.P.; Menicagli, M.; Campani, D. Granulocyte–macrophage colony-stimulating factor as an autocrine survival-growth factor in human gliomas. *Cytokine* **2012**, *57*, 347–359. [CrossRef] [PubMed]
33. Knowles, R.G.; Moncada, S. Nitric oxide synthases in mammals. *Biochem. J.* **1994**, *298*, 249–258. [CrossRef]
34. Schartner, J.M.; Hagar, A.R.; Van Handel, M.; Zhang, L.; Nadkarni, N.; Badie, B. Impaired capacity for upregulation of MHC class II in tumor-associated microglia. *Glia* **2005**, *51*, 279–285. [CrossRef]
35. Badie, B.; Schartner, J.; Prabakaran, S.; Paul, J.; Vorpahl, J. Expression of Fas ligand by microglia: Possible role in glioma immune evasion. *J. Neuroimmunol.* **2001**, *120*, 19–24. [CrossRef]
36. Liu, C.; Luo, D.; Reynolds, B.A.; Meher, G.; Katritzky, A.R.; Lu, B.; Gerard, C.J.; Bhadha, C.P.; Harrison, J.K. Chemokine receptor CXCR3 promotes growth of glioma. *Carcinog.* **2010**, *32*, 129–137. [CrossRef] [PubMed]
37. Cantanhede, I.G.; De Oliveira, J.R.M. PDGF Family Expression in Glioblastoma Multiforme: Data Compilation from Ivy Glioblastoma Atlas Project Database. *Sci. Rep.* **2017**, *7*, 15271. [CrossRef]
38. Liu, W.; Yan, M.; Liu, Y.; McLeish, K.R.; Coleman, W.G.; Rodgers, G.P. Olfactomedin 4 Inhibits Cathepsin C-Mediated Protease Activities, Thereby Modulating Neutrophil Killing ofStaphylococcus aureusandEscherichia coliin Mice. *J. Immunol.* **2012**, *189*, 2460–2467. [CrossRef]
39. Liu, W.; Yan, M.; Liu, Y.; Wang, R.; Li, C.; Deng, C.; Singh, A.; Coleman, W.G.; Rodgers, G.P. Olfactomedin 4 down-regulates innate immunity against Helicobacter pylori infection. *Proc. Natl. Acad. Sci. USA* **2010**, *107*, 11056–11061. [CrossRef]
40. Amirbeagi, F.; Thulin, P.; Pullerits, R.; Pedersen, B.; Andersson, B.A.; Dahlgren, C.; Welin, A.; Bylund, J. Olfactomedin-4 autoantibodies give unusual c-ANCA staining patterns with reactivity to a subpopulation of neutrophils. *J. Leukoc. Biol.* **2014**, *97*, 181–189. [CrossRef]
41. Gottfried-Blackmore, A.; Kaunzner, U.W.; Idoyaga, J.; Felger, J.C.; McEwen, B.S.; Bulloch, K. Acute in vivo exposure to interferon-enables resident brain dendritic cells to become effective antigen presenting cells. *Proc. Natl. Acad. Sci. USA* **2009**, *106*, 20918–20923. [CrossRef]
42. Facoetti, A.; Nano, R.; Zelini, P.; Morbini, P.; Benericetti, E.; Ceroni, M.; Campoli, M.; Ferrone, S. Human Leukocyte Antigen and Antigen Processing Machinery Component Defects in Astrocytic Tumors. *Clin. Cancer Res.* **2005**, *11*, 8304–8311. [CrossRef]
43. Zhu, J.; Petit, P.-F.; Van den Eynde, B.J. Apoptosis of tumor-infiltrating T lymphocytes: A new immune checkpoint mechanism. *Cancer Immunol. Immunother.* **2019**, *68*, 835–847. [CrossRef]
44. Hao, C.; Chen, G.; Zhao, H.; Li, Y.; Chen, J.; Zhang, H.; Li, S.; Zhao, Y.; Chen, F.; Li, W.; et al. PD-L1 Expression in Glioblastoma, the Clinical and Prognostic Significance: A Systematic Literature Review and Meta-Analysis. *Front. Oncol.* **2020**, *10*, 1015. [CrossRef]
45. Reardon, D.A.; Omuro, A.; Brandes, A.A.; Rieger, J.; Wick, A.; Sepulveda, J.; Phuphanich, S.; De Souza, P.; Ahluwalia, M.S.; Lim, M.; et al. OS10.3 Randomized Phase 3 Study Evaluating the Efficacy and Safety of Nivolumab vs Bevacizumab in Patients With Recurrent Glioblastoma: CheckMate 143. *Neuro-Oncology* **2017**, *19*, iii21. [CrossRef]
46. Zhao, J.; Chen, A.; Gartrell, R.D.; Silverman, A.M.; Aparicio, L.; Chu, T.; Bordbar, D.; Shan, D.; Samanamud, J.; Mahajan, A.; et al. Immune and genomic correlates of response to anti-PD-1 immunotherapy in glioblastoma. *Nat. Med.* **2019**, *25*, 462–469. [CrossRef]
47. Cloughesy, T.F.; Mochizuki, A.Y.; Orpilla, J.R.; Hugo, W.; Lee, A.H.; Davidson, T.B.; Wang, A.C.; Ellingson, B.M.; Rytlewski, J.A.; Sanders, C.M.; et al. Neoadjuvant anti-PD-1 immunotherapy promotes a survival benefit with intratumoral and systemic immune responses in recurrent glioblastoma. *Nat. Med.* **2019**, *25*, 477–486. [CrossRef] [PubMed]

48. Lamano, J.B.; Lamano, J.B.; Li, Y.D.; DiDomenico, J.D.; Choy, W.; Veliceasa, D.; Oyon, D.E.; Fakurnejad, S.; Ampie, L.; Kesavabhotla, K.; et al. Glioblastoma-Derived IL6 Induces Immunosuppressive Peripheral Myeloid Cell PD-L1 and Promotes Tumor Growth. *Clin. Cancer Res.* **2019**, *25*, 3643–3657. [CrossRef]
49. Sloin, H.E.; Ruggiero, G.; Rubinstein, A.; Storz, S.S.; Foulkes, N.S.; Gothilf, Y. Intera ctions between the circadian clock and TGF-β signaling pathway in zebrafish. *PLoS ONE* **2018**, *13*, e0199777. [CrossRef] [PubMed]
50. Kon, N.; Hirota, T.; Kawamoto, T.; Kato, Y.; Tsubota, T.; Fukada, Y. Activation of TGF-β/activin signalling resets the circadian clock through rapid induction of Dec1 transcripts. *Nat. Cell Biol.* **2008**, *10*, 1463–1469. [CrossRef]
51. Toedebusch, R.; Grodzki, A.C.; Dickinson, P.J.; Woolard, K.; Vinson, N.; Sturges, B.; Snyder, J.; Li, C.-F.; Nagasaka, O.; Consales, B.; et al. Glioma-associated microglia/macrophages augment tumorigenicity in canine astrocytoma, a naturally occurring model of human glioma. *Neuro-Oncol. Adv.* **2021**, *3*, vdab062. [CrossRef] [PubMed]

International Journal of *Molecular Sciences*

MDPI

Communication

Decoding the Role of DVL1 in Intracranial Meningioma

Anja Bukovac [1,2,*,†], Katarina Dragičević [3,†], Anja Kafka [1,2], Darko Orešković [4], Sanja Cesarec-Augustinović [5] and Nives Pećina-Šlaus [1,2]

1 Laboratory of Neurooncology, Croatian Institute for Brain Research, School of Medicine, University of Zagreb, 10000 Zagreb, Croatia; anja.kafka@mef.hr (A.K.); nina@mef.hr (N.P.-Š.)
2 Department of Biology, School of Medicine, University of Zagreb, 10000 Zagreb, Croatia
3 Biotech Research & Innovation Centre, University of Copenhagen, DK-2200 Copenhagen, Denmark; katarina.dragicevic@bric.ku.dk
4 Department of Neurosurgery, University Hospital Dubrava, 10000 Zagreb, Croatia; darkoreskov@gmail.com
5 "Ljudevit Jurak" Department of Pathology and Cytology, Clinical Hospital Center "Sestre milosrdnice", 10000 Zagreb, Croatia; sanja.cesarec@kbcsm.hr
* Correspondence: anja.bukovac@mef.hr
† Equally contributed.

Citation: Bukovac, A.; Dragičević, K.; Kafka, A.; Orešković, D.; Cesarec-Augustinović, S.; Pećina-Šlaus, N. Decoding the Role of DVL1 in Intracranial Meningioma. *Int. J. Mol. Sci.* **2021**, *22*, 11996. https://doi.org/10.3390/ijms222111996

Academic Editor: Anuska V. Andjelkovic

Received: 28 September 2021
Accepted: 2 November 2021
Published: 5 November 2021

Publisher's Note: MDPI stays neutral with regard to jurisdictional claims in published maps and institutional affiliations.

Abstract: In the search for molecular candidates for targeted meningioma therapies, increasing attention has been paid to the role of signaling pathways in the development and progression of intracranial meningiomas. Although it is well known that the Wnt signaling pathway is involved in meningioma progression, the role of its central mediator, DVL1, is still unclear. In order to investigate the influence of *DVL1* gene alterations on the progression of human intracranial meningioma, we focused on its central PDZ domain, which is responsible for DVL interaction with the Fzd receptor and the phosphorylation of DVL mediated through the casein kinases CK1 and CK2. A genetic analysis of genomic instability revealed the existence of microsatellite instability in 9.09% and the loss of heterozygosity in 6.06% of the samples. The sequencing of the PDZ gene region showed repetitive deletions of two bases located in intron 7 and exon 8, and a duplication in intron 8 in most samples, with different outcomes on the biological function of the DVL1 protein. Immunohistochemistry revealed that the nuclear expression of DVL1 was significantly correlated with a higher expression of active β-catenin (p = 0.029) and a higher meningioma grade (p = 0.030), which leads to the conclusion that it could be used as biomarker for meningioma progression and the activation of the Wnt signaling pathway.

Keywords: DVL1; PDZ domain; β-catenin; Wnt signaling pathway; intracranial meningioma

1. Introduction

Intracranial meningioma, with their mostly benign nature and slow growth, can still progress and show malignant characteristics that can lead to poor outcome for patients. According to statistics from the latest CBTRUS report (The Central Brain Tumor Registry of the United States) from 2012 to 2016 [1], the percentage of meningioma in the total sample of brain tumors is 37.1%, and among all benign brain tumors it is at 53.1%. The abovementioned numbers suggest that meningiomas are currently the most common primary tumors of the central nervous system [2], making the discovery of mechanisms involved in their development and progression highly valuable.

Molecular mechanisms underlying meningioma progression are usually connected to aberrant signaling pathways such as PI3K-AKT-mTOR, Ras-Raf-MEK, Rac-PAK-JNK, TGFβ-SMAD, RB, p53, Hedgehog, Notch and canonical Wnt [3–9]. This study concentrates on the canonical Wnt pathway, whose regulation of cell growth, development, and survival has great importance in tumorigenesis.

In tumors, Wnt signaling can play different roles; for instance, allowing them to reprogram their metabolism or promote chronic inflammation and oxidative stress, or enabling resistance to immunotherapy [10]. It has been shown that the activation of Wnt signaling can induce the malignant transformation of neural stem cells and thus contribute to the development of primary brain tumors [11–16]. The Wnt signaling pathway in meningiomas is known to be activated and plays a role in progression [17,18]. Activation occurs when Wnt ligands bind to membrane receptors called Frizzled. This binding is followed by the activation of a specific member of the Disheveled (Dvl) protein family, which is then pulled to the membrane where it promotes the phosphorylation and disassembly of the destruction complex, which targets β-catenin. The destruction of the central oncogenic molecule β-catenin is thus impaired, leading to its accumulation in the cytoplasm, followed by the translocation to the nucleus where it binds to the TCF/LEF factors and stimulates the transcription of other targeted oncogenes [19–22]. It has also been shown that DVL has nuclear activity by binding to the TCF factor together with the β-catenin and by activating the transcription of Wnt target genes [23].

The Disheveled family is a highly conserved, multifunctional group of proteins with three human homologues—DVL1, DVL2, and DVL3. The excessive expression of Disheveled proteins has been proven to cause increased activation of Wnt signaling in many different types of tumors [19]. All members of the DVL protein family contain three basic conserved domains—DIX, DEP, and PDZ. On the amino terminal side resides the DIX domain, which represents an important binding site for the AXIN [23]. On the carboxyl terminal side of the DVL protein is the DEP domain, which is key in DVL protein interactions with DAAM1 (Disheveled associated activator of morphogenesis 1) [23]. Between these two domains lies the central PDZ domain (Postsynaptic Density 95, Discs Large, Zonula Occludens-1) composed of 73 amino acids. The PDZ domain plays a role in both the canonical and non-canonical form of Wnt signaling and has the utmost importance in mediating the different interactions that DVL proteins carry out. The PDZ domain is essential for DVL interaction with the carboxyl terminal domain of the Fzd receptor, and also plays a role in DVL phosphorylation mediated through the casein kinases CK1 and CK2 [23].

The aim of this study was to determine genetic alterations in the *DVL1* gene and its central PDZ domain, and their role in the progression of intracranial meningioma. The results of this study could contribute to the discovery of new prognostic biomarkers and targets for personalized therapies.

2. Results

Out of 33 samples of intracranial meningiomas, 22 samples (66.67%) belonged to female and 11 samples (33.33%) to male patients. The age of the patients ranged from 23 to 85 years, with a mean of 61.12 and a median of 67 years.

Moreover, out of 33 samples, 16 were classified as grade I, 12 as grade II, and 5 as grade III. Using the Kruskal–Wallis test, there was no statistically significant difference between the incidence of grade with respect to the age of the patients ($p = 0.259$). Using the Fisher test, a statistically significant difference was found between the grade and sex of the patients ($p = 0.026$), where men showed a greater tendency to develop higher grade meningiomas and women the lower ones.

2.1. Genetic Instability of the DVL1 Gene Recorded with the D1S468 Marker

The *DVL1* microsatellite marker D1S468 proved to be highly informative, with a heterozygosity present in all samples. Out of 33 intracranial meningiomas analyzed, three samples (9.09%) showed microsatellite instability (MSI) and two samples (6.06%) demonstrated the loss of heterozygosity (LOH). The total number of samples in which the *DVL1* gene was altered was 15.15%. Out of the five samples in which one of these changes was found, four belonged to female patients. Based on the results of the Fisher test, no statistically significant difference was found between the sex and the genetic alteration ($\rho = 1.000$, $p = 0.542$), nor between the grade and the genetic alteration ($\rho = 0.053$, $p = 1.000$). On the contrary, Pearson's test showed a statistically significant difference between the incidence of genetic alteration and patients' age (r = −0.485, p = 0.004), indicating that younger patients harbored more alterations. Examples of MSI and LOH changes are shown in Figure 1.

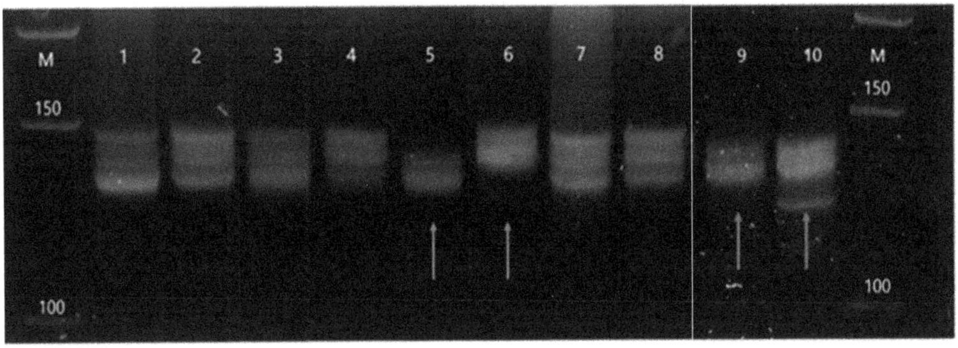

Figure 1. Examples of genetic changes—MSI and LOH, found in intracranial meningioma and detected on Spreadex gels. Legend: M—marker; odd numbers—tumor samples; even numbers—blood samples; green arrows—example of MSI in a tumor sample; red arrows—example of LOH in a tumor sample.

2.2. Mutations in the PDZ Region of the DVL1 Gene

Out of a total of 33 samples, the high-resolution melting method (HRM) revealed observable curve deviations in most samples; out of these, 10 samples (30.30%) showed large deviations of the tumor curve from the control blood curve. A total of 26 samples of tumor DNA were available for Sanger sequencing. A comparison of tumor sequences with the reference sequence revealed the existence of nucleotide changes in all of the sequenced samples, which are systematized in Table 1 and shown in Figure 2.

Table 1. Types of mutations detected by Sanger sequencing.

Location of Mutation	Type of Mutation	Number of Samples with This Type of Mutation
Intron 7	NG_008048.2: g.13921delT	7
Intron 7	NG_008048.2: g.13921T>C	4
Exon 8	NG_008048.2: g.13998delA	1
Exon 8	NG_008048.2: g.14004delA	9
Exon 8	NG_008048.2: g.14004A>T	1
Intron 8	NG_008048.2: g.14228_14267dup	25
Intron 8	NG_008048.2: g.14248G>T	1
Intron 8	NG_008048.2: g.14300G>T	1
Exon 9	NG_008048.2: g.14329G>C	1

Altogether, nine different types of mutations were found (Table 1). Twelve patients harbored one mutation, seven had two mutations, five had three mutations, one patient harbored four mutations, and another one had five different mutations. The mutation with the highest frequency was the g.14228_14267dup found in 25 samples (96.15%), followed by the deletion g.14004delA found in nine samples (34.61%) and g.13921delT found in seven samples (26.92%). The most pronounced mutation was the intron 8 duplication, which is 39 bp long (g.14228_14267dup), as shown in Figure 2a. This duplication was not registered in the reference sequence available at the National Center for Biotechnology Information (NCBI) database, but was found in 25 out of our 26 analyzed samples. The effect of the duplication on the biological function of the DVL1 protein could not be determined by the PROVEAN program.

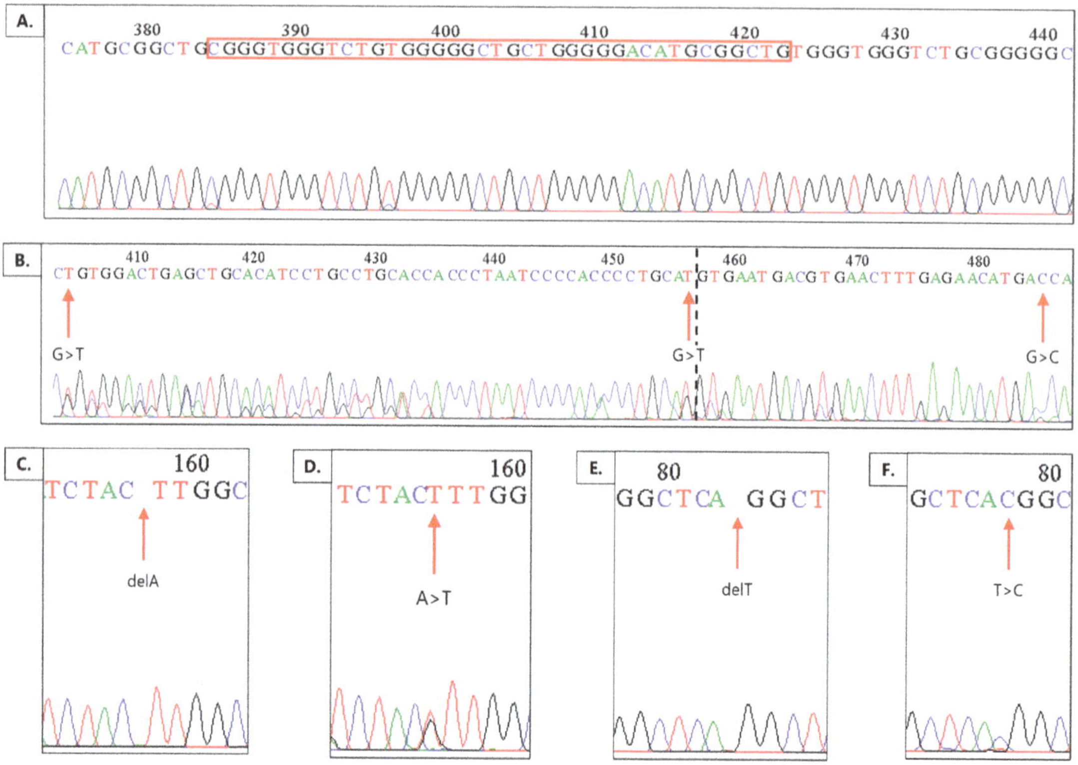

Figure 2. Mutations detected by the Sanger sequencing method. (**A**)—g.14228_14267dup (red frame); (**B**)—g.14248G>T, g.14300G>T, g.14329G>C; (**C**)—g.14004delA; (**D**)—g.14004A>T; (**E**)—g.13921delT; (**F**)—g.13921T>C. The dashed line indicates the pre-mRNA splicing site.

Intron 7 showed a particular location, g.13921, where two different mutations occurred in 11 samples—the deletion of nucleotide T (g.13921delT) (Figure 2e) and the substitution of the same nucleotide T into C (g.13921T>C) (Figure 2f). The effect of both mutations on the biological function of the DVL1 protein was not determined by the PROVEAN program. Our further analysis showed that both mutations were linked to the lower expression and H-score value of the DVL1 protein ($r = -0.478$, $p = 0.038$). Furthermore, the deletion of nucleotide T was significantly more tied to lower grade than to the higher grades ($\chi = -0.434$, $p = 0.027$), with no incidence observed in malignant grades. On the other hand, the substitution of the same nucleotide, was mostly present in malignant grades and a significant correlation to higher grades was established ($\chi = 0.514$, $p = 0.032$). Moreover, the deletion of T was found exclusively in female patients ($\chi = -0.442$, $p = 0.024$).

Exon 8 showed two locations hit by three different mutations in 11 samples. Two deletions of nucleotide A in exon 8 (g.13998delA and g.14004delA) caused harmful frameshift mutations. The more frequent deletion of nucleotide A (g.14004delA) (Figure 2c) is located at the codon ATT, which codes for the amino acid isoleucine. The change in the reading frame caused by this deletion introduced a downstream stop codon and caused the formation of the truncated DVL1 protein, with altered activity and functionality. The substitution of nucleotide A into T (g.14004A>T) (Figure 2d) at the same location caused the codon change from ACC to TCC, and consequently the amino acid change from Ile to Ser. Using a PROVEAN tool to analyze the effect of protein variations, this mutation was flagged as harmful for the DVL1 function. This substitution was found in a single sample that belonged to the meningothelial subtype of grade I meningioma. However, this sample showed the highest expression of active β-catenin (H-score = 262) and the highest percentage of nuclei with the DVL1 expression (90%).

The only sample that did not harbor the aforementioned duplication had two mutations in the intron 8—g.14248G>T and g.14300G>T. Mutation g.14300G>T is located at the splicing site between intron 8 and exon 9, and g.14248G>T is also located nearby. Due to their locations, these changes are potential splice site mutations that may change the length of the DVL1 protein (Figure 2b).

In exon 9, the substitution of g.14329G>C (Figure 2b) caused a change in the amino acid Ser to Thr. However, the analysis by PROVEAN showed that this change caused no disruption to the tertiary and quaternary structures of the DVL1 protein, nor a change in its biological function.

2.3. Protein Expression and Localization of the DVL1 and Active β-Catenin Form

The protein expression evaluation using the H-score revealed the presence of DVL1 in all inspected samples, mostly localized in cytoplasm but also in the nuclei (Figure 3). The H-score mean for DVL1 expression was 169.74. The majority of samples (16/23) showed a moderate signal (70%), while six samples showed a strong signal (26%). One sample did not show cytoplasmic expression but had a pronounced nuclear expression. A total of 48% of the samples did not express DVL1 in the nuclei, or expression was in less than 5% of the nuclei in the field of view. In 7/23 samples (30%), DVL1 was expressed in 10–50% of nuclei, while 5/23 samples (22%) showed the frequent nuclear expression of DVL1 in more than 50% of nuclei in the field of view. Spearman's test revealed that DVL1 H-score values dropped with the age of the patients ($\rho = -0.752$, $p = 0.000$). Moreover, the highest values of the DVL1 H-score (>200) were correlated with samples without any mutation in the PDZ domain ($r = -0.517$, $p = 0.023$), while samples comprised of mutations in the PDZ domain expressed less DVL1 protein product. The duplication g.14228_14267dup was omitted from calculation since it appeared in 96% of the analyzed meningiomas. Furthermore, the nuclear expression of the DVL1 protein was significantly correlated with the higher grade ($\rho = 0.453$, $p = 0.030$) and expression of the active form of β-catenin ($\rho = 0.456$, $p = 0.029$). On the contrary, Pearson's test showed no significant correlation between active β-catenin and cytoplasmic DVL1 H-score values ($r = 0.101$, $p = 0.647$).

Active or non-phosphorylated β-catenin was present in all samples and was localized in the cytoplasm of tumor cells. The mean H-score value of all samples was 100.17. The most pronounced was the low expression of active β-catenin, which was observed in 12/23 samples (52%). In 9/23 samples (39%) active beta-catenin had a moderate signal, and two samples (9%) showed a strong signal. No nuclear expression was found. Higher expressions of active β-catenin were significantly correlated with higher grades ($\rho = 0.580$, $p = 0.004$).

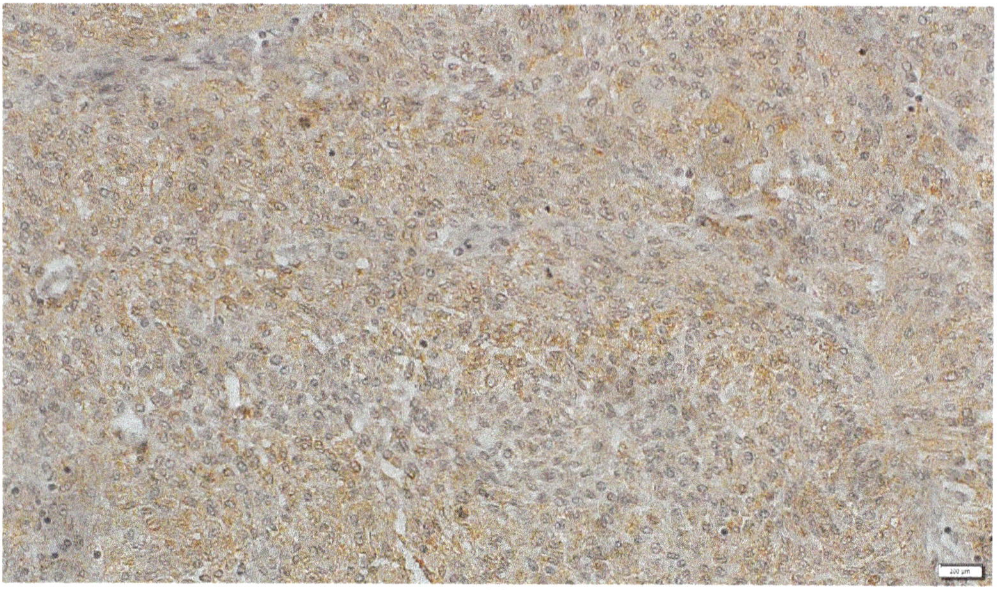

(**a**)

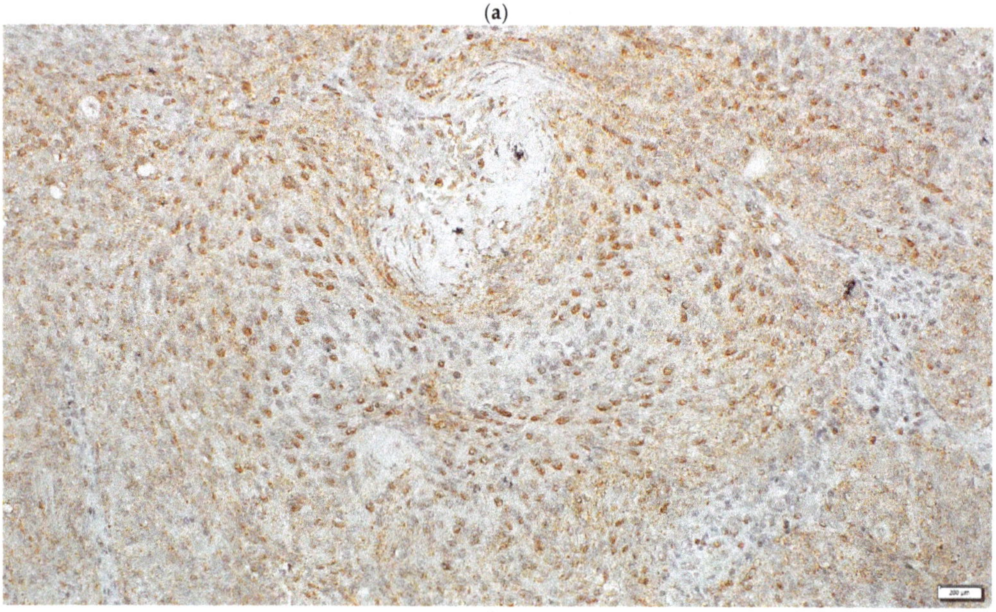

(**b**)

Figure 3. (**a**) Cytoplasmic localization of DVL1 in intracranial meningioma, with moderate to high expression. The specimen was taken from a 70-year-old female patient who was diagnosed with atypical (grade II) meningioma. (**b**) Nuclear localization of DVL1 in intracranial meningioma, with high expression. The specimen was taken from a 58-year-old female patient who was diagnosed with atypical (grade II) meningioma. Magnification: 200×, scale bar: 200 μm.

3. Discussion

Our findings concerning alterations of the *DVL1* gene in the intracranial meningioma set showed that 15.15% of the samples harbored changes. Microsatellite instability was present in 9.09% of the samples, and the loss of heterozygosity was observed in 6.06%. The genomic changes were not significantly associated with tumor grade or the sex of the patients. However, MSI was most frequently found in grade III and may be linked to the rise of mutational burden. On the contrary, genetic alterations were associated with patients younger than 60 years ($r = -0.485$, $p = 0.004$). When considering LOH, similar findings were described by Nagahata et al. [24], who studied *DVL1* in breast cancer and found that 10% of the samples harbored LOH of the *DVL1* gene, suggesting that *DVL1* has a role in breast cancer progression. Our previous work on *DVL1* in astrocytoma [25] showed the high frequency of MSI in all grades of astrocytomas, while LOH was detected only in glioblastomas (in 8.6% of samples). Our present results showed that the LOH of *DVL1* was not tied to the malignant grade and may represent early event in meningioma progression. Kafka et al. [25] investigated two different microsatellite markers, one of which was D1S468 and demonstrated higher frequency of genomic instability. Therefore, in future meningioma studies, samples could be tested with two *DVL1* gene microsatellite markers to show if a higher frequency of genomic instability is present.

A more detailed genetic analysis of the functionally crucial PDZ domain using Sanger sequencing showed a high mutation rate in the investigated sample. The most prominent was the duplication in intron 8, g.14228_14267dup, for which predictive software could not determine a significant impact on the DVL1 protein. Since the duplication was also found in the blood sequence of some patients, this may suggest a potential polymorphism or the germline existence of this variation. Such a 39 bp long duplication located within the intronic sequence may affect splicing signals. Many studies have shown that indel and other point mutations in introns, as well as changes at splicing sites, may all have an impact on proper splicing mechanisms [26–29]. Therefore, a duplication of 39 bp could lead to changes in secondary pre-mRNA structures affecting proper splicing. We can theorize that the new secondary structure causes the retention of the existing intron, resulting in a longer protein. On the other hand, duplication can cause the RNA loop that consists of an intron, but also includes one of the adjacent exons, resulting in a shorter protein. As described in a paper written by Lin et al. [30], structural stems at splicing sites can cause the formation of mRNA isoforms specific to some diseases, often tumors. The stem structures can also sterically impair the binding of splicing enhancers. It has also been shown that the higher percentage of GC base pairs positively correlates with the percentage of alternatively spliced exons. This fact is especially interesting in the context of studying the PDZ domain because in this region, a high percentage of GC base pairs naturally occurs, which is further increased by the duplication g.14228_14267dup riddled with GC.

The effect of two mutations at the same locus in intron 7, g.13921delT and g.13921T>C, were not determined by the PROVEAN program. Our study revealed that mutations at these loci were linked to the lower expression of the DVL1 protein ($r = -0.478$, $p = 0.038$), with the deletion of nucleotide T more tied to the lower grade ($\chi = -0.434$, $p = 0.027$), no incidence in the malignant grades, and the substitution of the same nucleotide more frequent in the highest grade ($\chi = 0.514$, $p = 0.032$). Likewise, the deletion of T was found exclusively in female patients ($\chi = -0.442$, $p = 0.024$). These results prove that mutations at locus g.13921 downregulate DVL1 expression, and depending on the type of mutation, this effect will manifest in different grades. Additionally, the substitution of T (g.13921T>C) could be a marker of low DVL1 expression in higher grade meningioma patients.

The mutation with the highest potential effect is the deletion of nucleotide A in exon 8, g.14004delA. The analysis found that this mutation causes a frameshift and the consequent truncated version of the DVL1 protein. Such a premature stop codon in exon 8 of the region encoding the PDZ domain may act in favor of tumor formation and development. In their study, Brennan et al. [31] found that truncated versions of the LRP5/6 co-receptor, involved in the Wnt signaling pathway, may have a protooncogenic role. Truncated forms of the co-receptor stabilize β-catenin independently of other membrane proteins involved in Wnt signal transduction, and are resistant to degradation and endocytosis. In addition to LRP5/6, the role of the truncated form of the APC protein in colorectal tumor was also established by Schneikert and Behrens [32]. According to their research, the truncated form of the APC protein stimulates the migration of colorectal tumor cells and promotes the development of chromosomal instability. These studies suggest that the truncated form of the DVL1 protein caused by the g.14004delA mutation in 34.61% of samples could be characteristic of meningiomas and enhance Wnt signaling, thus stimulating tumorigenesis. The substitution at the same locus (g.14004A>T) was associated with the highest expression of active β-catenin (H-score = 262) and the highest percentage of nuclear DVL1 expression (90%). These findings are also conclusive with regard to the activation of Wnt signaling. Since it was found in only one sample, this hypothesis should be tested with a higher number of tumor samples harboring this mutation.

The only sample that did not harbor the duplication showed five other different mutations, three of which were the substitution of nucleotide G. The substitutions g.14248G>T and g.14300G>T in intron 8 were assigned as mutations of the splicing site. Due to these two substitutions, a longer DVL1 protein retaining part of intron 8 is possible. The third substitution, g.14329G>C, causes the change from amino acid Ser to Thr, which have similar biochemical properties and therefore probably have no impact on the structure and function of the DVL1 protein, as stated in the paper by Castro-Chavez [33]. However, if there is a change at the splicing site due to the first two mutations in intron 8, there is a possibility of a new reading frame that affects the substitution in exon 9, leading to the formation of a protein with an altered amino acid composition and a new structure.

Although the PDZ domain was severely mutated, the protein expression of DVL1 was found in all samples, mostly with a moderate signal (70%). However, a strong signal was less present in 26% of the samples, while one sample showed a lack of cytoplasmic expression but had a pronounced nuclear expression. Furthermore, we established that the highest values of the DVL1 H-score (>200) were correlated with samples without any mutation in the PDZ domain ($r = -0.517$, $p = 0.023$), indicating that samples containing mutations in the PDZ domain expressed significantly less DVL1 product (the duplication g.14228_14267dup was omitted from the calculation). In 48% of the analyzed meningiomas, the nuclear expression of DVL1 was missing; however, 22% of the samples still showed DVL1 expression in more than 50% of the nuclei. Although studies rarely report on the nuclear expression of DVL1 in different tumors [34,35], Sharma et al. [36] demonstrated that the acetylation of the conserved lysines (K69 and K285), which are present in the DIX and PDZ domains, not only promoted the nuclear localization of DVL1, but also influenced its promoter binding and the regulation of genes implicated in cancer. Furthermore, our study showed that the cytoplasmic expression of DVL1 was not correlated with meningioma grade or the expression of active cytoplasmic β-catenin ($r = 0.101$, $p = 0.647$). However, this was not the case with nuclear DVL1 expression, which was associated with higher grades ($\rho = 0.453$, $p = 0.030$) and higher expressions of β-catenin's active form ($\rho = 0.456$, $p = 0.029$). This may suggest that the nuclear expression of DVL1 could promote Wnt signaling activation and potentially serve as a biomarker of meningioma progression. Similar findings on the influence of DVL1 progression in other tumors were also reported. For instance, in a study by Karin-Kujundzic et al. [37], the active involvement of DVL1 and significantly higher DVL1 expressions in serous ovarian carcinomas as compared to normal ovarian tissue were reported. Mizutani et al. [38] found that expressions of DVL1 and β-catenin are correlated, and that DVL1 expression increases with grade in prostate cancer. On the other hand, Ameli et al. [35] concluded that in invasive ductal and lobular breast carcinoma, DVL1 does not correlate with grade; however, they did not find a nuclear expression of the protein. Wei et al. [39] showed that the expression levels of DVL1 were higher in non-small-cell lung cancer metastases and correlated to β-catenin expression, while Zhang et al. [40] demonstrated that DVL1 increases the accumulation of β-catenin in ovarian cancer cells. They also showed that DVL1 was responsible for the nuclear translocation of β-catenin, which was not compatible with our study. However, similar findings to ours were recorded by the Kafka et al. [41]. They showed that most brain metastases (45.2%) had moderate DVL1 expression levels, with nuclear staining in

54.8% of the cases. In addition, they reported a correlation between the nuclear expression of β-catenin and upregulated DVL1 expression.

Interestingly, in our study, DVL1 H-score values dropped with the age of the patient ($\rho = -0.752$, $p = 0.000$). This finding is consistent with our previous study on the involvement of the Disheveled protein family in astrocytoma malignancy grades [25], which showed that younger patients had a stronger DVL1 expression than the older ones. The same study indicated that high-grade tumors had a lower expression of DVL1, suggesting that it may be an early event, which is contrary to our present findings on meningioma.

Active β-catenin, which is distinctive for activated Wnt signaling in tumor cells, was present in all samples and was localized in cytoplasm. The mean H-score value was 100.17. Low expression levels of active β-catenin were observed in 52% of meningiomas, while in 39% of the investigated samples of active β-catenin had a moderate signal, while two samples (9%) showed a strong signal. Although nuclear expression was not found, higher expression levels of active β-catenin were significantly correlated with higher grades ($\rho = 0.580$, $p = 0.004$) and nuclear DVL1 expression ($\rho = 0.456$, $p = 0.029$). This is in alignment with our previous study [18], indicating β-catenin involvement in meningioma progression.

4. Materials and Methods

Our study consisted of 33 samples of intracranial meningioma with different malignancy grades. All tumor samples were classified by pathologists according to the criteria of the World Health Organization [42]. Tumor samples, as well as 3–5 mL of autologous blood, were collected from patients with no prior radiation or chemotherapy treatments who were scheduled for operation. In addition to the pathohistological diagnosis and tumor grade, parameters such as the sex, location, and age of the patient were collected for each sample.

DNA was isolated from tumor tissue with the standard phenol/chloroform method [43], and from blood with the standard salting out method [44].

Samples were tested for genomic alternations—microsatellite instability (MSI) and loss of heterozygosity (LOH) of the *DVL1* gene by comparing DNAs from the tumor and blood of the same patient. Detection of MSI and LOH was performed on Spreadex gels EL400 Mini (Elchrom Scientific, AL-Labortechnik, AL-Diagnostic GmbH, Amstetten, Austria) using the microsatellite maker D1S468 (5′-TTAACCGTTTTGGTCCTACC-3′ and 5′-CTCTGACCAGCATTAAAGATTC-3′), with a high percentage of heterozygosity in the population.

For further investigation, to amplify the region encoding the PDZ domain, primers were designed using the NCBI database [45] and the primer design tool—Example Blast [46]. The primers (5′-TAACCGACTCCACCATGTCC-3′, 5′-GAAACGATCTCCCGCAGCA-3′) cover part of intron 7, whole exon 8 and intron 8, and part of exon 9 of the PDZ domain. Optimal PCR conditions for amplifications of the *DVL1* D1S468 microsatellite marker and the PDZ regions are shown in Table 2.

Table 2. PCR conditions for the *DVL1* microsatellite marker D1S468 and the PDZ genetic regions.

	(Pre) Denaturation	Denaturation	Annealing	Extending	No. of Cycles
D1S468	94 °C/5 min	94 °C/30 s	60 °C/30 s	72 °C/30 s	40
PDZ	94 °C/5 min	94 °C/35 s	58.6 °C/35 s	72 °C/35 s	35

Mutations in the PDZ domain were detected using the high-resolution melting method (HRM) and LightCycler® 480 High-Resolution Melting Master kit on the Roche LightCycler® Nano System. By comparing DNA sequences from the tumor and blood of the same patient, potential mutations were revealed. All samples that were suspected to harbor mutation were sequenced using the standard Sanger sequencing method and BigDyeTerminator v3.1 Cycle Sequencing kit on ABI 3730XL (Applied Biosystems, Foster City, CA, USA). The obtained tumor sequences of the PDZ domain of the *DVL1* gene were compared with the sequence of the paired blood sample and with the sequence available at the NCBI database [47]. (Verification of the impact of detected mutations on the biological function of the DVL1 protein was performed using the publicly available PROVEAN tool [48] and immunohistochemistry.

Immunohistochemistry was performed on 4 μm thick paraffin-embedded sections of 23 available meningioma samples collected during surgery. The sections were collected during a period of four years. For DVL1 detection, we used a polyclonal rabbit anti-Disheveled/Dvl1 antibody: ab233003 (Abcam, Cambridge, UK), diluted 1:200; its recombinant fragment (His-T7-tag) corresponded to Human Dvl1 aa 150–300 and partly covered the central PDZ domain. We also tested the expression and localization of the active form of β-catenin using the monoclonal antibody non-Phospho beta-catenin (Ser33-37/Thr41) (D131A1) Rabbit mAb #8814 (Cell Signalling Technology, Danvers, MA, USA), diluted 1:800. For visualization, we used DAB chromogen (EnVision^{TM}, Dako REAL^{TM}). Slides with antibody-labelled tissue were analyzed using bright-field microscopy on the Olympus BX53 microscope. The expression of the protein was observed in tumor hot spots, where at least 200 cells were counted. Immunopositivity was quantified using the H-score (Equation (1)), with a range of protein expression values on a scale of 0–300 [49]:

$$H = [1 \times (\% \text{ of stations } 1+) + 2 \times (\% \text{ of stations } 2+) + 3 \times (\% \text{ of stations } 3+)] \quad (1)$$

where 1+ indicates weak immunopositivity—yellowish / light brown color, 2+ indicates moderate immunopositivity—light brown, and 3+ indicates strong immunopositivity—dark brown. Depending on the obtained H-score value, samples were categorized into 3 groups: 0–100 = samples with no signal/weak signal (0/1+), 101–200 = samples with moderate signal (2+), and 201–300 = samples with strong signal (3+).

The results of the genetic and protein investigations were further analyzed and correlated using the publicly available program RStudio [50], including the R-package ggplot2, ggpubr, and plotly from the official R repository CRAN [51]. A significance level of $p < 0.05$ was used to process the results. The normality of the distribution was checked using the Shapiro–Wilk test.

5. Conclusions

We have shown that the central PDZ domain is highly mutated, with different outcomes on the biological function of the DVL1 protein. The samples containing mutations in the PDZ domain expressed significantly less DVL1 protein product, and the nuclear expression of DVL1 could potentially represent a good biomarker for meningioma progression and the activation of the Wnt signaling pathway. The results of this study contribute to a better understanding of the role of DVL1 in human intracranial meningiomas and point out molecules useful for diagnostics and the treatment of patients.

Author Contributions: Conceptualization, A.B. and N.P.-Š.; methodology, A.B. and K.D.; software, K.D.; validation, A.K., S.C.-A., D.O. and N.P.-Š.; formal analysis, A.B., K.D. and S.C.-A.; investigation, A.B., K.D., A.K., S.C.-A. and D.O.; resources, S.C.-A. and D.O.; data curation, A.B., K.D., A.K., S.C.-A. and D.O.; writing—original draft preparation, A.B. and K.D.; writing—review and editing, A.K., D.O., S.C.-A. and N.P.-Š.; visualization, A.B., K.D. and A.K.; supervision, N.P.-Š.; project administration, N.P.-Š.; funding acquisition, N.P.-Š. All authors have read and agreed to the published version of the manuscript.

Funding: This research was funded by the Scientific Centre of Excellence for Basic, Clinical, and Translational Neuroscience (project "Experimental and clinical research of hypoxic-ischemic damage in perinatal and adult brain"; GA KK01.1.1.01.0007, funded by the European Union through the European Regional Development Fund).

Institutional Review Board Statement: The study was conducted according to the guidelines of the Declaration of Helsinki and approved by the Ethics Committee of the School of Medicine, University of Zagreb (Case number: 380-59-10106-17-100/98; Class: 641-01/17-02/01, 23 March 2017), the Ethics Committee of the University Hospital Center Zagreb (number 02/21/AG, class:

8.1-16/215-2, 2 February 2017), the Ethics Committee of the University Hospital Center "Sisters of Charity" (number EP-5429/17-5, 23 March 2017), and the Ethics Committee of the University Hospital Dubrava (17 May 2017).

Informed Consent Statement: Informed consent was obtained from all subjects involved in the study.

Data Availability Statement: Data supporting the reported results are contained within the article. Some of the data presented in this study are available on request from the corresponding author. The data are not publicly available due to privacy issues.

Acknowledgments: We thank all medical personnel from the University Hospital Center Zagreb, University Hospital Center "Sisters of Charity", and University hospital Dubrava who were involved in the collection of tumor tissues.

Conflicts of Interest: The authors declare no conflict of interest. The funders had no role in the design of the study; in the collection, analyses, or interpretation of data; in the writing of the manuscript, or in the decision to publish the results.

References

1. The Central Brain Tumor Registry of the United States (CBTRUS). Available online: https://cbtrus.org/ (accessed on 30 January 2021).
2. Ostrom, Q.T.; Cioffi, G.; Gittleman, H.; Patil, N.; Waite, K.; Kruchko, C.; Barnholtz-Sloan, J.S. CBTRUS statistical report: Primary brain and other central nervous system tumors diagnosed in the United States in 2012–2016. *Neuro-Oncology* **2019**, *21*, v1–v100. [PubMed]
3. Pèrez-Magán, E.; Campos-Martín, Y.; Mur, P.; Fiaño, C.; Ribalta, T.; García, J.F.; Rey, J.A.; De Lope, A.R.; Mollejo, M.; Melendez, B. Genetic alterations associated with progression and recurrence in meningiomas. *J. Neuropathol. Exp. Neurol.* **2012**, *71*, 882–893.
4. Pećina-Šlaus, N.; Kafka, A.; Lechpammer, M. Molecular genetics of intracranial meningiomas with emphasis on canonical Wnt signalling. *Cancers* **2016**, *8*, 67. [CrossRef]
5. Galani, V.; Lampri, E.; Varouktsi, A.; Alexiou, G.; Mitselou, A.; Kyritsis, A.P. Genetic and epigenetic alterations in meningiomas. *Clin. Neurol. Neurosurg.* **2017**, *158*, 119–125. [CrossRef]
6. Nasser, M.M.; Mehdipour, P. Exploration of involved key genes and signaling diversity in brain tumors. *Cell. Mol. Neurobiol.* **2017**, *38*, 393–419. [CrossRef]
7. Preusser, M.; Brastianos, P.; Mawrin, C. Advances in meningioma genetics: Novel therapeutic opportunities. *Nat. Rev. Neurol.* **2018**, *14*, 106–115. [CrossRef]
8. Wang, N.; Osswald, M. Meningiomas: Overview and new directions in therapy. *Semin. Neurol.* **2018**, *38*, 112–120.
9. Pereira, B.J.A.; Oba-Shinjo, S.M.; de Almeida, A.N.; Marie, S. Molecular alterations in meningiomas: Literature review. *Clin. Neurol. Neurosurg.* **2019**, *176*, 89–96.
10. Vallée, A.; LeCarpentier, Y.; Vallée, J.-N. Targeting the canonical WNT/β-catenin pathway in cancer treatment using non-steroidal anti-inflammatory drugs. *Cells* **2019**, *8*, 726.
11. Kahn, M. Can we safely target the WNT pathway? *Nat. Rev. Drug Discov.* **2014**, *13*, 513–532. [CrossRef] [PubMed]
12. Noelanders, R.; Vleminckx, K. How Wnt signaling builds the brain: Bridging development and disease. *Neuroscientist* **2016**, *23*, 314–329. [CrossRef] [PubMed]
13. Matsui, W.H. Cancer stem cell signaling pathways. *Medicine* **2016**, *95*, S8–S19. [CrossRef] [PubMed]
14. Oliva, C.A.; Montecinos-Oliva, C.; Inestrosa, N.C. Wnt signaling in the central nervous system: New insights in health and disease. *Prog. Mol. Biol. Transl. Sci.* **2018**, *153*, 81–130.
15. El-Sahli, S.; Xie, Y.; Wang, L.; Liu, S. Wnt signaling in cancer metabolism and immunity. *Cancers* **2019**, *11*, 904. [CrossRef] [PubMed]
16. Ng, L.F.; Kaur, P.; Bunnag, N.; Suresh, J.; Sung, I.C.H.; Tan, Q.H.; Gruber, J.; Tolwinski, N.S. WNT signaling in disease. *Cells* **2019**, *8*, 826.
17. Pham, M.H.; Zada, G.; Mosich, G.M.; Chen, T.C.; Giannotta, S.L.; Wang, K.; Mack, W.J. Molecular genetics of meningiomas: A systematic review of the current literature and potential basis for future treatment paradigms. *Neurosurg. Focus* **2011**, *30*, E7. [CrossRef]
18. Bukovac, A.; Kafka, A.; Raguž, M.; Brlek, P.; Dragičević, K.; Müller, D.; Pećina-Šlaus, N. Are we benign? What can Wnt signaling pathway and epithelial to mesenchymal transition tell us about intracranial meningioma progression. *Cancers* **2021**, *13*, 1633.
19. Kafka, A.; Bašić-Kinda, S.; Pećina-Šlaus, N. The cellular story of dishevelleds. *Croat. Med. J.* **2014**, *55*, 459–46667. [CrossRef]
20. Yu, J.; Virshup, D.M. Updating the Wnt pathways. *Biosci. Rep.* **2014**, *34*, e00142. [CrossRef]
21. Wiese, K.; Nusse, R.; Van Amerongen, R. Wnt signalling: Conquering complexity. *Development* **2018**, *145*, 165902. [CrossRef]
22. Jackstadt, R.; Hodder, M.C.; Sansom, O.J. WNT and β-catenin in cancer: Genes and therapy. *Annu. Rev. Cancer Biol.* **2020**, *4*, 177–196.

23. Sharma, M.; Castro-Piedras, I.; Simmons, G.; Pruitt, K. Dishevelled: A masterful conductor of complex Wnt signals. *Cell. Signal.* **2018**, *47*, 52–64. [CrossRef] [PubMed]
24. Nagahata, T.; Shimada, T.; Harada, A.; Nagai, H.; Onda, M.; Yokoyama, S.; Shiba, T.; Jin, E.; Kawanami, O.; Emi, M. Amplification, up-regulation and over-expression of DVL-1, the human counterpart of the *Drosophila disheveled* gene, in primary breast cancers. *Cancer Sci.* **2003**, *94*, 515–518. [CrossRef] [PubMed]
25. Kafka, A.; Bačić, M.; Tomas, D.; Žarković, K.; Bukovac, A.; Njirić, N.; Mrak, G.; Krsnik, Z.; Pećina-Šlaus, N. Different behaviour of DVL1, DVL2, DVL3 in astrocytoma malignancy grades and their association to TCF1 and LEF1 upregulation. *J. Cell. Mol. Med.* **2018**, *23*, 641–655. [CrossRef]
26. Ars, E.; Serra, E.; Garcia, J.; Kruyer, H.; Gaona, A.; Lázaro, C.; Estivill, X. Mutations affecting mRNA splicing are the most common molecular defects in patients with neurofibromatosis type 1. *Hum. Mol. Genet.* **2000**, *9*, 237–247. [CrossRef] [PubMed]
27. Buratti, E.; Baralle, F. Influence of RNA secondary structure on the pre-mRNA splicing process. *Mol. Cell. Biol.* **2004**, *24*, 10505–10514. [CrossRef]
28. Rogic, S.; Montpetit, B.; Hoos, H.; Mackworth, A.; Ouellette, B.; Hieter, P. Correlation between the secondary structure of pre-mRNA introns and the efficiency of splicing in Saccharomyces cerevisiae. *BMC Genom.* **2008**, *9*, 355. [CrossRef] [PubMed]
29. Rubtsov, P. Role of pre-mRNA secondary structures in the regulation of alternative splicing. *Mol. Biol.* **2016**, *50*, 823–830. [CrossRef]
30. Lin, C.; Taggart, A.; Fairbrother, W.G. RNA structure in splicing: An evolutionary perspective. *RNA Biol.* **2016**, *13*, 766–771. [PubMed]
31. Brennan, K.; Sancho, J.M.G.; Castelo-Soccio, L.A.; Howe, L.; Mc Brown, A. Truncated mutants of the putative Wnt receptor LRP6/Arrow can stabilize β-catenin independently of Frizzled proteins. *Oncogene* **2004**, *23*, 4873–4884. [CrossRef]
32. Schneikert, J.; Behrens, J. Truncated APC is required for cell proliferation and DNA replication. *Int. J. Cancer* **2006**, *119*, 74–79. [CrossRef]
33. Castro-Chavez, F. The rules of variation: Amino acid exchange according to the rotating circular genetic code. *J. Theor. Biol.* **2010**, *264*, 711–721. [CrossRef] [PubMed]
34. Huang, M.Y.; Yen, L.C.; Liu, H.C.; Liu, P.P.; Chung, F.Y.; Wang, T.N.; Wang, J.Y.; Lin, S.R. Significant overexpression of DVL1 in Taiwanese colorectal cancer patients with liver metastasis. *Int. J. Mol. Sci.* **2013**, *14*, 20492–20507. [CrossRef]
35. Ameli, F.; Rose, I.M.; Masir, N. Expression of DDR1 and DVL1 in invasive ductal and lobular breast carcinoma does not correlate with histological type, grade and hormone receptor status. *Asian Pac. J. Cancer Prev.* **2015**, *16*, 2385–2390. [CrossRef] [PubMed]
36. Sharma, M.; Molehin, D.; Castro-Piedras, I.; Martinez, E.G.; Pruitt, K. Acetylation of conserved DVL-1 lysines regulates its nuclear translocation and binding to gene promoters in triple-negative breast cancer. *Sci. Rep.* **2019**, *9*, 16257. [CrossRef]
37. Karin-Kujundzic, V.; Kardum, V.; Sola, I.M.; Paic, F.; Skrtic, A.; Skenderi, F.; Serman, A.; Nikuseva-Martic, T.; Vranic, S.; Serman, L. Dishevelled family proteins in serous ovarian carcinomas: A clinicopathologic and molecular study. *APMIS* **2020**, *128*, 201–210. [CrossRef] [PubMed]
38. Mizutani, K.; Miyamoto, S.; Nagahata, T.; Konishi, N.; Emi, M.; Onda, M. Upregulation and overexpression of DVL1, the human counterpart of the *Drosophila dishevelled* gene, in prostate cancer. *Tumori* **2005**, *91*, 546–551. [CrossRef] [PubMed]
39. Wei, Q.; Zhao, Y.; Yang, Z.Q.; Dong, Q.Z.; Dong, X.J.; Han, Y.; Zhao, C.; Wang, E.H. Dishevelled family proteins are expressed in non-small cell lung cancer and function differentially on tumor progression. *Lung Cancer* **2008**, *62*, 181–192.
40. Zhang, K.; Song, H.; Yang, P.; Dai, X.; Li, Y.; Wang, L.; Du, J.; Pan, K.; Zhang, T. Silencing dishevelled-1 sensitizes paclitaxel-resistant human ovarian cancer cells via AKT/GSK-3β/β-catenin signalling. *Cell Prolif.* **2015**, *48*, 249–258.
41. Kafka, A.; Tomas, D.; Beroš, V.; Pećina, H.I.; Zeljko, M.; Pećina-Šlaus, N. Brain metastases from lung cancer show increased expression of DVL1, DVL3 and beta-catenin and down-regulation of E-cadherin. *Int. J. Mol. Sci.* **2014**, *15*, 10635–10651. [CrossRef]
42. Louis, D.N.; Perry, A.; Wesseling, P.; Brat, D.J.; Cree, I.A.; Figarella-Branger, D.; Hawkins, C.; Ng, H.K.; Pfister, S.M.; Reifenberger, G.; et al. The 2021 WHO classification of tumors of the central nervous system: A summary. *Neuro-Oncology* **2021**, *23*, 1231–1251. [CrossRef] [PubMed]
43. Green, M.R.; Sambrook, J. *Molecular Cloning—A Laboratory Manual*, 4th ed.; Cold Spring Harbor Laboratory Press: New York, NY, USA, 2012.
44. Miller, S.A.; Dykes, D.D.; Polesky, H.F. A simple salting out procedure for extracting DNA from human nucleated cells. *Nucleic Acids Res.* **1988**, *16*, 1215. [CrossRef] [PubMed]
45. National Center for Biotechnology Information—NCBI. Available online: https://www.ncbi.nlm.nih.gov (accessed on 25 October 2020).
46. Ye, J.; Coulouris, G.; Zaretskaya, I.; Cutcutache, I.; Rozen, S.; Madden, T. Primer-BLAST: A tool to design target-specific primers for polymerase chain reaction. *BMC Bioinform.* **2012**, *13*, 134.
47. National Center for Biotechnology Information—NCBI. Available online: https://www.ncbi.nlm.nih.gov/nucleotide/NG_008048?report=genbank&log$=nuclalign&blast_rank=1&RID=0 (accessed on 25 January 2021).
48. Protein Variation Effect Analyzer—PROVEAN. Available online: http://provean.jcvi.org/index.php (accessed on 30 January 2021).
49. Mazières, J.; Brugger, W.; Cappuzzo, F.; Middel, P.; Frosch, A.; Bara, I.; Klingelschmitt, G.; Klughammer, B. Evaluation of EGFR protein expression by immunohistochemistry using H-score and the magnification rule: Re-analysis of the SATURN study. *Lung Cancer* **2013**, *82*, 231–237. [CrossRef] [PubMed]
50. RStudio. Available online: https://rstudio.com/ (accessed on 15 March 2021).
51. CRAN. Available online: https://cran.r-project.org/ (accessed on 15 March 2021).

International Journal of
Molecular Sciences

MDPI

Review

Molecular Biomarkers in Glioblastoma: A Systematic Review and Meta-Analysis

Heena Sareen [1,2,*], Yafeng Ma [1,2], Therese M. Becker [1,2,3], Tara L. Roberts [1,2,3], Paul de Souza [2,3,4] and Branka Powter [1]

1. Centre for Circulating Tumour Cell Diagnostics and Research, Ingham Institute for Applied Medical Research, Liverpool, NSW 2170, Australia
2. South-Western Clinical School, University of New South Wales, Liverpool, NSW 2170, Australia
3. School of Medicine, Western Sydney University, Campbelltown, NSW 2560, Australia
4. Liverpool Hospital, Liverpool, NSW 2170, Australia

* Correspondence: h.sareen@student.unsw.edu.au; Tel.: +61-0406937108

Abstract: Background: Glioblastoma (GBM) is a highly aggressive cancer with poor prognosis that needs better treatment modalities. Moreover, there is a lack of reliable biomarkers to predict the response and outcome of current or newly designed therapies. While several molecular markers have been proposed as potential biomarkers for GBM, their uptake into clinical settings is slow and impeded by marker heterogeneity. Detailed assessment of prognostic and predictive value for biomarkers in well-defined clinical trial settings, if available, is scattered throughout the literature. Here we conducted a systematic review and meta-analysis to evaluate the prognostic and predictive significance of clinically relevant molecular biomarkers in GBM patients. Material and methods: A comprehensive literature search was conducted to retrieve publications from 3 databases (Pubmed, Cochrane and Embase) from January 2010 to December 2021, using specific terms. The combined hazard ratios (HR) and confidence intervals (95% CI) were used to evaluate the association of biomarkers with overall survival (OS) in GBM patients. Results: Twenty-six out of 1831 screened articles were included in this review. Nineteen articles were included in the meta-analyses, and 7 articles were quantitatively summarised. Fourteen studies with 1231 GBM patients showed a significant association of *MGMT* methylation with better OS with the pooled HR of 1.66 (95% CI 1.32–2.09, $p < 0.0001$, random effect). Five studies including 541 GBM patients analysed for the prognostic significance of *IDH1* mutation showed significantly better OS in patients with *IDH1* mutation with a pooled HR of 2.37 (95% CI 1.81–3.12; $p < 0.00001$]. Meta-analysis performed on 5 studies including 575 GBM patients presenting with either amplification or high expression of EGFR gene did not reveal any prognostic significance with a pooled HR of 1.31 (95% CI 0.96–1.79; $p = 0.08$). Conclusions: *MGMT* promoter methylation and *IDH1* mutation are significantly associated with better OS in GBM patients. No significant associations were found between *EGFR* amplification or overexpression with OS.

Keywords: glioblastoma; prognostic biomarkers; systematic review; meta-analysis

Citation: Sareen, H.; Ma, Y.; Becker, T.M.; Roberts, T.L.; de Souza, P.; Powter, B. Molecular Biomarkers in Glioblastoma: A Systematic Review and Meta-Analysis. *Int. J. Mol. Sci.* **2022**, *23*, 8835. https://doi.org/10.3390/ijms23168835

Academic Editors: Ivana Jovčevska and Nives Pećina-Šlaus

Received: 9 July 2022
Accepted: 4 August 2022
Published: 9 August 2022

1. Introduction

Glioblastoma (GBM), the most common and aggressive form of brain cancer, has an overall 5-year survival of only 7% [1]. Based on clinical presentation, GBM is classified into two different categories: Primary GBM accounts for approximately 90% of GBM cases, arises de-novo, and is more common in elderly patients [2]. It is characterised by distinct molecular alterations that include gene amplification of epidermal growth factor receptor (*EGFR*), overexpression of EGFR protein and loss of the tumour suppressor gene phosphatase and tensin homologue (*PTEN*) [2]. Secondary GBM accounts for approximately 10% of cases, is associated with younger patient age, and arises from lower grade precursors. Secondary GBM has better prognosis and typically carries mutations in isocitrate dehydrogenase 1 (*IDH1*) and Tumour protein 53 (*TP53*) genes [2].

Standard GBM treatment involves surgical resection, where the extent of resectable tumour is dictated by risk to the patient, as the tumour often infiltrates essential parts of the brain. Following surgical treatment, patients are treated with radiation and concomitant temozolomide (TMZ), then adjuvant TMZ to target remaining tumour cells [3]. Recurrence is, on average, observed 7–10 months post treatment [4]. The median overall survival (OS) of GBM patients is 12–14 months, even after treatment with TMZ and radiation, primarily due to the invasive nature of the cancer and resistance to therapies [4].

To date, few molecular biomarkers have been discovered. These include O^6-methylguanine DNA methyltransferase (*MGMT*) promoter methylation [5], Isocitrate dehydrogenase 1 (*IDH1*) mutation [6], mutations in the promoter region of the telomerase reverse transcriptase (*TERT*) gene [7], and amplification and/or overexpression of EGFR [8]. These markers have shown potential to predict the survival outcomes and treatment response in GBM patients. *MGMT* promoter methylation is known as a positive prognostic biomarker for patients treated with alkylating agents such as TMZ [5]. The *MGMT* gene encodes a DNA repair protein, which reverses DNA alkylation [9]. *MGMT* promoter methylation reduces its expression, thereby rendering cells more vulnerable to alkylating agents [9]. *IDH1* mutation (R132H) is also considered a favourable prognostic biomarker for GBM patients. It is more common in younger patients (18–45 years) and more frequent in secondary GBM (~73%), while rare in primary GBM (~3.7%) [10,11]. *EGFR* amplification and overexpression have been implicated as prognostic and predictive biomarkers [8]. Various EGFR targeting tyrosine kinase inhibitors (TKIs), such as gefitinib, erlotinib, and afatinib, have been trialled as targeted therapies for GBM [12].

These potential molecular biomarkers have value for GBM patient management or are informative in the context of standard of care TMZ and radiation treatment. However, for a subset of patients, the outcome is not well predicted by these markers, and may be comparably better or worse than predicted. This highlights the need to investigate other biomarkers associated with prognosis and response to treatment, particularly for newer treatment modalities. For instance, VEGF is proposed to drive angiogenesis and tumourigenesis due to its aberrant expression in GBM patients [13], and is therefore an attractive therapeutic target. Bevacizumab, a humanised antibody to vascular endothelial growth factor (VEGF), approved for recurrent GBM in many countries, was also trialled in newly diagnosed patients. However, no survival benefit was reported in the newly diagnosed GBM patients receiving bevacizumab in addition to standard of care [14]. Many studies have proposed various pharmacodynamic, prognostic and predictive biomarkers to preselect the patients that are more likely to receive survival benefits from anti-angiogenic therapies and to limit side effects.

An array of anti-angiogenic biomarkers including soluble vascular endothelial growth factor receptor 1 (sVEGFR1), soluble vascular endothelial growth factor receptor 2 (sVEGFR2), placental growth factor (PlGF) and VEGF are considered potential pharmacodynamic biomarkers. Their dynamics in peripheral blood samples are proposed to be associated with response to treatment and duration of survival [15].

To determine the value of these biomarkers, we were interested if reported data from GBM clinical trials could be evaluated for biomarkers of response to standard of care therapy or other treatment regimens trialled in the clinic. We therefore conducted a systematic review of key molecular biomarkers that have been investigated for their predictive value in recent GBM clinical trials and performed meta-analyses of such markers where statistical power (reported association with response) was sufficient.

2. Materials and Methods

2.1. Protocol and Registration

This review was registered in PROSPERO (registration number CRD42021238962) and was designed and carried out using Preferred Reporting Items for Systematic Reviews and Meta-analysis (PRISMA) formatting and guidelines [16].

2.2. Study Design and Search Strategy

A comprehensive literature search was conducted using three electronic databases: PubMed, Cochrane library and Embase databases for recent articles published between January 2010 to December 2021.The search strategy was deliberately broad and based on combination of keywords. The search terms used were "brain cancer biomarkers", "glioblastoma biomarkers" and "glioma biomarkers". Clinical studies published in English language in the last 10 years until December 2021 involving human subjects only were searched. Additional filters to include only clinical trials and randomised controlled trials (RCT) were applied in Pubmed and Embase. Included articles were screened for additional relevant studies cited for inclusion in our analysis if meeting criteria. The studies were then imported into the Rayyan Qatar Computing Research Institute (QCRI) systematic review application for further evaluation [17].

2.3. Study Selection and Criteria

In the screening process, two reviewers (H.S. and B.P.) independently screened all the imported publications in Rayyan. Studies were included if they evaluated histopathological confirmed GBM; patient number was more than 35; contained response evaluation of biomarkers; had OS/PFS/response rate and association of biomarker with OS/PFS; were an original study (RCT, cohort study or observational study). Publications were excluded if they were duplicates, reviews, letters, comments, clinical trial protocols or conference abstracts. Upon completion of inclusion and exclusion, any disagreements were resolved by consensus between the two reviewers. Included studies were inspected for duplication of patient cohorts or part of cohorts and if found to be duplicated the one with the most up to date data were included to avoid that the same data for identical cohorts was not included more than once.

2.4. Data Extraction

Ultimately a subset of 26 publications were included for data extraction and analysis and uploaded to Covidence for data extraction and quality assessment using the data extraction tool adapted for the current study. Extracted data included: general information (study title, lead author details), characteristics of included studies (study design, biomarkers tested, intervention and treatment outcomes (OS and PFS) associated with biomarkers, histopathology of tumour, total number of participants. Publications were included in meta-analyses if the hazard ratio and confidence intervals (HR and 95% CI) for the biomarkers affecting OS and PFS were given or were reliably calculated from provided Kaplan–Meier curves. For biomarkers, where number of studies or patient number did not warrant meta-analysis descriptive qualitative analyses was included. After detailed evaluation and discussion between two reviewers, 19 out of 26 studies were included in the meta-analyses while biomarkers of 7 studies underwent descriptive qualitative analysis.

2.5. Quality Assessment

Quality assessment was performed on all 26 included studies by two blinded reviewers using the Covidence Quality in Prognosis Studies (QUIPS) tool amended for the current study [18].

We assessed risk of bias across the six domains: study participation, study attrition, prognostic factor measurements, outcome measurement, study confounding, and statistical analysis and reporting [18]. Study participation was assessed for GBM histology, inclusion and exclusion criteria, adequate study participation (cohort size greater than 35), baseline characteristics (stage, grade, previous and current treatments). Study attrition included proportion of baseline samples available for biomarker analysis, reasons for not assessing samples (loss of follow up), attempts to collect information of non-assessed samples. Retrospective studies were not assessed for this domain. The prognostic factor measurements domain assessed whether the publication reported clear definition of prognostic factor. controls and methods for biomarker detection were valid and reliable. Method of measurement of prognostic factors is same for all the samples and is measured in an adequate proportion of study sample. The outcome measurement domain assessed whether the clear definition of outcome is provided and determined prior to biomarker analysis. Method of outcome measurement is reliable and valid. Outcome is assessed in adequate proportion of study sample and with the same method. The study confounding measurement domain assessed confounders measurements, including the previous and current treatments in relation to biomarkers, measured dose and duration of treatment. Statistical analysis and reporting assessed statistical tests used for biomarker expression in relation to survival outcomes. Appropriateness of the statistical tests for the data was assessed and description of the association of prognostic factors with the outcomes was reported.

2.6. Statistical Analysis and Data Analysis

Data retrieved from published reports underwent both quantitative and qualitative analysis. Statistical Analysis was performed using Review Manager (Review Manager–RevMan, 2020) and represented graphically. Random effect model based on the logarithm of the hazard ratio (HR) weighted by the inverse of the variance was used for combining results from the individual data. HR and CIs were used to evaluate the association of biomarkers with the OS. Statistical heterogeneity of included studies was assessed by the I^2 statistics and chi-square test, and I^2 value > 50% or $H_{eterogeneity}$, 0.05 indicated substantial heterogeneity.

HR and CIs of multivariate analysis were selected preferentially if both univariate and multivariate analysis data was specified in the publication. In some cases, where HR and CIs were not given in the publications, they were calculated from the Kaplan-Meier curves using Enguage Digitizer software with reported methods [19].

3. Results

Of 1831 screened publications (1827 from database searches, 4 from in-publication citations), 26 studies were identified as eligible for inclusion in this review and analyses. The process of search, inclusion and exclusion of studies is presented in Figure 1 [16]. 26 studies met inclusion criteria reporting predictive and prognostic role of molecular biomarkers in GBM patients (Table 1). Meta-analysis was performed on clinically relevant biomarker information available for GBM patients in included publications. The main biomarkers analysed here included *MGMT* methylation (14 studies), *IDH1* mutation (5 studies) and EGFR expression/amplification (5 studies). Due to limited data for meta-analyses, association with OS of GBM patients was qualitatively evaluated for seven "circulatory biomarker" studies as well as one study with "cytokine and immune signature biomarkers".

3.1. Risk of Bias Assessment and Sensitivity Analysis

The risk of bias quality assessment using QUIP tools is summarised in Table 2. Studies that have more than one domain assessed as high risk of bias were not included in the meta-analysis. Of 19 studies included in the meta-analysis, one study was assessed as high risk of bias for the study participation domain for not defining the inclusion criteria. This was still included in meta-analysis, as this was considered of low impact on analyses outcome. One study was assessed as high risk of bias for the study attrition domain. The included study was assessed as high risk of bias due to a smaller patient cohort size available for biomarker analyses (MGMT methylation was assessed for only 28 patients out of 53 included in that study). Studies included in the meta-analysis were either assessed as low risk of bias, moderate or unclear for the prognostic factor measurement domain and the confounding factors measurement domain. Six studies with high risk of bias for the outcome measurement domain were included in meta-analysis after carefully extracting the OS data and its association with biomarkers, while the other 4 studies did not have enough survival data for inclusion in the meta-analysis and are described qualitatively. All the studies included in the meta-analysis were assessed as low risk of bias for the statistical analyses' domain.

Sensitivity analysis was performed manually in RevMan by taking out one study at a time to determine the effect of that study on the overall association of biomarkers with OS.

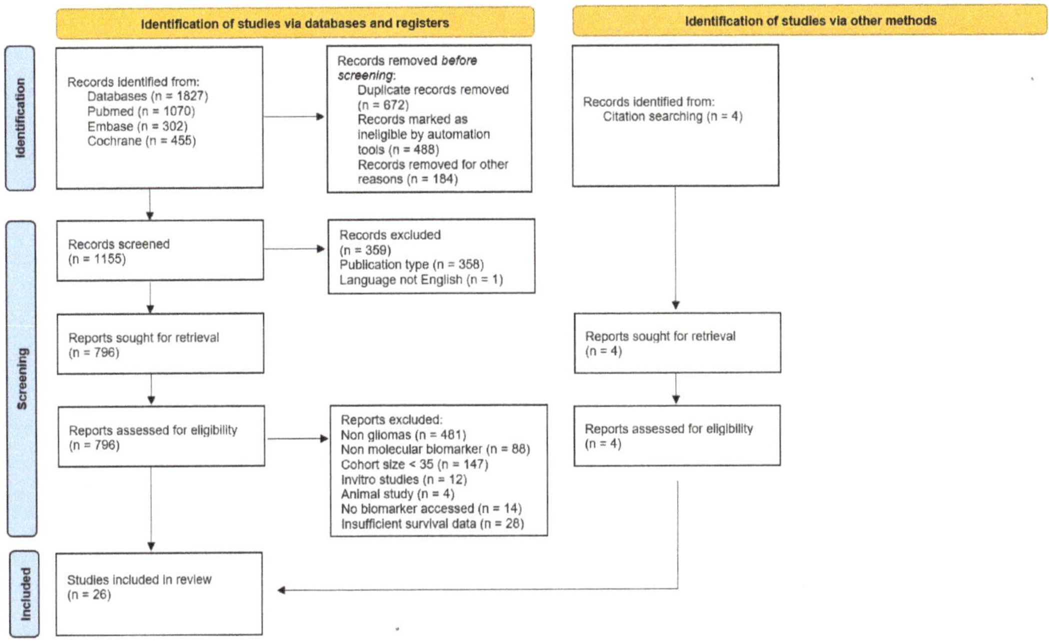

Figure 1. A PRISMA flow diagram of literature screening and exclusion criteria.

Table 1. The characteristics of included studies.

Study	Published Year	Histology	Study Design	Treatment	Median Age	No. of Patients	Endpoint/ Outcome	Biomarker Analysed
Abdullah et al. [20]	2015	Newly diagnosed GBM	R	Adjuvant chemotherapy $^{\alpha}$ + Radiotherapy	83	58	OS	EGFR, TP53
Accomando et al. [21]	2020	Recurrent GBM	R	Retroviral treatment Toca 511 + Toca FC	55	56	OS	Tumour immune signature and cytokine signature
Batchelor et al. [22]	2013	Newly diagnosed GBM	RCT	TKI (cediranib) + chemoradiotherapy	57	46	OS	EGFR, PDGFRA, MET and circulatory biomarkers
Batchelor et al. [23]	2017	Recurrent GBM	Clinical trial	TKI (tandutinib)	56	56	OS	circulatory biomarkers
Beije et al. [24]	2015	Recurrent GBM	P	TKI (bev/lomustine)	57	141	OS	CECs (circulatory epithelial cells)

Table 1. *Cont.*

Study	Published Year	Histology	Study Design	Treatment	Median Age	No. of Patients	Endpoint/ Outcome	Biomarker Analysed
Bloch et al. [25]	2017	Newly diagnosed GBM	RCT	Immunotherapy (HSPPC-96Prophage) + chemoradiotherapy	58	46	OS	MGMT, PDL1
Butowski et al. [26]	2011	Newly diagnosed GBM	RCT	TKI (enzastaurin) + chemoradiotherapy	57	66	OS	MGMT
Carvalho et al. [27]	2021	Recurrent GBM	R	TKI (bev + irinotecan)	59	40	OS	c-MET, VEGFR2
Cloughesy et al. [28]	2017	Recurrent GBM	RCT	TKI Arm 1 = (onartuzumab + bev) Arm 2 = (Pla + bev)	Arm1 = 57 Arm2 = 55	Arm1 = 64 Arm 2 = 65	OS	MGMT
Collins et al. [29]	2014	Recurrent GBM	R	Alkylating agents (TMZ/PVC)	53	309	OS	IDH1
Erdem-Eraslan et al. [30]	2016	Recurrent GBM	R	TKI (lomustine/bev)	57	148	OS	MGMT, IDH1
Galanis et al. [31]	2013	Recurrent GBM	Clinical trial	TKI (bev/sorafenib)	55	54	OS	Circulatory biomarkers, CECS
Gerstner et al. [32]	2015	Recurrent GBM	Cohort study	TKI (cediranib maleate + cilengitide)	54	45	OS	Circulatory Biomarkers
Han et al. [33]	2014	Recurrent GBM	Cohort study	Alkylating agents (TMZ)	53	60	OS	MGMT
Jan et al. [34]	2018	Newly diagnosed GBM	Cohort study	Immunotherapy (ADCTA vaccine) + chemoradiotherapy	51.8 *	ADCTA = 27 Reference = 20	OS	MGMT, IDH1
Lotsch et al. [35]	2013	Newly diagnosed GBM	R	NA	60 *	100	OS	MGMT, IDH1
Lee et al. [36]	2015	Newly diagnosed GBM	RCT	TKI (vandatinib) + chemoradiotherapy	Arm1 = 55 Arm2 = 59	Arm1 = 36 Arm 2 = 70	OS	Circulatory biomarkers
Michaelsen et al. [37]	2013	Newly diagnosed GBM	P	chemoradiotherapy	59.2	225	OS	MGMT, EGFR, TP53
Omuro et al. [38]	2014	Newly diagnosed GBM	Clinical trial	TKI (bev)+ chemoradiotherapy	55	40	OS	MGMT
Reardon et al. [15]	2018	Recurrent GBM	Cohort study	TKI (trebananib/bev)	Cohort 1 = 61.9 Cohort 2 = 63.1	Cohort1 = 11 Cohort 2 = 37	OS	Circulatory biomarkers, MGMT, IDH1
Reardon et al. [39]	2020	Recurrent GBM	RCT	TKI (nivolumab/bev)	Arm 1 = 55.5 Arm 2 = 55	Arm1 = 184 Arm 2 = 185	OS	MGMT
Roodakker et al. [40]	2016	Newly diagnosed GBM	R	Chemoradiotherapy	N1 = 57 * N2 $\geq$ 60 N3 $\leq$ 60	N1 = 86 N2 = 174 N3 = 80	OS	MGMT
Srividya et al. [41]	2010	Newly diagnosed GBM	P	Chemoradiotherapy	47	140	OS	EGFR
Tini et al. [42]	2015	NA	R	Chemoradiotherapy	63	144	OS	EGFR, MGMT

Table 1. *Cont.*

Study	Published Year	Histology	Study Design	Treatment	Median Age	No. of Patients	Endpoint/ Outcome	Biomarker Analysed
Weller et al. [43]	2015	Recurrent GBM	RCT	Alkylating agents (TMZ)	Arm 1 = 58 Arm 2 = 56	Arml = 52 Arm 2 = 53	OS	MGMT
Wirsching et al. [44]	2018	Newly diagnosed GBM	Clinical trial	TKI (bev) + rad	70	75	OS	MGMT

Studies are labelled as the last name of the first author and presented in alphabetical order. Abbreviations: Toca 511 = Vocimagene amiretrorepvector; Toca FC = 5-fluorocytosine; TMZ = Temozolomide, rad = radiation therapy, TKI = Tyrosine kinase inhibitors, bev = bevacizumab, Pla = Placebo, PVC = (procarbazine, CCNU (1-(2-chloroethyl)-3-cyclohexyl-1-nitrosourea) and vincristine, ADCTA = autologous dendritic cell tumour antigen vaccine, chemoradiotherapy = radiation therapy + chemotherapy with TMZ; R = Retrospective study, P = prospective study, RCT = Randomised control trial, OS = Overall survival; * = mean age; # = mean + STD DEV; N1 = screening cohort, N2 and N3 = Validation Cohort. $^{\alpha}$ = Chemotherapeutic drug not specified. NA = Treatment modality not given in the study.

Table 2. Risk of bias assessment.

Study ID	1.5 Summary of Study Participation	2.4 Summary Study Attrition	3.4 Summary of Prognostic Factor Measurement	4.4 Outcome Measurement Summary	5.3 Summary of Confounding Factors	6.4 Statistical Analysis and Reporting Summary
Abdullah 2015 [20]	Low	NA	Low	High	Moderate	Low
Accomando 2020 [21]	Low	NA	High	High	High	High
Batchelor 2013 [22]	Low	Low	Low	High	Low	Low
Batchelor 2017 [23]	Low	Unclear	Low	High	High	Low
Beije 2015 [24]	Low	Unclear	Low	Low	High	Low
Bloch 2017 [25]	Low	Low	Low	Low	Low	Low
Butowski 2011 [26]	Low	Low	Unclear	Low	Low	Low
Carvalho 2021 [27]	Low	NA	Low	Low	Low	Low
Cloughesy 2017 [28]	Low	Low	Low	High	Low	Low
Collins 2014 [29]	Low	NA	Low	Low	Low	Low
Erdem-Eraslan 2016 [30]	Low	NA	Low	Low	Low	Low
Galanis 2013 [31]	Low	Low	Low	Low	Low	Low
Gerstner 2015 [32]	Low	Low	Low	Low	Low	Low
Han 2014 [33]	Low	High	Low	Low	Low	Low
Jan-18 [34]	Low	Low	Low	Low	Low	Low
LÃ¶tsch 2013 [35]	High	NA	Low	Low	Low	Low
Lee 2015 [36]	Low	Low	Low	High	Low	Low
Michaelsen 2013 [37]	Low	Low	Low	Low	Low	Low
Omuro 2014 [38]	Low	Low	Low	High	Low	Low
Reardon 2018 [15]	Low	Low	Low	Low	Low	Low
Reardon 2020 [39]	Low	Low	High	Low	Low	Low
Roodakker 2016 [40]	Low	NA	Low	High	Low	Low
Srividya 2010 [41]	Low	Low	Low	Low	Low	Low
Tini 2015 [42]	Low	NA	Low	High	Low	Low
Weller 2015 [43]	Low	Low	Low	Low	Low	Low
Wirsching 2018 [44]	Low	Low	Unclear	High	Low	Low

Risk of bias accessed by QUIPS tool. NA = not applicable (domain not accessed for retrospective studies).

3.1.1. Quantitative Analysis

MGMT Methylation

MGMT methylation data from fourteen studies, involving a total of 1231 patients with differing treatment regimens were included in the analysis for association of OS and MGMT status. The MGMT methylation status was determined in 10 out of 14 studies by methylation specific PCR [15,26,28,30,33–35,38,42–44]. Pyrosequencing was used in one study [40], and 3 studies did not report the methodology of *MGMT* methylation assessment [15,25,39].

Overall, *MGMT* methylation showed a significant association with better OS in GBM patients with a combined HR ratio of 1.66 (95% CI 1.32–2.09, $p < 0.0001$, random effect; Figure 2). Since the therapeutic intervention varied for the 14 studies, sub-group analysis based on therapy was also performed to evaluate differential association of *MGMT* promoter methylation with OS (Figure 2).

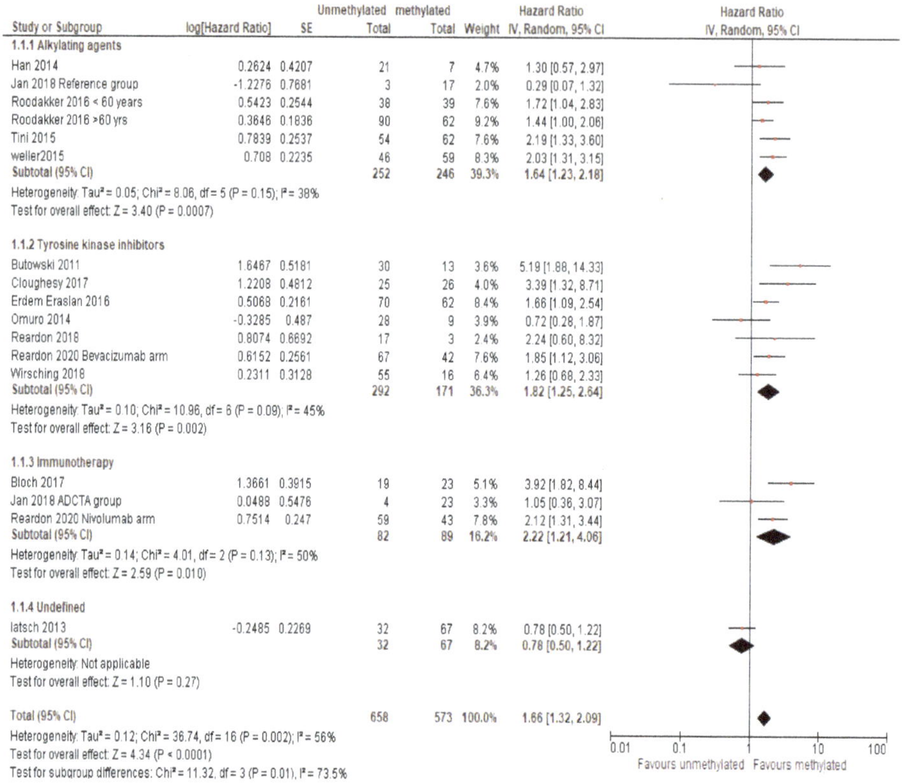

Figure 2. A forest plot demonstrating the association of *MGMT* methylation status with OS [15,25,26,28,30,33–35,38–40,42–44]. Abbreviations: SE: standard error; CI: confidence interval, bev= bevacizumab, niv= nivolumab < 60= < 60 years, > 60= >60 years. Size of the red square indicates the relative weight of the study as it contributes to the results of the overall comparison. The diamond at the bottom of the forest plot shows the result when all the individual studies are combined and averaged. The effect measure used was HR, where values greater than 1.0 indicate that patients with *MGMT* methylation has low risk of mortality than patients with unmethylated *MGMT* and vice versa for values less than 1.0.

As expected, in patients treated with alkylating agents, there was a significant association of *MGMT* methylation with better OS, with a pooled HR ratio of 1.64 (95% CI 1.23–2.18; $p = 0.0007$). Another subgroup of patients was treated with TKIs (with or without alkylating agent in combination) also revealed significant association of *MGMT* methylation with OS, with a pooled HR ratio of 1.82 (95% CI 1.25–2.64; $p = 0.002$). Similar results were observed in the subgroup of patients receiving immunotherapy with or without alkylating combination, with a pooled HR ratio of 2.22 (95% CI 1.21–4.06; $p = 0.01$), (Figure 2). Sensitivity analysis was performed for two different treatment types (alkylating agents and tyrosine kinase inhibitors) by removing one study at a time. There was no change found in the overall significance of association of biomarker with overall survival (Supplementary Table S1).

IDH1 Mutation

Five studies investigated *IDH1* status in 541 patients (480 with *IDH1* wildtype and 61 with *IDH1* mutation) [15,29,30,34,35]. Treatments in this cohort included alkylating agents [29], TKIs [15,30] and immunotherapy in combination with alkylating agent [34]. One study did not specify the treatment [35]. *IDH1* mutation was significantly associated with longer OS in GBM patients irrespective of the therapeutic intervention. The pooled HR ratio was 2.37 (95% CI 1.81–3.12; $p < 0.00001$) (Figure 3). No significant effect on data outcome was observed after performing a sensitivity analysis (Supplementary Table S2).

Study or Subgroup	log[Hazard Ratio]	SE	IDH1 WT Total	IDH1 MT Total	Weight	Hazard Ratio IV, Random, 95% CI
Collins 2014	0.8416	0.1527	191	41	83.0%	2.32 [1.72, 3.13]
Erdem Eraslan 2016	0.7885	0.5423	119	8	6.6%	2.20 [0.76, 6.37]
Jan 2018 ADCTA group	-2.9759	2.5086	24	3	0.3%	0.05 [0.00, 6.96]
Jan 2018 Reference group	2.0128	1.4325	19	1	0.9%	7.48 [0.45, 124.02]
Iatsch 2013	1.5239	0.5916	94	6	5.5%	4.59 [1.44, 14.63]
Reardon 2018	0.5365	0.7261	33	2	3.7%	1.71 [0.41, 7.10]
Total (95% CI)			**480**	**61**	**100.0%**	**2.37 [1.81, 3.12]**

Heterogeneity: Tau² = 0.00; Chi² = 4.47, df = 5 (P = 0.48); I² = 0%
Test for overall effect: Z = 6.21 (P < 0.00001)

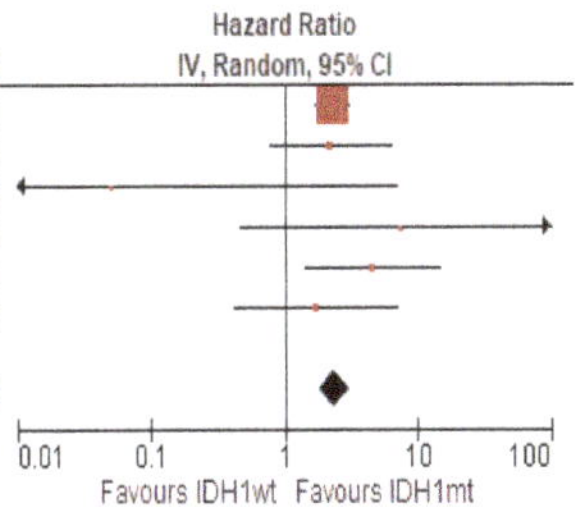

Figure 3. The association of OS with IDH1 mutation status [15,29,30,34,35]. Abbreviations; WT = wild type, MT = mutant. Size of the red square indicates the relative weight of the study as it contributes to the results of the overall comparison. The diamond at the bottom of the forest plot shows the result when all the individual studies are combined and averaged. The effect measure used was HR, where values greater than 1.0 indicate that patients with IDH1 MT has low risk of mortality than patients with IDH1 WT and vice versa for values less than 1.0.

EGFR Amplification or Overexpression of EGFR Protein

Five studies reported *EGFR* amplification and/or high expression of EGFR protein in a total of 575 patients [20,22,37,41,42]. Four studies included in the analysis investigated the association of high expression of EGFR [20,37,41,42] with OS and one study investigated the association of *EGFR* amplification with OS [22]. Treatment in this cohort included chemoradiotherapy (TMZ and radiotherapy) in 3 studies [37,41,42] and TKI with chemoradiotherapy in one study [22]. Treatment modality was not clearly defined in one study [20]. OS was not significantly associated with EGFR status, with a combined HR ratio of 1.31 (95% CI 0.96–1.79; $p = 0.08$) (Figure 4), possibly due to inadequate statistical power. Sensitivity analysis demonstrates the significant effect of one study [41] on the overall outcome on the association of EGFR with OS (Supplementary Table S3).

Study or Subgroup	log[Hazard Ratio]	SE	EGFR -ve Total	EGFR +ve Total	Weight	Hazard Ratio IV, Random, 95% CI
*Batchelor 2013	0.5596	0.3599	24	21	11.5%	1.75 [0.86, 3.54]
Abdullah 2015	0.678	0.3114	27	20	13.6%	1.97 [1.07, 3.63]
Michaelson 2013	0.0488	0.1531	54	145	22.9%	1.05 [0.78, 1.42]
Srividya 2010	-0.0408	0.0217	50	90	29.2%	0.96 [0.92, 1.00]
Tini 2015	0.5128	0.1558	51	93	22.8%	1.67 [1.23, 2.27]
Total (95% CI)			**206**	**369**	**100.0%**	**1.31 [0.97, 1.78]**

Heterogeneity: Tau² = 0.08; Chi² = 20.41, df = 4 (P = 0.0004); I² = 80%
Test for overall effect: Z = 1.74 (P = 0.08)

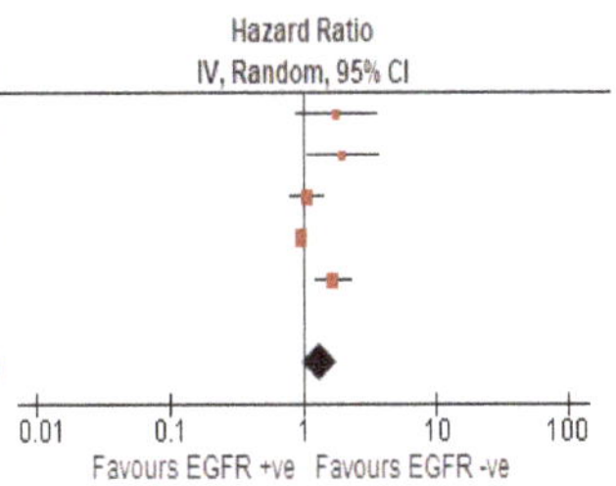

Figure 4. The association of OS with *EGFR* amplification or EGFR overexpression [20,22,37,41,42]. * Treatment type = chemoradiotherapy + TKI; Size of the red square indicates the relative weight of the study as it contributes to the results of the overall comparison. The diamond at the bottom of the forest plot shows the result when all the individual studies are combined and averaged. The effect measure used was HR, where values greater than 1.0 indicate that patients with *EGFR* amplification or EGFR overexpression has high risk of mortality than patients with no *EGFR* amplification or EGFR overexpression and vice versa for values less than 1.0. Note: expression of EGFR was determined by immunohistochemistry.

3.1.2. Qualitative Analysis

Our broad search for molecular biomarkers in GBM produced a set of candidates that may have value in specific trial treatment settings. However, data were insufficient for a meta-analysis, and is summarised in Table 3 and briefly discussed below.

Table 3. The association of other biomarkers with treatment response in GBM patients.

Study	Treatment	Biomarker	Outcome
Batchelor et al. 2013 [22]	Chemoradiation + cediranib	sVEGFR1	High plasma sVEGFR1 at treatment cycle 2/day 1: poor PFS & OS ($p < 0.05$)
Batchelor et al. 2017 [23]	tanutinib	sVEGFR1, plasma PlGF	1. Decrease in sVEGFR1 at treatment cycle 2/day 1: longer PFS & OS ($p = 0.05$; 0.01 respectively) 2. Decrease in plasma PlGF at day 10: longer PFS ($p = 0.04$)
Lee et al. 2015 [36]	Chemoradiation + vandatinib	sVEGFR1, plasma PlGF	1. Longitudinal sVEGFR1 increase: poor OS ($p < 0.05$) 2. Longitudinal PlGF increase: poor OS ($p < 0.05$)
Gerstner et al. 2015 [32]	cediranib maleate + cilengitide	Plasma PlGF	Early PIGF increase (at day 2): longer PFS ($p = 0.03$)
Reardon2018 [15]	trebananib/bevacizumab	Plasma VEGF and Interleukin-8 (IL-8) levels	1. High plasma VEGF: poor PFS & OS ($p < 0.005$) 2. High plasma IL-8: shorter OS ($p < 0.05$)
Beije et al. 2015 [24]	bevacizumab (avastin)/bevacizumab and lomustine/lomustine.	Circulatory endothelial cells (CECs)	For single agent lomustine treated patients with higher absolute CEC numbers after 4 and 6 weeks of treatment: longer OS ($p = 0.03$, $p = 0.004$ respectively) Absolute CEC numbers in patients receiving bevacizumab plus lomustine or bevacizumab single agent: no OS effect
Galanis et al. 2013 [31]	bevacizumab/sorafenib	Circulatory endothelial cells (CECs)	No correlation of baseline CEC values and 6 months PFS
Carvalho et al. [27]	bevacizumab	c-Met, VEGFR2	1. c-MET overexpression: TTP ($p = 0.05$) 2. VEGFR2 overexpression: Shorter TTP ($p = 0.009$) 3. Concomitant overexpression of c-Met and VEGFR2: worse TTP ($p = 0.001$) 4. Concomitant overexpression of c-Met and VEGFR2: worse OS ($p = 0.025$)
Accomando et al. [21]	Retroviral treatment Toca 511 + Toca FC	Pre-treatment tumour immune signature (in tumour microenvironment), post treatment Cytokine signature (in plasma)	1. Tumour immune signature was found to be higher in responders than non-responders ($p < 0.001$) 2. High cytokine signature: improved survival ($p < 0.05$)

Abbreviations: TTP = time to progression; Toca 511 = Vocimagene amiretrorepvector; Toca FC = 5-fluorocytosine; Tumour immune signature = Activated memory CD4 T cells * M1 macrophages/1 + Resting NK cells * M0 macrophages; Cytokine signature = E-selectin$_{max}$ * MIP-1β$_{max}$/1 + IL6$_{max}$; Max = maximum value of the 3 cytokines.

Tumour Immune Signature and Cytokine Signature

One study reported the molecular biomarkers associated with response to retroviral immunotherapy. Vocimagene amiretrorepvector (Toca 511) is a cancer selective, retroviral replicating vector that encodes cytosine deaminase. When administered, extended release 5-fluorocytosine (Toca FC) is converted by cytosine deaminase into the potent, short lived, chemotherapeutic agent, 5-fluorouracil, which diffuses into the tumour microenvironment from Toca 511–infected cells. Biomarkers that predicted the better clinical response to treatment in the TOCA 511/FC treated GBM patients were tumour immune signature and cytokine signature. Toca 511 and Toca FC cancer treatment has a putative mechanism of action that includes T cell–mediated antitumour immune activity, so the tumour immune signature based on the immune composition of the tumour micro-environment can potentially predict the clinical response in high grade glioma patients. Higher values of this signature indicate that more activated memory $CD4^+$ T cells, more M1 macrophages, fewer resting Natural killer cells (NK cells), and fewer M0 macrophages were detected in patient tumour tissue. This signature was found to be higher in responders than in non-responders (Wilcoxon rank-sum test, $p < 0.001$) [21].

The anti-tumour immune activity of TOCA 511/FC treatment can also be measured by cytokine levels from the patient's plasma samples. Accomando and colleagues measured a cytokine signature incorporating three cytokines (soluble E-selectin, Macrophage Inflammatory protein-1β and Interleukin-6) that were associated with the response to therapy and OS [21]. Increasing values of this cytokine signature indicate higher peak E-selectin, higher peak MIP-1β, and lower peak Interleukin-6 (IL-6) in peripheral blood during and after Toca 511 and Toca FC treatment [21]. A higher value of the signature was associated with improved survival ($p < 0.001$).

Circulatory Biomarkers

Eight studies included in this review reported the trial results of tyrosine kinase inhibitor therapies and the molecular biomarkers associated with response to treatment, including circulatory biomarkers. Circulatory biomarkers such as sVEGFR1, plasma PlGF and VEGF levels, and CECs are proposed as potential prognostic and predictive biomarkers in anti-VEGF therapies (Table 3). For the management of GBM which is characterised by high vascularisation and aberrantly high levels of VEGF expression, anti-VEGF therapies are being trialled [45]. sVEGFR1 is implicated as a negative regulator of the VEGF pathway and proposed as a resistance biomarker to anti-VEGF therapies in other solid cancers [46].

PlGF is another member of VEGF family, and its dynamics are now being considered as a potential pharmacodynamic biomarker to anti-VEGF therapy [47,48]. Overexpression of PlGF in preclinical models promotes tumour growth, which makes it an attractive therapeutic target [49].

Circulatory endothelial cells (CECs) are mature endothelial cells shed off the blood vessels as a result of vascular damage. Increased plasma levels of CECs are reported in cancer patients that corelate with VEGF levels. CECs may serve as a surrogate marker of anti-angiogenic activity that reflect the disease status and response to anti-angiogenic treatment [50].

Pharmacodynamics of blood based sVEGFR1, sVEGFR2, PlGF, VEGF, cytokine signature, and CECs may also be useful to monitor the target effect, tumour response and treatment outcome in response to anti-VEGF therapies and immunotherapy [15,21–24,31,32,36]. If these biomarkers indeed guide decision making to continue or terminate treatment in the early phases of a trial, benefit may be maximised.

Immunotherapies targeting the PDL1-PD1 axis have entered standard clinical practice for various solid cancers including (non-small cell lung cancer, gastric cancer, urothelial cancer, cervical cancer, and melanoma) [51–54]. Recent studies have shown the direct association of PDL1 expression with survival in GBM patients [55–57], although more studies are needed to evaluate benefit of immunotherapy in GBM.

4. Discussion

GBM is the most common and aggressive type of brain cancer and treatment options have not notably improved for decades. There are different molecular subtypes within GBM and, conceivably, targeting driver pathways or molecular "weaknesses" may lead to better patient outcomes. *MGMT* and *IDH1* are widely accepted biomarkers in the clinical context to provide prognostic or predictive information and their utilities are linked to the standard of care therapy. Here, we were interested in not only re-evaluating the utility of *MGMT* and *IDH1* but also other possible candidate biomarkers for their association with GBM patient OS in the setting of clinical trials using standard of care and other treatment modalities. Such biomarkers may add benefit to future clinical trials and better GBM patient management. Yet, perhaps not surprisingly, the best studied biomarkers, even in the clinical trials context, remain *MGMT, IDH1,* and EGFR.

MGMT methylation, as a prognostic and predictive biomarker of GBM, has been comprehensively studied previously [5,58]. Initially Stupp et al. provided evidence of association of *MGMT* promoter methylation to outcome in GBM patients treated with TMZ and radiation therapy versus radiation alone [3]. Further trials involving 206 GBM patients confirmed better survival outcomes in those with *MGMT* promoter methylation when treated with TMZ and radiation [59]. A previous meta-analysis which analysed 30 studies with the total of 2986 patients demonstrated *MGMT* methylation status as a prognostic factor in GBM patients showing significant association with better OS and progression free survival (PFS) for patient treatment with alkylating agents [5]. In our systematic review focusing on recent clinical trials (conducted in the last 10 years), we included 14 studies/1231 patients and investigated the association of *MGMT* promoter methylation with OS outcomes in GBM patients, irrespective of therapeutic intervention. Our analysis of *MGMT* methylation in GBM agrees with previous findings, manifesting a significant association of *MGMT* methylation with good OS in GBM patients. Interestingly, the survival benefit is not limited to patients treated with alkylating agents but was observed in all the GBM patients irrespective of treatment.

However, substantial heterogeneity was observed in the overall analysis of association of *MGMT* methylation with OS for the 14 included studies (I^2 = 56%), while this was smaller (I^2 = 38%) for studies focusing on alkylating agent treatments. We were also able to perform subgroup analysis based on the treatment type, and still found significant OS association with *MGMT* methylation. This observation is intriguing and suggests that while the close functional link between MGMT and alkylating agents would predict such a relationship, there may be more biological significance to MGMT methylation resulting in clinical benefits from other agents. Of the 7 studies included for the TKIs treatment group, 2 studies investigated newly diagnosed GBM patients who received TKIs therapy together with standard of care alkylating agents [26,38] and one study investigated newly diagnosed GBM patients treated with TKIs and radiation therapy [44]. The other 4 studies investigated the prognostic value of *MGMT* methylation in patients at first or second recurrence after standard therapy (chemotherapy with TMZ and radiation). In these 4 studies, patients were treated with either bevacizumab alone or in combination with other drugs [15,28,30,39]. The prognostic significance of MGMT methylation for progressive GBM patients treated with bevacizumab has been reported previously [60,61]. Wick et al. reported MGMT methylation as positive prognostic biomarker in the recurrent GBM patients treated with either bevacizumab or combination of bevacizumab and lomustine (HR: 0.48; $p < 0.001$) [60]. Similar findings were reported by Gleeson et al. with better OS observed in patients with MGMT methylated tumours as compared to those with unmethylated tumours (HR:0.61, $p = 0.027$) [61].

Three studies were included in the subgroup analysis of prognostic significance of *MGMT* methylation in patients receiving immunotherapy [25,34,39]. Two of these studies were conducted on newly diagnosed patients who also received standard of care along with immunotherapy [25,34] and one study enrolled patients at their first recurrence after standard treatment with TMZ and radiation [39]. This study compared the OS survival benefit in patients treated with nivolumab (PD-1 immune check point inhibitor) vs bevacizumab. No statistical difference was observed in the risk of death between groups (HR, 1.04; 95% CI, 0.83–1.30, $p = 0.76$). However, *MGMT* methylation status was prognostic in both groups. Taken together, these findings suggest MGMT methylation as strong prognostic biomarker in both newly diagnosed and recurrent GBM patients regardless of treatment intervention. However, the association of MGMT methylation with survival may still be functionally linked to alkylating agents and radiotherapy received either in parallel or prior to the trial.

IDH1/2 catalyses the reversible oxidation of isocitrate to yield α-ketoglutarate with simultaneous reduction of NADP+ to NADPH. This NADPH produced by the cells provides a cellular defence against intracellular oxidative damage [62]. *IDH1* mutations are found in approximately 12% of GBM patients [10]. Mutation in IDH1 is favourable for OS and an independent prognostic GBM biomarker [63]. Our analysis adds support to these findings [15,29,30,34,35].

EGFR, a receptor tyrosine kinase, upstream of central signalling pathways such as PI3K/AKT and RAS/RAF/MEK/MAPK pathways, is often altered in cancer [64]. Alterations and overexpression of EGFR are often linked with oncogenesis in GBM and are widely investigated in this context [65,66]. *EGFR* amplification and/or overexpression is observed in 50–60% of GBM [67,68]. Past studies which explored the prognostic significance of *EGFR* mutations, amplification and/or overexpression in GBM reported conflicting results [8,67,69–71]. While some studies found association of EGFR overexpression and amplification with poor prognosis [8,67], others did not find prognostic value of EGFR in GBM [69,71]. In addition, EGFR is also considered a potential target for newer therapies in GBM. However, the results from clinical trials targeting EGFR through various small kinase inhibitors (erlotinib, gefitinib, afatinib, and lapatinib) were disappointing, at least in part due to poor drug penetrance through the blood–brain barrier. However, adaptive reliance on redundant pathways to overcome EGFR inhibition has been proposed [72] in line with observations in other cancers treated with EGFR inhibitors.

In our meta-analysis, we included 5 studies with 575 patients and did not find significant association of *EGFR* amplification and or overexpression with OS in GBM patients. However, substantial heterogeneity was found among the included studies ($I^2 = 81\%$). Factors that may contribute to the heterogeneity include methods of determination of EGFR expression and amplification, therapeutic intervention, first diagnosis vs recurrence, median age, ethnic diversity and experimental design. Four studies included in this review assessed EGFR overexpression by immunohistochemistry [20,37,41,42], while amplification of the *EGFR* gene was assessed by fluorescence in-situ hybridisation (FISH) in another study [22]. Among the included studies, 4 studies investigated the overexpression of EGFR and its prognostic value in the patients receiving standard of care [20,37,41,42]. Results were diverging with some showing strong association of EGFR overexpression with worse survival while others produced no or limited association with survival [20,37,41,42].

Of note, one study investigated the association of EGFR amplification with OS in patients receiving standard of care treatment in combination with anti-VEGF TKI cediranib [22]. This study demonstrated improved survival in a subset of newly diagnosed GBM patients with improved tumour blood perfusion after receiving standard 6 weeks of fractionated radiation along with daily temozolomide and cediranib. They found an interesting correlation of EGFR amplification with lack of increase in perfusion after treatment. EGFR amplification was thus a negative prognostic factor for the patients treated with this combination therapy. Further detailed investigation is needed to determine whether EGFR is merely a poor prognostic variable or if it is associated with the vascular function after anti-VEGF therapies.

While our data based on limited patient numbers suggests no statistical association of EGFR amplification/overexpression, more homogeneous studies and larger patient cohorts are needed to clarify the prognostic and predictive significance of EGFR in GBM.

Other important biomarkers discussed in the review include the circulatory biomarkers (sVEGFR1, sVEGFR2, PlGF, VEGF, cytokine signature, and CECs). These pharmacodynamic biomarkers can be used to examine the target effect, tumour response and treatment outcome for drugs targeting tyrosine kinase receptors [47]. The closer examination of these biomarkers in the early phases of trials may be helpful in directing management decisions.

5. Conclusions

In conclusion, our meta-analysis confirms the positive prognostic significance of MGMT methylation and IDH1 mutation in GBM patients regardless of treatment type. The prognostic significance of EGFR amplification and overexpression still needs clarification. We also highlighted potential biomarkers, especially easily accessible circulating blood-based markers, which, however, need thorough future evaluation of their prognostic and/or predictive utility for GBM in certain therapy settings. This study also highlights the key knowledge gap in the literature which did not produce sufficient data to perform meta-analysis on the biomarkers associated with novel therapies.

6. Limitations

This review has several limitations that need to be considered. Firstly, we deliberately used broad search terms to retrieve all the studies evaluating prognostic and predictive biomarkers with standard of care or novel treatment modalities for GBM patients. Although considerable numbers of studies were identified in our search, a large proportion of studies were excluded due to their small cohort size (n < 35) and inclusion of patients with brain metastases. Secondly, insufficient data on novel biomarkers precluded a meta-analysis and we are therefore unable to provide evidence for their prognostic or predictive value.

The low number of studies included in the meta-analysis of EGFR as biomarker was another limitation of our review, so we combined EGFR amplification and overexpression to increase sample size. Further analysis with a greater number of studies and homogeneous biomarker detection is required to clarify evidence towards the prognostic significance of EGFR in GBM.

Another limitation was the inclusion of clinical trials that showed no survival benefits of trial drugs over standard of care. Thus, biomarkers evaluated in this context hold no value for prediction of response to the trial treatment over standard of care.

Finally, the variation in methodologies for molecular investigation could confound any statistical associations, either in favour of or against the trial hypothesis. Conceivably, the use of 'better' methods of determining molecular alterations, and optimised tissues (biopsy vs circulating) in carefully conducted trials with rigorous sampling and storage conditions, and sufficient follow-up with many longitudinal samples, even if not of large size, can provide good evidence of predictive and prognostic significance.

Supplementary Materials: The supporting information can be downloaded at: https://www.mdpi.com/article/10.3390/ijms23168835/s1.

Author Contributions: H.S., B.P., T.M.B., P.d.S. and T.L.R. conceptualised the study. H.S. and B.P. performed the initial screening. H.S. extracted and interpreted the data. H.S. and B.P. performed quality assessment using QUIPS tool. H.S. and Y.M. performed the statistical analysis. H.S. headed the writing of the manuscript in discussion with other authors. All authors have read and agreed to the published version of the manuscript.

Funding: H.S. is recipient of a Research Training Program (RTP) scholarship. T.L.R is an Irene and Arnold Vitocco Cancer Research Fellow. B.P. is recipient of a Sydney Partnership for Health, Education, Research and Enterprise (SPHERE) Cancer Clinical Academic Group, ECR Brain Cancer Seed Grant.

Institutional Review Board Statement: Not applicable.

Informed Consent Statement: Not applicable.

Data Availability Statement: Not applicable.

Conflicts of Interest: The authors declare no conflict of interest.

References

1. Miller, K.D.; Ostrom, Q.T.; Kruchko, C.; Patil, N.; Tihan, T.; Cioffi, G.; Fuchs, H.E.; Waite, K.A.; Jemal, A.; Siegel, R.L.; et al. Brain and other central nervous system tumor statistics, 2021. *CA A Cancer J. Clin.* **2021**, *71*, 381–406. [CrossRef] [PubMed]
2. Ohgaki, H.; Kleihues, P. The definition of primary and secondary glioblastoma. *Clin. Cancer Res.* **2013**, *19*, 764–772. [CrossRef] [PubMed]
3. Stupp, R.; Mason, W.P.; van den Bent, M.J.; Weller, M.; Fisher, B.; Taphoorn, M.J.; Belanger, K.; Brandes, A.A.; Marosi, C.; Bogdahn, U.; et al. Radiotherapy plus concomitant and adjuvant temozolomide for glioblastoma. *N. Engl. J. Med.* **2005**, *352*, 987–996. [CrossRef] [PubMed]
4. Roy, S.; Lahiri, D.; Maji, T.; Biswas, J. Recurrent Glioblastoma: Where we stand. *S. Asian J. Cancer* **2015**, *4*, 163–173. [CrossRef] [PubMed]
5. Zhang, K.; Wang, X.-Q.; Zhou, B.; Zhang, L. The prognostic value of MGMT promoter methylation in Glioblastoma multiforme: A meta-analysis. *Fam. Cancer* **2013**, *12*, 449–458. [CrossRef]
6. Christians, A.; Adel-Horowski, A.; Banan, R.; Lehmann, U.; Bartels, S.; Behling, F.; Barrantes-Freer, A.; Stadelmann, C.; Rohde, V.; Stockhammer, F.; et al. The prognostic role of IDH mutations in homogeneously treated patients with anaplastic astrocytomas and glioblastomas. *Acta Neuropathol. Commun.* **2019**, *7*, 156. [CrossRef] [PubMed]
7. Powter, B.; Jeffreys, S.A.; Sareen, H.; Cooper, A.; Brungs, D.; Po, J.; Roberts, T.; Koh, E.-S.; Scott, K.F.; Sajinovic, M.; et al. Human TERT promoter mutations as a prognostic biomarker in glioma. *J. Cancer Res. Clin. Oncol.* **2021**, *147*, 1007–1017. [CrossRef] [PubMed]
8. Li, J.; Liang, R.; Song, C.; Xiang, Y.; Liu, Y. Prognostic significance of epidermal growth factor receptor expression in glioma patients. *Onco Targets Ther.* **2018**, *11*, 731–742. [CrossRef]
9. Feldheim, J.; Kessler, A.F.; Monoranu, C.M.; Ernestus, R.-I.; Löhr, M.; Hagemann, C. Changes of O(6)-Methylguanine DNA Methyltransferase (MGMT) Promoter Methylation in Glioblastoma Relapse-A Meta-Analysis Type Literature Review. *Cancers* **2019**, *11*, 1837. [CrossRef] [PubMed]
10. Han, S.; Liu, Y.; Cai, S.J.; Qian, M.; Ding, J.; Larion, M.; Gilbert, M.R.; Yang, C. IDH mutation in glioma: Molecular mechanisms and potential therapeutic targets. *Br. J. Cancer* **2020**, *122*, 1580–1589. [CrossRef]
11. Ferguson, S.D.; Xiu, J.; Weathers, S.-P.; Zhou, S.; Kesari, S.; Weiss, S.E.; Verhaak, R.G.; Hohl, R.J.; Barger, G.R.; Reddy, S.K.; et al. GBM-associated mutations and altered protein expression are more common in young patients. *Oncotarget* **2016**, *7*, 69466–69478. [CrossRef] [PubMed]
12. An, Z.; Aksoy, O.; Zheng, T.; Fan, Q.-W.; Weiss, W.A. Epidermal growth factor receptor and EGFRvIII in glioblastoma: Signaling pathways and targeted therapies. *Oncogene* **2018**, *37*, 1561–1575. [CrossRef] [PubMed]
13. Xu, C.; Wu, X.; Zhu, J. VEGF promotes proliferation of human glioblastoma multiforme stem-like cells through VEGF receptor 2. *Sci. World J.* **2013**, *2013*, 417413. [CrossRef]
14. Chinot, O.L.; Wick, W.; Mason, W.; Henriksson, R.; Saran, F.; Nishikawa, R.; Carpentier, A.F.; Hoang-Xuan, K.; Kavan, P.; Cernea, D.; et al. Bevacizumab plus Radiotherapy–Temozolomide for Newly Diagnosed Glioblastoma. *N. Engl. J. Med.* **2014**, *370*, 709–722. [CrossRef] [PubMed]
15. Reardon, D.A.; Lassman, A.B.; Schiff, D.; Yunus, S.A.; Gerstner, E.R.; Cloughesy, T.F.; Lee, E.Q.; Gaffey, S.C.; Barrs, J.; Bruno, J.; et al. Phase 2 and biomarker study of trebananib, an angiopoietin-blocking peptibody, with and without bevacizumab for patients with recurrent glioblastoma. *Cancer* **2018**, *124*, 1438–1448. [CrossRef] [PubMed]
16. Page, M.J.; McKenzie, J.E.; Bossuyt, P.M.; Boutron, I.; Hoffmann, T.C.; Mulrow, C.D.; Shamseer, L.; Tetzlaff, J.M.; Akl, E.A.; Brennan, S.E.; et al. The PRISMA 2020 statement: An updated guideline for reporting systematic reviews. *BMJ* **2021**, *10*, 89. [CrossRef]
17. Ouzzani, M.; Hammady, H.; Fedorowicz, Z.; Elmagarmid, A. Rayyan—a web and mobile app for systematic reviews. *Syst. Rev.* **2016**, *5*, 210. [CrossRef] [PubMed]
18. Hayden, J.A.; van der Windt, D.A.; Cartwright, J.L.; Côté, P.; Bombardier, C. Assessing bias in studies of prognostic factors. *Ann. Intern. Med.* **2013**, *158*, 280–286. [CrossRef]
19. Tierney, J.F.; Stewart, L.A.; Ghersi, D.; Burdett, S.; Sydes, M.R. Practical methods for incorporating summary time-to-event data into meta-analysis. *Trials* **2007**, *8*, 16. [CrossRef] [PubMed]
20. Abdullah, K.G.; Ramayya, A.; Thawani, J.P.; Macyszyn, L.; Martinez-Lage, M.; O'Rourke, D.M.; Brem, S. Factors Associated with Increased Survival after Surgical Resection of Glioblastoma in Octogenarians. *PLoS ONE* **2015**, *10*, e0127202. [CrossRef]
21. Accomando, W.P.; Rao, A.R.; Hogan, D.J.; Newman, A.M.; Nakao, A.; Alizadeh, A.A.; Diehn, M.; Diago, O.R.; Gammon, D.; Haghighi, A.; et al. Molecular and Immunologic Signatures are Related to Clinical Benefit from Treatment with Vocimagene Amiretrorepvec (Toca 511) and 5-Fluorocytosine (Toca FC) in Patients with Glioma. *Clin. Cancer Res.* **2020**, *26*, 6176–6186. [CrossRef] [PubMed]
22. Batchelor, T.T.; Gerstner, E.R.; Emblem, K.E.; Duda, D.G.; Kalpathy-Cramer, J.; Snuderl, M.; Ancukiewicz, M.; Polaskova, P.; Pinho, M.C.; Jennings, D.; et al. Improved tumor oxygenation and survival in glioblastoma patients who show increased blood perfusion after cediranib and chemoradiation. *Proc. Natl. Acad. Sci. USA* **2013**, *110*, 19059–19064. [CrossRef] [PubMed]

23. Batchelor, T.T.; Gerstner, E.R.; Ye, X.; Desideri, S.; Duda, D.G.; Peereboom, D.; Lesser, G.J.; Chowdhary, S.; Wen, P.Y.; Grossman, S.; et al. Feasibility, phase I, and phase II studies of tandutinib, an oral platelet-derived growth factor receptor-β tyrosine kinase inhibitor, in patients with recurrent glioblastoma. *Neuro Oncol.* **2017**, *19*, 567–575. [CrossRef] [PubMed]
24. Beije, N.; Kraan, J.; Taal, W.; van der Holt, B.; Oosterkamp, H.M.; Walenkamp, A.M.; Beerepoot, L.; Hanse, M.; van Linde, M.E.; Otten, A.; et al. Prognostic value and kinetics of circulating endothelial cells in patients with recurrent glioblastoma randomised to bevacizumab plus lomustine, bevacizumab single agent or lomustine single agent. A report from the Dutch Neuro-Oncology Group BELOB trial. *Br. J. Cancer* **2015**, *113*, 226–231. [CrossRef] [PubMed]
25. Bloch, O.; Lim, M.; Sughrue, M.E.; Komotar, R.J.; Abrahams, J.M.; O'Rourke, D.M.; D'Ambrosio, A.; Bruce, J.N.; Parsa, A.T. Autologous Heat Shock Protein Peptide Vaccination for Newly Diagnosed Glioblastoma: Impact of Peripheral PD-L1 Expression on Response to Therapy. *Clin. Cancer Res. Off. J. Am. Assoc. Cancer Res.* **2017**, *23*, 3575–3584. [CrossRef]
26. Butowski, N.; Chang, S.M.; Lamborn, K.R.; Polley, M.-Y.; Pieper, R.; Costello, J.F.; Vandenberg, S.; Parvataneni, R.; Nicole, A.; Sneed, P.K.; et al. Phase II and pharmacogenomics study of enzastaurin plus temozolomide during and following radiation therapy in patients with newly diagnosed glioblastoma multiforme and gliosarcoma. *Neuro Oncol.* **2011**, *13*, 1331–1338. [CrossRef]
27. Carvalho, B.; Lopes, J.M.; Silva, R.; Peixoto, J.; Leitão, D.; Soares, P.; Fernandes, A.C.; Linhares, P.; Vaz, R.; Lima, J. The role of c-Met and VEGFR2 in glioblastoma resistance to bevacizumab. *Sci. Rep.* **2021**, *11*, 6067. [CrossRef] [PubMed]
28. Cloughesy, T.; Finocchiaro, G.; Belda-Iniesta, C.; Recht, L.; Brandes, A.A.; Pineda, E.; Mikkelsen, T.; Chinot, O.L.; Balana, C.; Macdonald, D.R.; et al. Randomized, Double-Blind, Placebo-Controlled, Multicenter Phase II Study of Onartuzumab Plus Bevacizumab Versus Placebo Plus Bevacizumab in Patients With Recurrent Glioblastoma: Efficacy, Safety, and Hepatocyte Growth Factor and O6-Methylguanine–DNA Methyltransferase Biomarker Analyses. *J. Clin. Oncol.* **2017**, *35*, 343–351. [CrossRef] [PubMed]
29. Collins, V.P.; Ichimura, K.; Di, Y.; Pearson, D.; Chan, R.; Thompson, L.C.; Gabe, R.; Brada, M.; Stenning, S.P. Prognostic and predictive markers in recurrent high grade glioma; results from the BR12 randomised trial. *Acta Neuropathol. Commun.* **2014**, *2*, 68. [CrossRef] [PubMed]
30. Erdem-Eraslan, L.; van den Bent, M.J.; Hoogstrate, Y.; Naz-Khan, H.; Stubbs, A.; van der Spek, P.; Böttcher, R.; Gao, Y.; de Wit, M.; Taal, W.; et al. Identification of Patients with Recurrent Glioblastoma Who May Benefit from Combined Bevacizumab and CCNU Therapy: A Report from the BELOB Trial. *Cancer Res.* **2016**, *76*, 525–534. [CrossRef] [PubMed]
31. Galanis, E.; Anderson, S.K.; Lafky, J.M.; Uhm, J.H.; Giannini, C.; Kumar, S.K.; Kimlinger, T.K.; Northfelt, D.W.; Flynn, P.J.; Jaeckle, K.A.; et al. Phase II study of bevacizumab in combination with sorafenib in recurrent glioblastoma (N0776): A north central cancer treatment group trial. *Clin. Cancer Res. Off. J. Am. Assoc. Cancer Res.* **2013**, *19*, 4816–4823. [CrossRef] [PubMed]
32. Gerstner, E.R.; Ye, X.; Duda, D.G.; Levine, M.A.; Mikkelsen, T.; Kaley, T.J.; Olson, J.J.; Nabors, B.L.; Ahluwalia, M.S.; Wen, P.Y.; et al. A phase I study of cediranib in combination with cilengitide in patients with recurrent glioblastoma. *Neuro Oncol.* **2015**, *17*, 1386–1392. [CrossRef] [PubMed]
33. Han, S.J.; Rolston, J.D.; Molinaro, A.M.; Clarke, J.L.; Prados, M.D.; Chang, S.M.; Berger, M.S.; DeSilva, A.; Butowski, N.A. Phase II trial of 7 days on/7 days off temozolmide for recurrent high-grade glioma. *Neuro Oncol.* **2014**, *16*, 1255–1262. [CrossRef]
34. Jan, C.-I.; Tsai, W.C.; Harn, H.J.; Shyu, W.C.; Liu, M.C.; Lu, H.M.; Chiu, S.C.; Cho, D.Y. Predictors of Response to Autologous Dendritic Cell Therapy in Glioblastoma Multiforme. *Front. Immunol.* **2018**, *9*, 727. [CrossRef]
35. Lötsch, D.D.; Ghanim, B.; Laaber, M.; Wurm, G.; Weis, S.; Lenz, S.; Webersinke, G.; Pichler, J.; Berger, W.; Spiegl-Kreinecker, S. Prognostic significance of telomerase-associated parameters in glioblastoma: Effect of patient age. *Neuro Oncol.* **2013**, *15*, 423–432. [CrossRef] [PubMed]
36. Lee, E.Q.; Kaley, T.J.; Duda, D.G.; Schiff, D.; Lassman, A.B.; Wong, E.T.; Mikkelsen, T.; Purow, B.W.; Muzikansky, A.; Ancukiewicz, M.; et al. A Multicenter, Phase II, Randomized, Noncomparative Clinical Trial of Radiation and Temozolomide with or without Vandetanib in Newly Diagnosed Glioblastoma Patients. *Clin. Cancer Res.* **2015**, *21*, 3610–3618. [CrossRef] [PubMed]
37. Michaelsen, S.R.; Christensen, I.J.; Grunnet, K.; Stockhausen, M.-T.; Broholm, H.; Kosteljanetz, M.; Poulsen, H.S. Clinical variables serve as prognostic factors in a model for survival from glioblastoma multiforme: An observational study of a cohort of consecutive non-selected patients from a single institution. *BMC Cancer* **2013**, *13*, 402. [CrossRef]
38. Omuro, A.A.; Beal, K.; Gutin, P.; Karimi, S.; Correa, D.D.; Kaley, T.J.; DeAngelis, L.M.; Chan, T.A.; Gavrilovic, I.T.; Nolan, C.; et al. Phase II study of bevacizumab, temozolomide, and hypofractionated stereotactic radiotherapy for newly diagnosed glioblastoma. *Clin. Cancer Res.* **2014**, *20*, 5023–5031. [CrossRef]
39. Reardon, D.A.; Brandes, A.A.; Omuro, A.; Mulholland, P.; Lim, M.; Wick, A.; Baehring, J.; Ahluwalia, M.S.; Roth, P.; Bähr, O.; et al. Effect of Nivolumab vs. Bevacizumab in Patients With Recurrent Glioblastoma: The CheckMate 143 Phase 3 Randomized Clinical Trial. *JAMA Oncol.* **2020**, *6*, 1003–1010. [CrossRef]
40. Roodakker, K.R.; Elsir, T.; Edqvist, P.H.D.; Hagerstrand, D.; Carison, J.; Lysiak, M.; Henriksson, R.; Ponten, F.; Rosell, J.; Soderkvist, P.; et al. PROX1 is a novel pathway-specific prognostic biomarker for high-grade astrocytomas; results from independent glioblastoma cohorts stratified by age and IDH mutation status. *Oncotarget* **2016**, *7*, 72431–72442. [CrossRef]
41. Srividya, M.R.; Thota, B.; Arivazhagan, A.; Thennarasu, K.; Balasubramaniam, A.; Chandramouli, B.A.; Hegde, A.S.; Santosh, V. Age-dependent prognostic effects of EGFR/p53 alterations in glioblastoma: Study on a prospective cohort of 140 uniformly treated adult patients. *J. Clin. Pathol.* **2010**, *63*, 687. [CrossRef]

42. Tini, P.; Cerase, A.; Cevenini, G.; Carbone, S.F.; Miracco, C.; Pirtoli, L. Epidermal Growth Factor Receptor Expression May Correlate with Survival Through Clinical and Radiological Features of Aggressiveness in Glioblastoma Treated with Radiochemotherapy. *Anticancer Res.* **2015**, *35*, 4117. [PubMed]
43. Weller, M.; Tabatabai, G.; Kästner, B.; Felsberg, J.; Steinbach, J.P.; Wick, A.; Schnell, O.; Hau, P.; Herrlinger, U.; Sabel, M.C.; et al. MGMT Promoter Methylation Is a Strong Prognostic Biomarker for Benefit from Dose-Intensified Temozolomide Rechallenge in Progressive Glioblastoma: The DIRECTOR Trial. *Clin. Cancer Res.* **2015**, *21*, 2057–2064. [CrossRef] [PubMed]
44. Wirsching, H.G.; Tabatabai, G.; Roelcke, U.; Hottinger, A.F.; Jörger, F.; Schmid, A.; Plasswilm, L.; Schrimpf, D.; Mancao, C.; Capper, D.; et al. Bevacizumab plus hypofractionated radiotherapy versus radiotherapy alone in elderly patients with glioblastoma: The randomized, open-label, phase II ARTE trial. *Ann. Oncol.* **2018**, *29*, 1423–1430. [CrossRef] [PubMed]
45. Miletic, H.; Niclou, S.P.; Johansson, M.; Bjerkvig, R. Anti-VEGF therapies for malignant glioma: Treatment effects and escape mechanisms. *Expert Opin. Ther. Targets* **2009**, *13*, 455–468. [CrossRef]
46. Duda, D.G.; Willett, C.G.; Ancukiewicz, M.; Tomaso, E.; Shah, M.; Czito, B.G.; Bentley, R.; Poleski, M.; Lauwers, G.Y.; Carroll, M.; et al. Plasma Soluble VEGFR-1 Is a Potential Dual Biomarker of Response and Toxicity for Bevacizumab with Chemoradiation in Locally Advanced Rectal Cancer. *Oncologist* **2010**, *15*, 577–583. [CrossRef]
47. Jain, R.K.; Duda, D.G.; Willett, C.G.; Sahani, D.V.; Zhu, A.X.; Loeffler, J.S.; Batchelor, T.T.; Sorensen, A.G. Biomarkers of response and resistance to antiangiogenic therapy. *Nat. Rev. Clin. Oncol.* **2009**, *6*, 327–338. [CrossRef]
48. Groot, J.F.D.; Lamborn, K.R.; Chang, S.M.; Gilbert, M.R.; Cloughesy, T.F.; Aldape, K.; Yao, J.; Jackson, E.F.; Lieberman, F.; Robins, H.I.; et al. Phase II Study of Aflibercept in Recurrent Malignant Glioma: A North American Brain Tumor Consortium Study. *J. Clin. Oncol.* **2011**, *29*, 2689–2695. [CrossRef]
49. de Groot, J.F.; Piao, Y.; Tran, H.; Gilbert, M.; Wu, H.-K.; Liu, J.; Bekele, B.N.; Cloughesy, T.; Mehta, M.; Robins, H.I.; et al. Myeloid Biomarkers Associated with Glioblastoma Response to Anti-VEGF Therapy with Aflibercept. *Clin. Cancer Res.* **2011**, *17*, 4872–4881. [CrossRef]
50. Michael, A.; Relph, K.; Pandha, H. Emergence of potential biomarkers of response to anti-angiogenic anti-tumour agents. *Int. J. Cancer* **2010**, *127*, 1251–1258. [CrossRef]
51. Hao, C.; Chen, G.; Zhao, H.; Li, Y.; Chen, J.; Zhang, H.; Li, S.; Zhao, Y.; Chen, F.; Li, W.; et al. PD-L1 Expression in Glioblastoma, the Clinical and Prognostic Significance: A Systematic Literature Review and Meta-Analysis. *Front. Oncol.* **2020**, *10*, 1015. [CrossRef] [PubMed]
52. Wu, C.; Zhu, Y.; Jiang, J.; Zhao, J.; Zhang, X.-G.; Xu, N. Immunohistochemical localization of programmed death-1 ligand-1 (PD-L1) in gastric carcinoma and its clinical significance. *Acta Histochem.* **2006**, *108*, 19–24. [CrossRef] [PubMed]
53. Faghfuri, E.; Faramarzi, M.A.; Nikfar, S.; Abdollahi, M. Nivolumab and pembrolizumab as immune-modulating monoclonal antibodies targeting the PD-1 receptor to treat melanoma. *Expert Rev. Anticancer Ther.* **2015**, *15*, 981–993. [CrossRef] [PubMed]
54. Reck, M.; Rodríguez-Abreu, D.; Robinson, A.G.; Hui, R.; Csőszi, T.; Fülöp, A.; Gottfried, M.; Peled, N.; Tafreshi, A.; Cuffe, S. Pembrolizumab versus chemotherapy for PD-L1–positive non–small-cell lung cancer. *N. Engl. J. Med.* **2016**, *375*, 1823–1833. [CrossRef]
55. Yang, T.; Kong, Z.; Ma, W. PD-1/PD-L1 immune checkpoint inhibitors in glioblastoma: Clinical studies, challenges and potential. *Hum. Vaccines Immunother.* **2021**, *17*, 546–553. [CrossRef] [PubMed]
56. Lee, K.S.; Lee, K.; Yun, S.; Moon, S.; Park, Y.; Han, J.H.; Kim, C.-Y.; Lee, H.S.; Choe, G. Prognostic relevance of programmed cell death ligand 1 expression in glioblastoma. *J. Neuro-Oncol.* **2018**, *136*, 453–461. [CrossRef]
57. Zeng, J.; Zhang, X.-K.; Chen, H.-D.; Zhong, Z.-H.; Wu, Q.-L.; Lin, S.-X. Expression of programmed cell death-ligand 1 and its correlation with clinical outcomes in gliomas. *Oncotarget* **2016**, *7*, 8944–8955. [CrossRef]
58. Kim, Y.S.; Kim, S.H.; Cho, J.; Kim, J.W.; Chang, J.H.; Kim, D.S.; Lee, K.S.; Suh, C.-O. MGMT Gene Promoter Methylation as a Potent Prognostic Factor in Glioblastoma Treated with Temozolomide-Based Chemoradiotherapy: A Single-Institution Study. *Int. J. Radiat. Oncol. Biol. Phys.* **2012**, *84*, 661–667. [CrossRef]
59. Hegi, M.E.; Diserens, A.C.; Gorlia, T.; Hamou, M.F.; de Tribolet, N.; Weller, M.; Kros, J.M.; Hainfellner, J.A.; Mason, W.; Mariani, L.; et al. MGMT gene silencing and benefit from temozolomide in glioblastoma. *N. Engl. J. Med.* **2005**, *352*, 997–1003. [CrossRef]
60. Wick, W.; Gorlia, T.; Bendszus, M.; Taphoorn, M.; Sahm, F.; Harting, I.; Brandes, A.A.; Taal, W.; Domont, J.; Idbaih, A.; et al. Lomustine and Bevacizumab in Progressive Glioblastoma. *N. Engl. J. Med.* **2017**, *377*, 1954–1963. [CrossRef]
61. Gleeson, J.P.; Keane, F.; Keegan, N.M.; Mammadov, E.; Harrold, E.; Alhusaini, A.; Harte, J.; Eakin-Love, A.; O'Halloran, P.J.; MacNally, S.; et al. Similar overall survival with reduced vs. standard dose bevacizumab monotherapy in progressive glioblastoma. *Cancer Med.* **2020**, *9*, 469–475. [CrossRef] [PubMed]
62. Lee, S.M.; Koh, H.-J.; Park, D.-C.; Song, B.J.; Huh, T.-L.; Park, J.-W. Cytosolic NADP+-dependent isocitrate dehydrogenase status modulates oxidative damage to cells. *Free Radic. Biol. Med.* **2002**, *32*, 1185–1196. [CrossRef]
63. Cheng, H.-B.; Yue, W.; Xie, C.; Zhang, R.-Y.; Hu, S.-S.; Wang, Z. IDH1 mutation is associated with improved overall survival in patients with glioblastoma: A meta-analysis. *Tumor Biol.* **2013**, *34*, 3555–3559. [CrossRef] [PubMed]
64. Seshacharyulu, P.; Ponnusamy, M.P.; Haridas, D.; Jain, M.; Ganti, A.K.; Batra, S.K. Targeting the EGFR signaling pathway in cancer therapy. *Expert Opin. Ther. Targets* **2012**, *16*, 15–31. [CrossRef] [PubMed]
65. Muñoz-Hidalgo, L.; San-Miguel, T.; Megías, J.; Monleón, D.; Navarro, L.; Roldán, P.; Cerdá-Nicolás, M.; López-Ginés, C. Somatic copy number alterations are associated with EGFR amplification and shortened survival in patients with primary glioblastoma. *Neoplasia* **2020**, *22*, 10–21. [CrossRef]

66. López-Ginés, C.; Muñoz-Hidalgo, L.; San-Miguel, T.; Megías, J.; Triviño, J.C.; Calabuig, S.; Roldán, P.; Cerdá-Nicolás, M.; Monleón, D. Whole-exome sequencing, EGFR amplification and infiltration patterns in human glioblastoma. *Am. J. Cancer Res.* **2021**, *11*, 5543–5558.
67. Shinojima, N.; Tada, K.; Shiraishi, S.; Kamiryo, T.; Kochi, M.; Nakamura, H.; Makino, K.; Saya, H.; Hirano, H.; Kuratsu, J.-I.; et al. Prognostic Value of Epidermal Growth Factor Receptor in Patients with Glioblastoma Multiforme. *Cancer Res.* **2003**, *63*, 6962–6970.
68. Hatanpaa, K.J.; Burma, S.; Zhao, D.; Habib, A.A. Epidermal growth factor receptor in glioma: Signal transduction, neuropathology, imaging, and radioresistance. *Neoplasia* **2010**, *12*, 675–684. [CrossRef]
69. Chen, J.-R.; Xu, H.-Z.; Yao, Y.; Qin, Z.-Y. Prognostic value of epidermal growth factor receptor amplification and EGFRvIII in glioblastoma: Meta-analysis. *Acta Neurol. Scand.* **2015**, *132*, 310–322. [CrossRef]
70. Tripathy, K.; Das, B.; Singh, A.K.; Misra, A.; Misra, S.; Misra, S.S. Prognostic Significance of Epidermal Growth Factor Receptor in Patients of Glioblastoma Multiforme. *J. Clin. Diagn. Res.* **2017**, *11*, EC05–EC08. [CrossRef]
71. Quan, A.L.; Barnett, G.H.; Lee, S.-Y.; Vogelbaum, M.A.; Toms, S.A.; Staugaitis, S.M.; Prayson, R.A.; Peereboom, D.M.; Stevens, G.H.J.; Cohen, B.H.; et al. Epidermal growth factor receptor amplification does not have prognostic significance in patients with glioblastoma multiforme. *Int. J. Radiat. Oncol. Biol. Phys.* **2005**, *63*, 695–703. [CrossRef] [PubMed]
72. Westphal, M.; Maire, C.L.; Lamszus, K. EGFR as a Target for Glioblastoma Treatment: An Unfulfilled Promise. *CNS Drugs* **2017**, *31*, 723–735. [CrossRef] [PubMed]

International Journal of
Molecular Sciences

MDPI

Review

Functional Precision Oncology: The Next Frontier to Improve Glioblastoma Outcome?

Dena Panovska and Frederik De Smet *

Department of Imaging and Pathology, Katholieke Universiteit Leuven, 3000 Leuven, Belgium; dena.panovska@kuleuven.be
* Correspondence: frederik.desmet@kuleuven.be; Tel.: +32-1637-2575

Abstract: Glioblastoma remains the most malignant and intrinsically resistant brain tumour in adults. Despite intensive research over the past few decades, through which numerous potentially druggable targets have been identified, virtually all clinical trials of the past 20 years have failed to improve the outcome for the vast majority of GBM patients. The observation that small subgroups of patients displayed a therapeutic response across several unsuccessful clinical trials suggests that the GBM patient population probably consists of multiple subgroups that probably all require a distinct therapeutic approach. Due to extensive inter- and intratumoral heterogeneity, assigning the right therapy to each patient remains a major challenge. Classically, bulk genetic profiling would be used to identify suitable therapies, although the success of this approach remains limited due to tumor heterogeneity and the absence of direct relationships between mutations and therapy responses in GBM. An attractive novel strategy aims at implementing methods for functional precision oncology, which refers to the evaluation of treatment efficacies and vulnerabilities of (ex vivo) living tumor cells in a highly personalized way. Such approaches are currently being implemented for other cancer types by providing rapid, translatable information to guide patient-tailored therapeutic selections. In this review, we discuss the current state of the art of transforming technologies, tools and challenges for functional precision oncology and how these could improve therapy selection for GBM patients.

Keywords: functional precision oncology; glioblastoma; drug sensitivity

Citation: Panovska, D.; De Smet, F. Functional Precision Oncology: The Next Frontier to Improve Glioblastoma Outcome? *Int. J. Mol. Sci.* **2022**, *23*, 8637. https://doi.org/10.3390/ijms23158637

Academic Editors: Ivana Jovčevska and Nives Pećina-Šlaus

Received: 29 June 2022
Accepted: 1 August 2022
Published: 3 August 2022

1. Introduction

Targeted therapies hold the promise to eradicate cancer cells through the inhibition of specific oncogenic proteins [1]. The efficiency of this approach largely depends on the dependency of the cancer cells to the targeted pathway, meaning that the identification of eligible patients is crucial to achieve clinical benefits. Current clinical practice uses a variety of diagnostic approaches through which disease-specific biomarkers are identified to select the most appropriate patients. For instance, the identification of HER2 amplified or estrogen receptor-positive breast cancer anticipates favorable response to HER2-targeted therapy or hormone therapy [2], EGFR mutation in lung cancer predicts response to EGFR-targeted therapeutic compounds [3], while imatinib in Philadelphia chromosome-positive leukemia predicts a favorable outcome in that context [4]. The completion of multiple cancer genome projects and the installation of better, faster and cheaper methods for genomic interrogations over the past 15 years has led to a better understanding of the pathogenic mutations that are connected to various cancer types, and fueled the concept of precision oncology [5]. Indeed, precision oncology aims at identifying effective therapeutic approaches based on properties (biomarkers) that are specific to each patient's tumor [6]. While the success stories highlighted above have now been around for more than a decade, the applicability of this one-on-one relationship between specific biomarkers and associated therapeutic responses has also faced many challenges and could only be exploited to a very limited extent across available cancer therapies. There are several reasons why the identification of a simple biomarker to predict therapy response is not trivial. For instance, inter- and intra-tumoral heterogeneity can greatly hinder the interpretation for a treating oncologist as multiple genotypes (with potentially divergent treatment sensitivities) can simultaneously populate the same tumor, thereby significantly affecting treatment efficacy. Moreover, the simultaneous presence of additional, potentially interfering genomic aberrations further complicates the interpretation of the relationship between a single biomarker and clinical outcome.

1.1. The Complexity of GBM

Glioblastoma (GBM) [7], still the most malignant primary brain cancer in adults [8], significantly suffers from the above described drawbacks. Already since 2005, the standard-of-care treatment of GBM includes a multidisciplinary approach combining surgery, ionizing radiation (RT) and chemotherapy. In spite of this aggressive approach, the median survival of GBM patients generally does not exceed 2 years [9]. This is caused by a combination of factors. (i) GBM is a highly infiltrative tumor, meaning that surgeons are commonly unable to resect the entire tumor, resulting in significant amounts of residual disease. In line with this, the extent of resection (EOR) has been identified as an important prognostic factor for GBM [10]. (ii) Targeting the residual tumor cells, primarily done by a combination of radiation therapy and temozolomide (TMZ), turns out to be extremely difficult: already in more than 50% of patients, progressive disease is radiologically observed even before finishing TMZ treatment (typically already within 3 months of therapy) [9]. This strongly suggests that large amounts of intrinsically unresponsive tumor cells were residing in the brain tissue even before starting therapy, which rapidly cause recurrence in GBM patients. Identifying more suitable and patient-tailored therapies that are accessible to the central nervous system (CNS) and are able to target a heterogeneous population of tumor cells therefore remains a major challenge in achieving clinical benefits.

To identify appropriate drug targets for GBM, large-scale sequencing programs were initiated to uncover disease causing genetic aberrations [11–13]. Over the past decade, several hundreds of GBM tumors have been sequenced within various consortia, uncovering complex and elaborate genetic alterations, including single nucleotide variants, focal or large chromosomal deletions and/or amplifications, and gene fusions [13]. For several of these genetic aberrations, drugs that target the affected cellular pathways have been developed, either in the context of GBM or other cancer types. Examples of such targets/pathways include receptor tyrosine kinases (e.g., EGFR, PDGFRA, VEGFR, MET) and various downstream intracellular signaling pathways (e.g., PI3K, AKT, mTOR, MEK/ERK) [14], hyperactive fusion proteins (e.g., TACC-FGFR and NTRK-fusions) [15], DNA repair (ATR/CHK1/CHK2, MDM2/4, PARP1, WEE) [16,17], and cell cycle regulation (CDK4/6) [11]. In spite of numerous clinical trials that were conducted to test the efficacy of these drugs against GBM, clinical results have been disappointing [8].

The failure of these trials could in part be explained by a lack of sufficiently precise selection procedures to enroll the appropriate patients that could actually benefit from the given therapy [18]. So far, such selection has primarily been based on bulk genetic analyses, where the presence of specific genetic aberrations was used as inclusion criteria for assigning appropriate therapy for each patient—an approach used for instance in the INSIGhT trial for GBM patients (NCT02977780). However, the complexity and interpatient heterogeneity of the genetic aberrations in GBM are so extensive that multiple interfering pathways are often simultaneously affected [11–13], making it largely unclear whether tumor cells of a particular patient would be responsive to a given therapy (even in the presence of the particular targetable mutation). On top of this, with the advent of single-cell sequencing methods, it turns out that the cellular composition of a GBM tumor is more complex than initially anticipated. Indeed, single-cell RNA sequencing (scRNAseq) studies showed that multiple of the TCGA-based tumor cell subtypes and a variety of stem cell-like states (i.e., neural progenitor-like, astrocyte-like, oligodendrocyte progenitor-like and mesenchymal-like) can be simultaneously present in a single GBM tumor [19–21] while containing multiple, often divergent genomic aberrations. Moreover, the various stem cell-like cellular states are plastic, meaning that they are interchangeable, a process that seems driven by stress factors caused by the environment of the cells. As such, stem cell-like cells are often more resistant to therapeutic perturbations. Therefore, instead of the initially anticipated subgrouping into 4 major subtypes [11,22], current insights suggest that GBM tumors harbor dozens of different tumor cell profiles, probably each requiring a specific therapeutic approach [7].

Finally, GBM tumor cells can also acquire de novo resistance upon therapy. Indeed, in an initially TMZ-responsive tumor cell population, resistance can easily be acquired by upregulating DNA-repairing enzymes such as MGMT or by inactivating the DNA mismatch repair (MMR) system, eventually leading to tumor recurrence [23]. At this point, a second surgical resection is often used as salvage therapy combined with other chemotherapeutic agents, such as lomustine/CCNU [24]. Additionally, in the recurrent setting, it would be highly beneficial to have better tools available to identify more suitable therapeutic options. All the above shows that identifying an appropriate therapy for each GBM patient, either in the newly diagnosed or recurrent setting, remains a daunting task. Being able to more precisely match particular therapies to the appropriate patients would not only significantly increase our ability to delay disease progression, it could also increase the success rates of clinical trials by more precisely identifying eligible patients.

1.2. Exceptional Responders across GBM Trials

In spite of the overall inability to treat GBM with durable clinical outcomes, clinical trials sometimes describe small groups of patients that did show a clinical response. For instance, in the multicentric INTELLANCE trial (NCT01800695), a small group of patients did experience clinical benefits from the treatment. Recurrent GBM patients that harboured an EGFR amplification were treated with a combination of TMZ and anti-EGFR monoclonal antibody coupled to a toxin (ABT-414; DEPATUX-M®) [25,26], These exceptional responders included one patient with a durable response beyond 40 months in addition to 4 and 9 patients with a reduction in tumor volume of 25–50% and 25%, respectively, out of a total of 60 patients that received this treatment [27]. The overall statistics of the trial were however insufficient to warrant approval by the regulatory agencies [27], but being able to identify those patients more carefully could have improved the outcome of the trial. Furthermore, molecular profiling of the patient samples was unable to identify an overall correlation between OS/PFS and EGFRvIII mutations, even though preclinical results from GBM cell lines and xenograft models showed high specificity and effectivity of the antibody–drug conjugate towards EGFRvIII and EGFR amplified tumor cells [28]. This is one of the many examples where a direct relationship between a genetic aberration and therapy response could not be confirmed in clinical practice, highlighting that more sophisticated assays may be required to achieve better therapy matching. In addition, the molecular analyses in this and most other trials remains largely confined to bulk analyses without taking tumor heterogeneity into account.

1.3. Functional Diagnostics: Evolving from a Static to a Dynamic Interrogation of Cancer Cells' Ability to Respond to Therapy

Major efforts are currently being put in matching specific (genetic) cancer features to drug responses [29]. However, in order to determine therapeutic efficacy across different patients and within a single tumor, as highlighted above, genetic information alone is often proven insufficient. Indeed, most studies only use baseline measurements in a 'static' setting (i.e., one snapshot prior to treatment), and intend to correlate the presence of specific genetic features to subsequent responsiveness to therapy. The simultaneous aberration of multiple cellular pathways, which can significantly interfere with each other, or for which multiple therapeutic options are sometimes available, make it difficult to predict the most suitable therapy. A *functional* interpretation [6] (e.g., what happens before and after cells are exposed to a certain therapy; what are the effects of the drug on the cellular state) on the other hand could provide *dynamic*, faster, more detailed and potentially longitudinal insights into the ability of cells to respond to therapy in a genotype agnostic way, although methods to do so remain difficult. In this light, approaches for assessing differential drug responses are gaining traction by which live tumor cells are ex vivo exposed to various therapeutic insults, while a chosen cellular response is carefully monitored—an approach coined *functional diagnostics, functional oncology or functional precision medicine* [6]. When monitoring the right features, such approach does not necessarily require a complete biological understanding, while still providing medically relevant insights (e.g., do cells respond to therapy or not, rather than why do they respond or not), allowing faster translation to a clinical setting.

Functional diagnostics is, however, not a novel approach. Such strategy has been widely applied in other biomedical domains, such as infectious diseases where antibiogram screens are used to select the most appropriate antibiotic in a patient-tailored way. Still, translating such functional diagnostic assays to a cancer setting requires further amelioration and validation in order to become medically applicable. In this regard, we endorse functional diagnostic insights as a complementary component to conventional genetic, imaging (i.e., MRI, CT scans) and baseline pathological (tissue) analyses. Indeed, coming to a proper patient-tailored interpretation will require that the different levels of information (imaging, pathology, genetic and functional assays) are integrated into an overarching framework to steer clinical decision making.

Overall, the goal of functional testing is to bring forward personalized medicine to patients diagnosed with complex disease entities, where treatment options are rather limited. In other words, functional tests ought to facilitate the matching of each patient to the most beneficial treatment. This being said, ex vivo drug exposure of freshly isolated tumor biopsies can directly inform on cell death, alterations in signaling networks, cellular phenotype and morphology or even tumor cell–tumor microenvironment (TME) crosstalk and adverse events in normal tissue. Certainly, the type of functional readout informing on tumor and non-malignant cellular vulnerabilities would largely depend on the mechanism of action of the given treatment. Typically, investigators rely on commonly available, FDA-approved therapies or drugs in clinical trials where dose-escalating studies where safety and tolerability of the therapy of interest has been already assessed and approved.

In the particular case of GBM, the functional screening method should not only be able to map each tumor in great detail—given GBM's high degree of intra-tumoral heterogeneity [30]—it should also be able to track molecular responses to drug treatments. Currently, every patient diagnosed with GBM is profiled using a uniformed diagnostic procedure, consisting of MRI scans and "static" measurements of pathological trademark alterations, such as chromosomal rearrangements, mutational patterns and MGMT promotor methylation (Figure 1). Functional testing gives the opportunity to directly evaluate therapy efficacy, either in dissociated GBM samples or tumor tissue slices [31]. In order to predict tumor cell behavior in such a complex and dynamic system as the GBM/brain, one must first familiarize with the baseline features (mutational status, transcripts, proteins/protein modifications, metabolites) and interactions between these components (gene–RNA, RNA–protein, protein–protein) across various cellular states [6,32]. Ex vivo drug sensitivity towards a panel of therapies can then be measured by monitoring the direction and strength of evolution of these interactive signaling events in all (non-) malignant cells. Miniaturizing the assay, for instance with chip technology would enable testing of multiple treatment conditions, while still providing sufficient multiparameter resolution on phenotypic and functional changes. These results could then be used by a medical board to integrate the functional finding (e.g., a ranked list of therapeutic options from most to least active in the tumor cells of the investigated patient) with baseline features and clinical parameters, such as tumor size, tumor location, extension of tumor resection and drug tolerability. Once all data is integrated, the most suitable drug/combination could be selected for each, individual patient. As anticipated, this procedure can be repeated once the GBM tumor recurs (Figure 1).

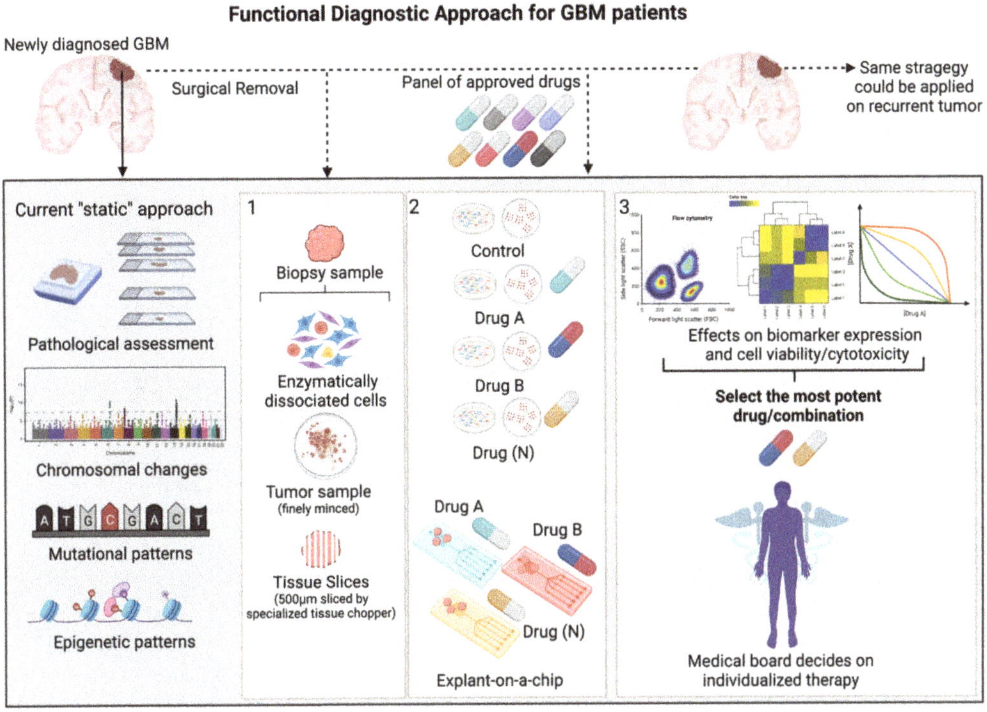

Figure 1. Schematic overview of functional diagnostic approach in GBM (Created with BioRender.com). During craniotomy, biopsy samples are routinely collected from newly diagnosed or recurrent GBM patients and pathologically assessed using standard clinical procedures, including immunohistochemical staining (IHC) of a handful of markers aiding histological grading, next generation sequencing and molecular analysis uncovering mutational patterns and epigenetic sequencing that measures the MGMT-promotor methylation status Although highly relevant, all these techniques offer only a single glance at the tumor's baseline features ("static" measurement) and do not completely capture the intra-tumoral heterogeneity and therapeutic vulnerability of the patient's tumor. To resolve this task, functional diagnostic is a personalized medicine strategy that makes use of live tumor samples derived from each individual patient. Panel 1: These biopsy samples can be enzymatically dissociated, minced or cut into fine tissue layers/slices. Panel 2: As such, these probes can be ex vivo treated with a panel of approved GBM-targeting therapies in cell culture flasks/plates or microfluidic chips. Panel 3: Various methods could be applied in order to optimally capture the effects of the given therapy on functional cellular features (cyto-toxic/-static events, various cellular states or cellular signaling pathways) relevant and corresponding to the given treatment. The output of these functional measurements would be a ranked list of most potent therapies, whereby a medical board could integrate this information together with histological, molecular measurements and clinical parameters. Finally, clinicians could decide on which therapy would be the most beneficial for each patient. This strategy could be applied on patients diagnosed with a recurrent tumor.

1.4. Tools and Methods for Functional Diagnostics

The development of functional diagnostic assays strongly depends on the availability of representative cancer models that maximally capture the genetic and phenotypic features of patients' tumor. So far, in vitro cancer research has been relying on so-called conventional cancer cell lines [33], which, although easy to use and representative in broader disease terms, have important limitations including: (i) lack of predictive value with regard to activity in clinical trials; and (ii) display of major and irreversible alterations in biological properties, such as gains and losses of genetic information, alterations in growth and invasive properties and loss of biomarker expression compared to the original tumor [33]. The growing body of evidence of heterogeneity, along with technological advances and platforms for drug development, steered pre-clinical research towards models derived from diseased individuals, such as patient-derived cell lines (PDCLs), patient-derived organoids (PDOs) and patient-derived xenografts (PDX). For GBM, an armamentarium of such models has been developed, although the installation of optimal readouts to assess drug activity in either of them still remains challenging (Figure 2).

FD platforms in GBM

Model	Advantages	Disadvantages	Readouts
2D conventional cell lines 	• Easy to expand and maintain • Disease-specific features • Large scale, high-throughput drug screens	• Limited predictive power, heterogeneity, microenvironment and patient-specific biomarkers • Homogeneous after long-term culturing	• Cell viability • Gene alteration and expression targeted drug screenings • Colony formation and migration assays • Biomarker expression
Patient-derived cell lines 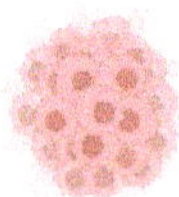	• Rapid, personalized drug readout • Closer resemblance of patients' tumor • Large scale, high-throughput drug screens	• Lack of microenvironment • Moderate initiation (~30% of patients) and growth rate (3-9 weeks) • Loss of heterogeneity; assay dependent	
Patient-derived organoids 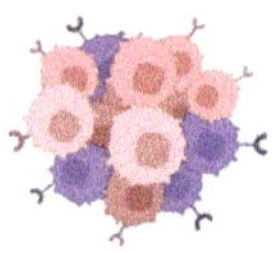	• 3D cellular arrangement and organization • Recapitulate tumor-of-origin biological features • Moderate drug throughput screening	• Patient-dependent growth potential (80%) • Do not expand beyond 2 weeks of culture • Lack of some microenvironment components, vasculature • Laborious, expensive	
Ex-vivo biopsies Dissociated cells 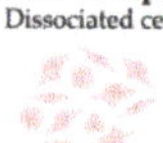 Tumor slices/ Minced tissue 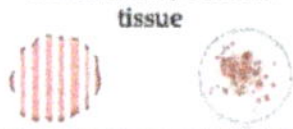	• Spatial organization preserved in tumor slices and minced tissues • Rapid readout • No need for in vitro propagation • Preserved TME & heterogeneity	• Limiting factors: sample quality and size • Only few drug conditions could be tested; low throughput	• Cell viability • Biomarker expression
Explants-on-a-chip 	• Extracellular environment, 3D structures • Dynamic testing of multiple conditions • Cost-effective	• Small sample size could be loaded and analyzed • Low throughput	• Cell viability • Morpholo gy, mass • Biomarker s
Patient-derived xenografts 	• In vivo environment • Blood-brain-barrier • Retain heterogeneity and patient-specific biomarkers	• Laborious • Time and cost inefficient • Patient-dependent engraftment capacity • Low throughput drug testing • Lack of major immune components	• Survival • Drug toxicity and potential side effects

Figure 2. Summary of pre-clinical models and platforms, which could be used for functional testing in GBM. Advantages and disadvantages of each model together with potential assay readout are outlined (Created with Biorender.com).

For the initial identification and evaluation of drug targets, compound design and efficacy testing, patient-derived cell cultures and organoids provide an excellent platform to preclinically explore and evaluate pharmacological responses across individual tumors. Given the fact that such models more faithfully recapitulate features of the tumor-of-origin, drug screening across cohorts of such models, offers the identification of therapeutic options that can be immediately linked to particular features present in the identified models [33]. As such, PDCLs/PDOs are compatible with large-scale pharmacogenomic platforms, such as Cancer Cell Line Encyclopedia [34], Genomics of Drug Sensitivity in Cancer [35], and Cancer Therapeutics Response Portal [36] and finally, the Connectivity Map [37], a project integrated in the CLUE platform (https://clue.io/about, accessed on 2 July 2022), comprising extensive and continuously expanding connectivity maps of protein, RNA expression and/or morphological changes in cancer cells, as a response to drug perturbation in addition to drug repurposing library of FDA-approved drugs, clinical trial drugs and pre-clinical compounds. All these large drug-screening resources are invaluable for exploring and understanding the mechanisms of various classes of compounds, drug repurposing and matching (combinations of) genomic mutations with functional responses over time [38]. However, reliable patient-derived in vitro models (PDCLs and organoids) may also take significant time to develop (from several weeks to months) and can typically only be generated from a subset of patients (for GBM, this ranges from 30–50%), thereby making them less suitable as generic tools to determine appropriate treatment regimens within acceptable time frames (from days to few weeks). Additionally, patient-derived in vitro models often lack the presence of an appropriate extracellular matrix (ECM) and (immune) tumor microenvironment (TME) which may also skew cellular behavior away from its original phenotype present in the patient [39]. Similarly, long-term culturing and expansion often leads to clonal selection and loss of heterogeneity [40], reducing their representative nature. PDX models emerged as patients' avatars—in vivo systems that closely mimic primary tumor biology and features. In this manner, PDX models are not only a powerful tool for preclinical drug development and testing, but also proven beneficial in providing clinically relevant information upon PDX clinical trials [41] and co-clinical trials. In co-clinical trials, mouse PDX models are established from tumor samples of each clinical trial participant and serve as personalized models for drug testing, from which the most appropriate therapy can subsequently be applied to the patient/donor [42–44].

Drug screening in PDX clinical trials were executed for various cancer types and solid tumors [41]. With this concept. it was confirmed that PDX models have the ability to predict trial responses, by evaluating predictive response biomarkers, map resistance mechanisms [41] and guide treatment decision making [33,42–44]. Patient-derived xenograft models for GBM are generated by direct transplantation of dissociated patient tumor material or tumor pieces. While a tendency for CNV-loss in heterotopic models has been suggested, orthotopic PDX models typically retain a close resemblance to the primary tumor [45]. Interestingly, studies confirm that the tumor-of-origin resemblance is highly dependent on the region from which the biopsy has been harvested, meaning that two PDX models generated from distinct regions of a single tumor could generate PDX models with dissimilar tumor subpopulations [45]. XENOGBM is a study currently evaluating the molecular analogy between the primary tumors of GBM patients and their corresponding PDX models (NCT02904525). PDX platforms are more advantageous over in vitro cultures as they retain 3D structural organization, clinical features, such as tumor invasiveness, vascularization, pseudopalisading necrosis and therapy-induced tumor evolution, and molecular features of the primary tumor, for instance crucial biomarkers such as EGFR expression, which is regularly suppressed by culturing conditions [46,47]. Furthermore, orthotopic PDX models provide the in vivo CNS environment enclosed behind the blood–brain barrier (BBB) allowing the direct evaluation of the penetration capacity and metabolomics of pharmacologic agents. Although seemingly superior over other models, PDXs still have several disadvantages for precision medicine in GBM. These models are laborious, time consuming and expensive in comparison to cell lines and organoids. The tumor take rate has been shown to be quite variable, meaning that PDX models would not be generated for all patients, or the number of models would be too limited in order to evaluate sufficient numbers of drug or drug combinations. Furthermore, the time between tumor engraftment and therapy decision may be too long for GBM patients. Finally, the use of immunodeficient mice largely hinders the interrogation of the role of the immune system in treatment responsiveness and general tumor biology.

To circumvent these issues and in line with the rapidly advancing organoid technology, organoid cultures were successfully established from patient specimens or through pluripotent stem cell reprogramming. In either case, organoids represent self-organizing, 3D systems which highly resemble the tissue from which they were derived. PDOs conserve tumor heterogeneity and TME components, tissue architecture, molecular and functional features. PDOs can be efficiently expanded over time while conserving patient-specific genomic features and intra-tumor heterogeneity, which could be reliably correlated with functional responses to therapeutics [48]. Non-malignant cells that are retained after 2 weeks of culture in the GBM organoid models include macrophage/microglia, T-cells, stromal cells and oligodendrocytes [49]. These features make organoids remarkable platforms for high-throughput drug screening, treatment evaluation in personalized chemo- and immunotherapies [39] and prediction of patient outcome. As such, a growing body of evidence shows clear correlation between organoid in vitro responses to long-term clinical responses of individual patient donors. Currently, these evaluations have been mostly performed on patients diagnosed with gastrointestinal cancer [50–52], colorectal cancer [53–58], breast cancer [59,60], pancreatic cancer [57,59,61], ovarian cancer [62,63] and esophageal adenocarcinoma [64]. PDO conceptualization for personalized treatment is lagging for CNS tumors and needs further validation. PDOs were successfully propagated for 24 chordoma patients, enabling the evaluation of the response rate of PDL1-postive and -negative organoids to decreasing concentrations of nivolumab after 72 h [65]. Another study using PDOs for retinoblastoma (RB) tumors confirmed the therapeutic efficacy of a combination treatment of topotecan and melphalan against recurrent retinal tumors and subretinal seeds, which was in line with previous reports. Importantly, these RB PDOs contained tumor stroma consisting of glial cells, which have a tumor supportive role [66], again showing the potential of these models in precision medicine trials. GBM PDOs were successfully generated from primary patient tissue and allowed an in-depth characterization which confirmed the close resemblance to the patient material, not only on the phenotypic but also on the functional level [67]. However, GBM tumors are incredibly heterogeneous at the spatial level, so that PDOs derived from a single patient and three different tumor regions (infiltrating edge, necrotic core and bulk tumor region containing necrosis, gliosis and putative treatment response) generated organoids with functionally distinct features, implying a wide range of cellular diversity between the organoids [67]. How these differences influence therapy responsiveness is yet to be interrogated. Technical and methodological efforts are continuously being put into the improvements of GBM-PDO generation and maintenance. Thus, 4D-printed self-programmable cell-culture arrays were fabricated to alter and shape their 3D environment as a response to external stimuli, whereby the fourth dimension is time. As such, these platforms have been extensively used for characterization and high-fidelity drug screening purposes [68]. In addition, the time of GBM-PDO generation has been radically reduced: 1–2 weeks for 4D-printed models [68] and 2–4 weeks after surgery in a novel method, whereby micro-dissected tumor pieces are applied in an optimized, specifically formulated medium for GBM-PDOs propagation and placed on an orbital shaker (instead of Matrigel) [49]. This method avoids tumor dissociation and, in turn, enabled successful generation, biobanking, in-depth characterization and co-culturing with CAR T-cells, proving the specificity and capacity of CAR T-cells in targeting EGFRvIII mutant tumor cells [49]. To date, only one report noted the applicability of GBM-PDOs in guiding therapy regime, which was performed for a single patient. In this report, everolimus was selected as the most potent therapeutic drug among a panel of FDA-approved mTOR inhibitors and led to tumor regression in the patient diagnosed with a recurrent GBM [69]. As all previously described in vitro models, PDOs have some shortcomings, such as the lack of complete TME and vascular network of endothelial cells. Additionally, to maximally capture tumor heterogeneity, the patient material should be sampled from different tumor foci consisting of highly viable and metabolically active cells, avoiding necrotic and hemorrhagic areas. Obtaining such tissue from recurrent tumors can be challenging, because of the abundance of low-quality, cell-sparse and necrotic areas and lack of proliferating cells [49].

In general, tumors including GBM, release cells and cellular content into the bloodstream or cerebrospinal fluid (CSF). These biomarkers are shed from the tumor residing site in form of circulating tumor cells (CTCs), proteins, cell-free nucleic acids and extracellular vesicles (EVs), accordingly systemized as liquid biopsies [70]. As such, liquid biopsies set ground for a rapid, noninvasive way for cancer diagnosis and prognostic markers [70]. Currently, liquid biopsies have gained clinical application for metastatic breast cancer [71], small cell lung cancer [72], prostate [73] and colorectal cancer [74] in the context of tumor diagnosis and longitudinal monitoring of therapy responses in both primary and metastasized tumors. Specifically, it has been shown that CTC count in peripheral blood correlates to therapy response. Advanced molecular profiling of these cells shows a high degree of concordance between genomic and transcriptomic profiles with the tumor of origin, making CTCs an excellent tool that could support personalized medicine approaches [75]. CTC-derived cell lines for various cancer types enabled CTC characterization and in vitro drug treatments, which may inform on the treatment susceptibility of the primary tumor and identify ways to inhibit metastasis [76]. This has been further corroborated by short-term ex vivo propagation of small cell lung cancer (SCLC) CTCs in culture from 23 patients. The CTC-derived cultures were in vitro treated with cisplatin and etopside and the results were correlated with individual responses from the respective patients. The results of this investigation showed correlation between response profiles of ex vivo expanded CTCs and three patients. Furthermore, this study highlights the ability of in vitro treated CTCs to accurately inform on innate and acquired chemo-resistance, based on patients' treatment history and clinical outcomes [72]. A similar report emphasized the predictive accuracy of in vitro-treated CTCs and two respective patients diagnosed with head and neck cancer and treated with cisplatin [77].

Current research is focused on refining methods for CTC isolation, ex vivo expansion and the establishment of CTC cell lines [78,79]. Additionally, two observational clinical trials, one in melanoma (EXPEVIVO-CTC; NCT03797053) and a second one in stage I-III lung cancer (CTMS 18-0056; NCT03655015), are anticipating the correlation of patients' response to ex vivo expanded and treated CTCs.

Owing to their location, brain tumors are challenging for surgical resection. Even when the tumor is accessible, the invasive surgery and biopsy collection present a risk of swelling and neurological damage. As patients receive an MRI scan within 12 weeks of treatment, contrast-enhancing lesions that are revealed on the images can indicate tumor progression, but might also be caused by post-radiotherapy edema, termed as pseudoprogression, which can spontaneously resolve [80,81]. At the moment, there are no methods that could reliably differentiate between glioma progression and pseudoprogression, or longitudinally monitor disease and treatment effects. The validation of biomarkers from liquid biopsies that could aid GBM prognosis is steadily progressing [81]. Liquid biopsies can be collected from cerebrospinal fluid (CSF), as it is in close contact with the CNS and accumulates tumor-specific markers, but CSF collection through lumbar punction is an invasive procedure [80,81]. In this light, the minimally invasive procedure to obtain liquid biopsies from GBM patients is through blood draw, but one must assume that the BBB is compromised at the tumor site. BBB disruption and permeability increases, as GBM tumors invade and progress into the surrounding tissue [81]. Therefore, CTC enumeration or EVs detection can potentially complement current strategies for more precise prediction of GBM progression. At present, methods for optimal CTC isolation and detection are advancing for GBM [70]. Unlike other epithelial-derived cancers where strong surface expression of EpCAM is detected (such as breast, prostate cancer, pancreatic, colorectal and hepatocellular), RNA sequencing of GBM-derived CTCs revealed Wnt-activated stemness and enrichment of mesenchymal features [81,82]. Alternative methods for CTCs detection in GBM include: GFAP labelling, telomerase-based assay, FISH detection of aneuploidy of chromosome 8, CTC-iChip microfluidic platform, recombinant VAR2CSA Malaria Protein and hTERT-specific oHSV1 expressing GFP [81]. All these studies point out the clinical utility of CTCs and liquid biopsies in real-time disease monitoring, prediction of progression and even functional measurements [83]. However, the number of CTCs is genuinely low (1–10 cells per 10 mL blood; 1 cell per 10^9 blood cells); therefore, efficient CTC isolation which recapitulates intra-tumoral heterogeneity and enables functional assessment is still far beyond the reach of GBM patients [70,80,81].

Hence, the ideal model for rapid functional assessment of drug sensitivity in GBM would be a system which maximally preservers the native cellular integrity [40] and interaction of the tumor cells with the microenvironment [46]. This includes ex-vivo drug treatment of tumor slices [31] or cellular suspensions of freshly dissociated patients' biopsies within hours post-surgery. Regarding GBM's extensive heterogeneity and invasiveness, one must consider sampling from distinct tumor regions in order to gain an "as close as possible" perspective of the therapeutic vulnerabilities of the invading cells that are remaining after tumor debulking. In a recent proof-of-concept study of a single GBM patient, tumor material was harvested and analyzed with single-cell RNA sequencing and scATAC-seq. The leftover patient material was orthotopically transplanted into mice, which were then treated with standard-of-care therapy (irradiation and temozolomide). Subsequently, the patient tumor was harvested and analyzed at recurrence. This framework provided mechanistic genetic and epigenetic insights into therapy-driven evolution and identified potential druggable targets, therefore providing an approach for designing therapeutic regimens for GBM [84]. Yet another proof-of-principle study demonstrated the efficacy of drug screening human breast cancer cell lines through imaging mass cytometry, assessing more than 40 markers [85]. All these methodologies are facing technological challenges, which need to be improved, upscaled and validated in order to meet the needs of routine clinical practice.

An auspicious high-throughput drug screening methodology has emerged with microfluidic devices [86]. The chip technology closely mimics the extracellular environment, which in turn is capable of generating 3D structures of cells. Such a device was designed to recapitulate the complex vasculature of the BBB and track the transport of nanoparticles to GBM spheroids. Analogous permeability measurements were performed on orthotopic xenografts through intravital imaging, which matched the in vitro model [87]. Microfluidic devices are automated and multiplexed platforms where the controlled environment offers a way to monitor drug effects, such as cell viability, changes in cellular mass accumulation rate upon treatment and morphology [88] at multiple timepoints [86,89].

1.5. Clinical Trials Implementing Functional Diagnostic Assays

Currently, numerous clinical trials are testing the efficacy of functional diagnostic methods in the prediction of patients' outcomes (Table 1). For instance, the EXALT-1 trial [90] showed that functional ex-vivo testing has the capacity to guide treatment and facilitate matching of patients with advanced hematological malignances to the right treatment. Strikingly, the progression-free survival in patients was prolonged 1.3-fold in comparison to the previously applied therapy. Briefly, patient material was obtained from biopsies, bone marrow aspirates or peripheral blood, dissociated (if necessary) and incubated with a drug library containing ~139 drugs at two different concentrations. After fixation, the cells were stained with antibodies against cancer cells and normal tissue, which allowed measuring the proportion of each population that remained alive following drug exposure. One of the greatest advantages of this approach was the short time between the testing and making treatment decisions [90]. A follow-up study, EXALT-2 is currently recruiting patients (NCT04470947) and in this three-arm study the treatment decision is going to be guided by genomic profiling, drug screening or the clinician's choice. Whether this approach would favor direct, acute cytotoxic agents over slower-acting but potentially also very effective therapeutics options remains one of the outstanding questions.

For GBM, several clinical centers are engaging into functional measurements and ex vivo tumor profiling. At the Oslo University Hospital, GBM patients recurring or progressing after first-line treatment are recruited in the ISM-GBM study, in which individual cancer stem cells (CSCs) are first propagated in vitro as PDCL models, and subsequently subjected to high-throughput screening (HTS) towards FDA/EMA approved drugs (NCT05043701). An individualized drug combination would be prescribed to each patient, based on the outcomes from functional tests [91,92]. Similarly, the safety and efficacy of HTS in CSCs will be evaluated by the Swedish Medical Center (NCT02654964). While highly valuable as approach, the ability to propagate CSCs from GBM patients typically remains limited to ~30–50%, making this approach only applicable to a subset of patients [93].

A handful of preclinical studies have been initiated for GBM, where drug vulnerabilities of organoids or PDCLs are going to be measured in in vitro assays (NCT04868396, NCT04180046, PRISM—NCT03336931). In the case of NCT04180046, primary GBM cell lines are going to be established from patient samples, in order to pathologically characterize the presence of GBM-related hallmarks (IDH1, GFAP, P53, ATRX and Ki67) and measure dose-response effects of natural and synthetic drugs [94]. Similarly, the PRISM trial will perform detailed tumor molecular profiling of pediatric brain tumors on several levels (proteogenomic, transcriptomic, methylation analysis), which ultimately would enable treatment tailoring. In parallel, individual PDCLs and PDX models ("mouse avatars") will be generated to facilitate the evaluation of the efficacy of the molecular-driven therapy within clinically acceptable timeframe [95]. The NCT04868396 study will, on the other hand, focus on the generation of organoid cultures initiated from tumor tissue collected by standard surgery. Here, the primary organoid library would be used to study mechanisms of aggressiveness and recurrence of GBM.

3D-PREDICT (NCT03561207) is a multicenter prospective study which among other cancer types (ovarian cancer, advanced cancers) is also enrolling patients diagnosed with adult high-grade glioma (anaplastic astrocytoma and GBM). Here, extensive molecular profiling and direct ex vivo drug testing of patient tumor materials are carried out, with the ultimate goal being to make personalized medicine recommendations. Thus, the primary outcome measure of this trial is correlation between patient outcomes and functional results. Initially, four patients have been included in this study generating PDCLs and organoids for these patients. This enables the characterization and comparison of patients' biopsies and individual tumor-derived models at genomic and transcriptomic level and further performing functional tests, such as clonogenic assays and 3D-PREDICT assays. The latter is practically a viability test of 3D spheroids treated with a mono-drug library at different concentrations for an adapted period of time, which extrapolates IC50 values and stratifies the response predictions as unresponsive, moderate and responsive. Based on these results, tumor spheroids derived from one of the four included patients (male, 24) diagnosed with GBM featuring ependymoma regions was found sensitive to JAK/STAT and mTOR inhibition. In this manner, clinicians opted to treat this patient with a combination therapy consisting of ruxolitinib and everolimus. After his seventh progression, the patient was classified as having stable disease for more than 4 months post-treatment [96]. A second study report from this trial enrolled 55 patients with newly diagnosed (ND) and recurrent high-grade glioma. In the case of ND patients, 71 patients were included at first. However, 13 patients had to be excluded because of model generation/assay failure. Then 15 patients were excluded because of premature enrollment in the study (<6 months); 9 patients decided not to take advantage of the trial and 1 patient progressed due to other events. As such, 33 patients were considered eligible for TMZ + RT treatment, where in-vitro TMZ response prediction was made 7 days post-surgery and subsequently compared to clinical OS after patients completed their treatment cycle. Of note, 20/33 patients had already progressed at the time this comparison was made. However, the median OS of assay responders was 11.6 months, as opposed to assay non-responders—5.9 months. Thus, 85% prediction accuracy was achieved. Interestingly, in the case of recurrent tumors, two remarkable observations were made. In some patients, PFS was exceeding the reported median PFS for carmustine and irinothecan and two patients were predicted as responders to BRAF inhibitors by the 3D Predict Glioma Assay, without harboring the targeted mutation, demonstrating the autonomy of in vitro/ex vivo tests in personalized medicine beyond NGS characterization. One of the patients diagnosed with GBM (IDH-WT), received combination of bevacizumab/dabrafenib and progressed after 4 months. The other patient had anaplastic astrocytoma and was prescribed dabrafenib for 12 months after which radiographical progression was noted. In both cases, at recurrence onset, patient tissue was collected by re-resection and screened in 3D Predict assay, recording a decrease in sensitivity towards BRAF inhibitors, which again was in line with the anticipated clinical outcomes [97].

An Ex Vivo DEtermiNed Cancer Therapy (EVIDENT) trial has been recently initiated (NCT05231655), which aims at determining the efficacy and feasibility of ex vivo screening in prediction of standard-of-care therapy outcome and novel therapy identification, including patients diagnosed with solid tumors (kidney, bladder, head and neck cancers, melanoma, sarcoma and GBM). This trial seems most prominent because solid tumor biopsies will be directly screened and the response would be quantified and correlated to patient clinical outcome.

Table 1. Clinical trials in GBM using functional diagnostic evaluation.

Identifier	Name	Title	Status	Models	Study Type	Purpose	Readout	Diagnosis (n = Number of Recruited Patients)	Ref.
NCT05043701	ISM-GBM	Individualized Systems Medicine Functional Profiling for Recurrent Glioblastoma (ISM-GBM)	Recruiting	PDCLs (CSCs from rGBM) *	Interventional	A personalized drug combination will be prescribed to each patient based on the functional drug screen	HTS FDA/EMA approved drugs; cell viability	rGBM (n = 15)	[91,92]
NCT02654964	/	Cancer Stem Cell High-Throughput Drug Screening Study	Unknown	PDCLs (CSCs from rGBM) *	Interventional	A personalized drug combination will be prescribed to each patient based on the functional drug screen	CSC/HTS viability assay of drugs/combinations	rGBM (n = 10)	/
NCT04868396	/	Patient-derived Glioma Stem Cell Organoids	Active, not recruiting	PDO	Observational	Baseline characterization	Mechanisms that contribute to aggressive tumor growth and treatment resistance in primary and recurrent GBM	ND-GBM & rGBM (n = 60)	/
NCT04180046	/	Utility of Primary Glioblastoma Cell Lines	Recruiting	PDCLs	Observational	Baseline characterization	Phenotypic, genetic (IDH-, MGMT- status) and IHC characterization	ND-GBM (n = 10)	[94]
NCT03336931	PRISM	PRecISion Medicine for Children With Cancer	Recruiting	PDCLs and PDX	Observational	Molecular profiling, drug testing, recommendation of personalized therapy	In vitro HTS testing; In vivo drug testing using PDX models; Liquid biopsies	Childhood solid tumors (n = 550)	[95]
NCT03561207	3D-PREDICT	3D Prediction of Patient-Specific Response	Recruiting	PDCLs and PDOs	Observational	Compare Assay results to reported patient outcomes	Cell viability	GBM, anaplastic astrocytoma/solid tumors (n = 570)	[96,97]
NCT05231655	EVIDENT	Ex VIvo DEtermiNed Cancer Therapy	Recruiting	Ex-vivo biopsies	Observational	High-throughput ex-vivo drug screen of cells processed directly from solid tumors to determine sensitivity/resistance profiles	Ex-vivo HTS of cells processed directly from solid tumors to determine sensitivity/resistance profiles	GBM, Solid tumors (n = 600)	/

* ND-GBM—newly diagnosed GBM tumor; rGBM—recurrent GBM tumor; HTS—high-throughput screening; CSC—cancer stem cells.

2. Concluding Remarks

Conventional genetic-based matching of patients and treatments may be beneficial for only a fraction of cancers, where the oncogene driver mutation is uniform and maintained at a stable level among the cancer cell population. Such examples are seen in HER2-positive breast cancers [2], Philadelphia chromosome in chronic myeloid leukemia [4], BRAF mutations in melanoma [98] and few other cancer types. However, this approach underestimates complex circuits of non-genetic mechanisms that define the pathological behavior of tumors. Therefore, most of the large-scale clinical trials matching targetable genetic alterations to inhibitors resulted in unsatisfying survival rates, widely accepting the fact that "one-size-fits-all" therapy approach is not beneficial for complex and heterogeneous diseases. From here, it became obvious that the personalization of cancer medicine is the way to tackle this disease. Although personalized chemo-sensitivity assays are thriving in the academic and commercial enterprises, still there are several hurdles that need to be addressed. Fundamentally, functional diagnostic assays require the availability of adequate tissue material to enable an efficient yield of viable tumor cells. This means that the hospital where the surgery is performed should include a department/laboratory, ensuring rapid transfer and minimal tissue manipulation before the functional diagnostic test is performed under strictly controlled conditions. Additionally, surgeons and clinicians should clearly communicate requirements and conditions for optimal tissue harvest and handling. However, in many cases, good-quality tumor samples cannot always be obtained, especially from metastatic and recurrent solid tumors [99]. Most commonly, core needle biopsies, fine needle aspirates and circulating tumor cells are collected, which are not sufficient for high-throughput ex-vivo drug screening or model establishment [100]. Next to sufficient viable cell yield, a key requirement for proper interpretation of functional diagnostic assays are treatment conditions. The diverse mechanisms of action of targeted inhibitors implies that concentrations and treatment duration should be optimized for each drug. One strategy to solve this is to evaluate various concentration and time ranges of each drug in representative cohorts of 2D patient-derived models/organoids and optimally validate a predictive biomarker correlating to response or direct measurement of tumor cell viability and fitness. Again, results from functional diagnostic assays should be routinely obtained within a clinically relevant timeframe. Considering GBM's nature, all of the before mentioned points should be well considered. Firstly, based on the tumor location, surgeons are not always able to provide sufficient material for all pathological (IHC, genomic) and functional evaluations, meaning that a functional model and platform might not be established for all patients. In spite of GBM's aggressive nature, the timeframe between the functional diagnostic readout to treatment selection should be well accounted. Finally, because of the vast spatiotemporal heterogeneity, biopsy materials sampled from distinct tumor regions might give rise in slightly biologically distinct models. Unfortunately, no current method can precisely profile remaining cell populations after tumor debulking, which eventually will invade the surrounding tissue and cause tumor recurrence. Efficiently targeting these cells remains a dreadful challenge for all oncology specialists. In summary, solemn genomic assessments do not identify obvious druggable targets and therapies for advanced and heterogeneous cancers. In this regard, functional diagnostic tests may provide a platform for exploring cytotoxicity profiles of cancer cells derived by affected individuals towards drug-and-drug combinations.

3. Future Directions

A growing appreciation of biobanking, the generation of living biobanks of patient-derived models and ex-vivo treatments have the potential to enhance the development of rationally selected combined therapies and guide prospective clinical trials. As the number of clinical trials and assessments of functional diagnostic platforms increase, we anticipate the implementation of this strategy in standard clinical oncology practice. Importantly, the integration of molecular characterization data, functional profiles, clinical parameters and patient follow-up from a multitude of individuals into a single database might even enable informing clinical decisions for patients from whom sufficient tumor material may not be available. Finally, while multiple endeavors are ongoing to implement functional diagnostics to select appropriate therapeutic options for GBM, it still remains to be seen which approach will prove to be the most predictive and clinically relevant.

Author Contributions: Conceptualization, F.D.S. and D.P.; writing—original draft preparation, D.P.; writing—review and editing, F.D.S.; visualization, D.P.; supervision, F.D.S.; funding acquisition, F.D.S. All authors have read and agreed to the published version of the manuscript.

Funding: This research was funded by KU Leuven Research grant; grant number C14/17/084 and FWO research grant; grant numbers G0I1118N, I007418N and G0B3722N.

Institutional Review Board Statement: Not applicable.

Informed Consent Statement: Not applicable.

Data Availability Statement: Not applicable.

Conflicts of Interest: The authors declare no conflict of interest.

References

1. Masui, K.; Gini, B.; Wykosky, J.; Zanca, C.; Mischel, P.S.; Furnari, F.B.; Cavenee, W.K. A Tale of Two Approaches: Complementary Mechanisms of Cytotoxic and Targeted Therapy Resistance May Inform next-Generation Cancer Treatments. *Carcinogenesis* **2013**, *34*, 725–738. [CrossRef] [PubMed]
2. Krishnamurti, U.; Silverman, J.F. HER2 in Breast Cancer: A Review and Update. *Adv. Anat. Pathol.* **2014**, *21*, 100–107. [CrossRef] [PubMed]
3. Castellanos, E.; Feld, E.; Horn, L. Driven by Mutations: The Predictive Value of Mutation Subtype in EGFR-Mutated Non–Small Cell Lung Cancer. *J. Thorac. Oncol.* **2017**, *12*, 612–623. [CrossRef]
4. Sampaio, M.M.; Santos, M.L.C.; Marques, H.S.; Gonçalves, V.L.d.S.; Araújo, G.R.L.; Lopes, L.W.; Apolonio, J.S.; Silva, C.S.; Santos, L.K.d.S.; Cuzzuol, B.R.; et al. Chronic Myeloid Leukemia-from the Philadelphia Chromosome to Specific Target Drugs: A Literature Review. *World J. Clin. Oncol.* **2021**, *12*, 69–94. [CrossRef]
5. Wheeler, D.A.; Wang, L. From Human Genome to Cancer Genome: The First Decade. *Genome Res.* **2013**, *23*, 1954. [CrossRef] [PubMed]
6. Friedman, A.A.; Letai, A.; Fisher, D.E.; Flaherty, K.T. Precision Medicine for Cancer with Next-Generation Functional Diagnostics. *Nat. Rev. Cancer* **2015**, *15*, 747–756. [CrossRef]
7. Yabo, Y.A.; Simone, P. Cancer Cell Heterogeneity and Plasticity: A Paradigm Shift in Glioblastoma. *Neuro-Oncology* **2021**, *24*, 669–682. [CrossRef] [PubMed]
8. Zanders, E.D.; Svensson, F.; Bailey, D.S. Therapy for Glioblastoma: Is It Working? *Drug Discov. Today* **2019**, *24*, 1193–1201. [CrossRef]
9. Stupp, R.; Mason, W.; van den Bent, M.; Weller, M.; Fischer, B.; Taphoorn, M.; Belanger, K.; Brandes, A.; Marosi, C.; Bogdahn, U.; et al. Radiotherapy plus Concomitant and Adjuvant Temozolomide for Glioblastoma. *J. Neurooncol.* **2005**, *352*, 987–989. [CrossRef]
10. Brown, T.J.; Brennan, M.C.; Li, M.; Church, E.W.; Brandmeir, N.J.; Rakszawski, K.L.; Patel, A.S.; Rizk, E.B.; Suki, D.; Sawaya, R.; et al. Association of the Extent of Resection with Survival in Glioblastoma. *JAMA Oncol.* **2016**, *2*, 1460. [CrossRef]
11. Brennan, C.W.; Verhaak, R.G.W.; McKenna, A.; Campos, B.; Noushmehr, H.; Salama, S.R.; Zheng, S.; Chakravarty, D.; Sanborn, J.Z.; Berman, S.H.; et al. The Somatic Genomic Landscape of Glioblastoma. *Cell* **2013**, *155*, 462–477. [CrossRef] [PubMed]
12. Ceccarelli, M.; Barthel, F.P.; Sabedot, T.S.; Salama, S.R.; Murray, B.A.; Morozova, O.; Newton, Y.; Pagnotta, S.M.; Anjum, S.; Wang, J.; et al. Molecular Profiling Reveals Biologically Discrete Subsets and Pathways of Progression in Diffuse Glioma. *Cell* **2016**, *164*, 550–563. [CrossRef] [PubMed]
13. Verhaak, R.G.W.; Hoadley, K.A.; Purdom, E.; Wang, V.; Qi, Y.; Wilkerson, M.D.; Miller, C.R.; Ding, L.; Golub, T.; Mesirov, J.P.; et al. Integrated Genomic Analysis Identifies Clinically Relevant Subtypes of Glioblastoma Characterized by Abnormalities in PDGFRA, IDH1, EGFR, and NF1. *Cancer Cell* **2010**, *17*, 98–110. [CrossRef] [PubMed]
14. Kim, G.; Ko, Y.T. Small Molecule Tyrosine Kinase Inhibitors in Glioblastoma. *Arch. Pharm. Res.* **2020**, *43*, 385–394. [CrossRef]
15. Di Stefano, A.L.; Fucci, A.; Frattini, V.; Labussiere, M.; Mokhtari, K.; Zoppoli, P.; Marie, Y.; Bruno, A.; Boisselier, B.; Giry, M.; et al. Detection, Characterization, and Inhibition of FGFR-TACC Fusions in IDH Wild-Type Glioma. *Clin. Cancer Res.* **2015**, *21*, 3307–3317. [CrossRef] [PubMed]
16. Qiu, Z.; Oleinick, N.L.; Zhang, J. ATR/CHK1 Inhibitors and Cancer Therapy. *Radiother. Oncol.* **2018**, *126*, 450–464. [CrossRef]
17. Carrassa, L.; Damia, G. DNA Damage Response Inhibitors: Mechanisms and Potential Applications in Cancer Therapy. *Cancer Treat. Rev.* **2017**, *60*, 139–151. [CrossRef] [PubMed]
18. Touat, M.; Idbaih, A.; Sanson, M.; Ligon, K.L. Glioblastoma Targeted Therapy: Updated Approaches from Recent Biological Insights. *Ann. Oncol.* **2017**, *28*, 1457–1472. [CrossRef] [PubMed]

19. Patel, A.P.; Tirosh, I.; Trombetta, J.J.; Shalek, A.K.; Shawn, M.; Wakimoto, H.; Cahill, D.P.; Nahed, B.V.; Curry, W.T.; Martuza, R.L.; et al. Single-Cell RNA-Seq Highlights Intratumoral Heterogeneity in Primary Glioblastoma. *Science* **2014**, *344*, 1396–1401. [CrossRef] [PubMed]
20. Darmanis, S.; Sloan, S.A.; Croote, D.; Mignardi, M.; Chernikova, S.; Samghababi, P.; Zhang, Y.; Neff, N.; Kowarsky, M.; Caneda, C.; et al. Single-Cell RNA-Seq Analysis of Infiltrating Neoplastic Cells at the Migrating Front of Human Glioblastoma. *Cell Rep.* **2017**, *21*, 1399–1410. [CrossRef]
21. Chen, X.; Wen, Q.; Stucky, A.; Zeng, Y.; Gao, S.; Loudon, W.G.; Ho, H.W.; Kabeer, M.H.; Li, S.C.; Zhang, X.; et al. Relapse Pathway of Glioblastoma Revealed by Single-Cell Molecular Analysis. *Carcinogenesis* **2018**, *16*, 872–876. [CrossRef]
22. Wang, L.; Babikir, H.; Müller, S.; Yagnik, G.; Shamardani, K.; Catalan, F.; Kohanbash, G.; Alvarado, B.; Di Lullo, E.; Kriegstein, A.; et al. The Phenotypes of Proliferating Glioblastoma Cells Reside on a Single Axis of Variation. *Cancer Discov.* **2019**, *9*, 1708–1719. [CrossRef] [PubMed]
23. Daniel, P.; Sabri, S.; Chaddad, A.; Meehan, B.; Jean-Claude, B.; Rak, J.; Abdulkarim, B.S. Temozolomide Induced Hypermutation in Glioma: Evolutionary Mechanisms and Therapeutic Opportunities. *Front. Oncol.* **2019**, *9*, 1–7. [CrossRef] [PubMed]
24. Wick, W.; Gorlia, T.; Bendszus, M.; Taphoorn, M.; Sahm, F.; Harting, I.; Brandes, A.A.; Taal, W.; Domont, J.; Idbaih, A.; et al. Lomustine and Bevacizumab in Progressive Glioblastoma. *N. Engl. J. Med.* **2017**, *377*, 1954–1963. [CrossRef] [PubMed]
25. van den Bent, M.; Gan, H.K.; Lassman, A.B.; Kumthekar, P.; Merrell, R.; Butowski, N.; Lwin, Z.; Mikkelsen, T.; Nabors, L.B.; Papadopoulos, K.P.; et al. Efficacy of Depatuxizumab Mafodotin (ABT-414) Monotherapy in Patients with EGFR-Amplified, Recurrent Glioblastoma: Results from a Multi-Center, International Study. *Cancer Chemother. Pharmacol.* **2017**, *80*, 1209–1217. [CrossRef] [PubMed]
26. Reardon, D.A.; Lassman, A.B.; Van Den Bent, M.; Kumthekar, P.; Merrell, R.; Scott, A.M.; Fichtel, L.; Sulman, E.P.; Gomez, E.; Fischer, J.; et al. Efficacy and Safety Results of ABT-414 in Combination with Radiation and Temozolomide in Newly Diagnosed Glioblastoma. *Neuro. Oncol.* **2017**, *19*, 965–975. [CrossRef] [PubMed]
27. Lassman, A.B.; Van Den Bent, M.J.; Gan, H.K.; Reardon, D.A.; Kumthekar, P.; Butowski, N.; Lwin, Z.; Mikkelsen, T.; Nabors, L.B.; Papadopoulos, K.P.; et al. Safety and Efficacy of Depatuxizumab Mafodotin + Temozolomide in Patients with EGFR -Amplified, Recurrent Glioblastoma: Results from an International Phase I Multicenter Trial. *Neuro. Oncol.* **2019**, *21*, 106–114. [CrossRef] [PubMed]
28. Phillips, A.C.; Boghaert, E.R.; Vaidya, K.S.; Mitten, M.J.; Norvell, S.; Falls, H.D.; DeVries, P.J.; Cheng, D.; Meulbroek, J.A.; Buchanan, F.G.; et al. ABT-414, an Antibody-Drug Conjugate Targeting a Tumor-Selective EGFR Epitope. *Mol. Cancer Ther.* **2016**, *15*, 661–669. [CrossRef] [PubMed]
29. Shin, S.H.; Bode, A.M.; Dong, Z. Precision Medicine: The Foundation of Future Cancer Therapeutics. *npj Precis. Oncol.* **2017**, *1*, 12. [CrossRef] [PubMed]
30. Soeda, A.; Hara, A.; Kunisada, T.; Yoshimura, S.I.; Iwama, T.; Park, D.M. The Evidence of Glioblastoma Heterogeneity. *Sci. Rep.* **2015**, *5*, 7979. [CrossRef] [PubMed]
31. Zhao, W.; Dovas, A.; Spinazzi, E.F.; Levitin, H.M.; Banu, M.A.; Upadhyayula, P.; Sudhakar, T.; Marie, T.; Otten, M.L.; Sisti, M.B.; et al. Deconvolution of Cell Type-Specific Drug Responses in Human Tumor Tissue with Single-Cell RNA-Seq. *Genome Med.* **2021**, *13*, 1–15. [CrossRef] [PubMed]
32. Neftel, C.; Laffy, J.; Filbin, M.G.; Hara, T.; Shore, M.E.; Rahme, G.J.; Richman, A.R.; Silverbush, D.; Shaw, M.L.; Hebert, C.M.; et al. An Integrative Model of Cellular States, Plasticity, and Genetics for Glioblastoma. *Cell* **2019**, *178*, 835–849.e21. [CrossRef] [PubMed]
33. Pauli, C.; Hopkins, B.D.; Prandi, D.; Shaw, R.; Fedrizzi, T.; Sboner, A.; Sailer, V.; Augello, M.; Puca, L.; Rosati, R.; et al. Personalized in Vitro and in Vivo Cancer Models to Guide Precision Medicine. *Cancer Discov.* **2017**, *7*, 462–477. [CrossRef] [PubMed]
34. Barretina, J.; Caponigro, G.; Stransky, N.; Venkatesan, K.; Margolin, A.A.; Kim, S.; Wilson, C.J.; Lehár, J.; Kryukov, G.V.; Sonkin, D.; et al. The Cancer Cell Line Encyclopedia Enables Predictive Modelling of Anticancer Drug Sensitivity. *Nature* **2012**, *483*, 603–607. [CrossRef] [PubMed]
35. Yang, W.; Soares, J.; Greninger, P.; Edelman, E.J.; Lightfoot, H.; Forbes, S.; Bindal, N.; Beare, D.; Smith, J.A.; Thompson, I.R.; et al. Genomics of Drug Sensitivity in Cancer (GDSC): A Resource for Therapeutic Biomarker Discovery in Cancer Cells. *Nucleic Acids Res.* **2013**, *41*, 955–961. [CrossRef]
36. Basu, A.; Bodycombe, N.E.; Cheah, J.H.; Price, E.V.; Liu, K.; Schaefer, G.I.; Ebright, R.Y.; Stewart, M.L.; Ito, D.; Wang, S.; et al. An Interactive Resource to Identify Cancer Genetic and Lineage Dependencies Targeted by Small Molecules. *Cell* **2013**, *154*, 1151–1161. [CrossRef] [PubMed]
37. Wang, Y.; Yella, J.; Jegga, A.G. Transcriptomic Data Mining and Repurposing for Computational Drug Discovery. *Methods Mol. Biol.* **2019**, *1903*, 73–95. [CrossRef] [PubMed]
38. McFarland, J.M.; Paolella, B.R.; Warren, A.; Geiger-Schuller, K.; Shibue, T.; Rothberg, M.; Kuksenko, O.; Colgan, W.N.; Jones, A.; Chambers, E.; et al. Multiplexed Single-Cell Transcriptional Response Profiling to Define Cancer Vulnerabilities and Therapeutic Mechanism of Action. *Nat. Commun.* **2020**, *11*, 4296. [CrossRef]
39. Klein, E.; Hau, A.C.; Oudin, A.; Golebiewska, A.; Niclou, S.P. Glioblastoma Organoids: Pre-Clinical Applications and Challenges in the Context of Immunotherapy. *Front. Oncol.* **2020**, *10*, 1–18. [CrossRef] [PubMed]

40. Rosenberg, S.; Verreault, M.; Schmitt, C.; Guegan, J.; Guehennec, J.; Levasseur, C.; Marie, Y.; Bielle, F.; Mokhtari, K.; Hoang-Xuan, K.; et al. Multi-Omics Analysis of Primary Glioblastoma Cell Lines Shows Recapitulation of Pivotal Molecular Features of Parental Tumors. *Neuro. Oncol.* **2017**, *19*, 219–228. [CrossRef] [PubMed]
41. Gao, H.; Korn, J.M.; Ferretti, S.; Monahan, J.E.; Wang, Y.; Singh, M.; Zhang, C.; Schnell, C.; Yang, G.; Zhang, Y.; et al. High-Throughput Screening Using Patient-Derived Tumor Xenografts to Predict Clinical Trial Drug Response. *Nat. Med.* **2015**, *21*, 1318–1325. [CrossRef] [PubMed]
42. Vargas, R.; Gopal, P.; Kuzmishin, G.B.; DeBernardo, R.; Koyfman, S.A.; Jha, B.K.; Mian, O.Y.; Scott, J.; Adams, D.J.; Peacock, C.D.; et al. Case Study: Patient-Derived Clear Cell Adenocarcinoma Xenograft Model Longitudinally Predicts Treatment Response. *NPJ Precis. Oncol.* **2018**, *2*, 14. [CrossRef] [PubMed]
43. Izumchenko, E.; Paz, K.; Ciznadija, D.; Sloma, I.; Katz, A.; Vasquez-Dunddel, D.; Ben-Zvi, I.; Stebbing, J.; McGuire, W.; Harris, W.; et al. Patient-Derived Xenografts Effectively Capture Responses to Oncology Therapy in a Heterogeneous Cohort of Patients with Solid Tumors. *Ann. Oncol.* **2017**, *28*, 2595–2605. [CrossRef]
44. Inoue, A.; Deem, A.K.; Kopetz, S.; Heffernan, T.P.; Draetta, G.F.; Carugo, A. Current and Future Horizons of Patient-Derived Xenograft Models in Colorectal Cancer Translational Research. *Cancers* **2019**, *11*, 1321. [CrossRef] [PubMed]
45. Haddad, A.F.; Young, J.S.; Amara, D.; Berger, M.S.; Raleigh, D.R.; Aghi, M.K.; Butowski, N.A. Mouse Models of Glioblastoma for the Evaluation of Novel Therapeutic Strategies. *Neuro-Oncol. Adv.* **2021**, *3*, 1–16. [CrossRef] [PubMed]
46. Dirkse, A.; Golebiewska, A.; Buder, T.; Nazarov, P.V.; Muller, A.; Poovathingal, S.; Brons, N.H.C.; Leite, S.; Sauvageot, N.; Sarkisjan, D.; et al. Stem Cell-Associated Heterogeneity in Glioblastoma Results from Intrinsic Tumor Plasticity Shaped by the Microenvironment. *Nat. Commun.* **2019**, *10*, 1–16. [CrossRef] [PubMed]
47. Schulte, A.; Günther, H.S.; Martens, T.; Zapf, S.; Riethdorf, S.; Wülfing, C.; Stoupiec, M.; Westphal, M.; Lamszus, K. Glioblastoma Stem-like Cell Lines with Either Maintenance or Loss of High-Level EGFR Amplification, Generated via Modulation of Ligand Concentration. *Clin. Cancer Res.* **2012**, *18*, 1901–1913. [CrossRef] [PubMed]
48. Zhou, Z.; Cong, L.; Cong, X. Patient-Derived Organoids in Precision Medicine: Drug Screening, Organoid-on-a-Chip and Living Organoid Biobank. *Front. Oncol.* **2021**, *11*, 1–16. [CrossRef]
49. Jacob, F.; Ming, G.-L.; Song, H. Generation and Biobanking of Patient-Derived Glioblastoma Organoids and Their Application in CAR T Cell Testing. *Nat. Protoc.* **2020**, *15*, 4000–4033. [CrossRef]
50. Vlachogiannis, G.; Hedayat, S.; Vatsiou, A.; Jamin, Y.; Fernández-mateos, J.; Khan, K.; Lampis, A.; Eason, K.; Huntingford, I.; Burke, R.; et al. Patient-derived organoids model treatment response of metastatic gastrointestinal cancers. *Science* **2018**, *359*, 920–926. [CrossRef]
51. Yan, H.H.N.; Siu, H.C.; Law, S.; Ho, S.L.; Yue, S.S.K.; Tsui, W.Y.; Chan, D.; Chan, A.S.; Ma, S.; Lam, K.O.; et al. A Comprehensive Human Gastric Cancer Organoid Biobank Captures Tumor Subtype Heterogeneity and Enables Therapeutic Screening. *Cell Stem Cell* **2018**, *23*, 882–897.e11. [CrossRef] [PubMed]
52. Steele, N.G.; Chakrabarti, J.; Wang, J.; Biesiada, J.; Holokai, L.; Chang, J.; Nowacki, L.M.; Hawkins, J.; Mahe, M.; Sundaram, N.; et al. An Organoid-Based Preclinical Model of Human Gastric Cancer. *Cell Mol. Gastroenterol. Hepatol.* **2019**, *7*, 161–184. [CrossRef] [PubMed]
53. Ganesh, K.; Wu, C.; O'Rourke, K.P.; Szeglin, B.C.; Zheng, Y.; Sauvé, C.E.G.; Adileh, M.; Wasserman, I.; Marco, M.R.; Kim, A.S.; et al. A Rectal Cancer Organoid Platform to Study Individual Responses to Chemoradiation. *Nat. Med.* **2019**, *25*, 1607–1614. [CrossRef] [PubMed]
54. Ooft, S.N.; Weeber, F.; Dijkstra, K.K.; McLean, C.M.; Kaing, S.; van Werkhoven, E.; Schipper, L.; Hoes, L.; Vis, D.J.; van de Haar, J.; et al. Patient-Derived Organoids Can Predict Response to Chemotherapy in Metastatic Colorectal Cancer Patients. *Sci. Transl. Med.* **2019**, *11*, eaay2574. [CrossRef] [PubMed]
55. Yao, Y.; Xu, X.; Yang, L.; Zhu, J.; Wan, J.; Shen, L.; Xia, F.; Fu, G.; Deng, Y.; Pan, M.; et al. Patient-Derived Organoids Predict Chemoradiation Responses of Locally Advanced Rectal Cancer. *Cell Stem Cell* **2020**, *26*, 17–26.e6. [CrossRef] [PubMed]
56. Narasimhan, V.; Wright, J.A.; Churchill, M.; Wang, T.; Rosati, R.; Lannagan, T.R.M.; Vrbanac, L.; Richardson, A.B.; Kobayashi, H.; Price, T.; et al. Medium-Throughput Drug Screening of Patient-Derived Organoids from Colorectal Peritoneal Metastases to Direct Personalized Therapy. *Clin. Cancer Res.* **2020**, *26*, 3662–3670. [CrossRef] [PubMed]
57. Pasch, C.A.; Favreau, P.F.; Yueh, A.E.; Babiarz, C.P.; Gillette, A.A.; Sharick, J.T.; Karim, M.R.; Nickel, K.P.; DeZeeuw, A.K.; Sprackling, C.M.; et al. Patient-Derived Cancer Organoid Cultures to Predict Sensitivity to Chemotherapy and Radiation. *Clin. Cancer Res.* **2019**, *25*, 5376–5387. [CrossRef]
58. Arena, S.; Corti, G.; Durinikova, E.; Montone, M.; Reilly, N.M.; Russo, M.; Lorenzato, A.; Arcella, P.; Lazzari, L.; Rospo, G.; et al. A Subset of Colorectal Cancers with Cross-Sensitivity to Olaparib and Oxaliplatin. *Clin. Cancer Res.* **2020**, *26*, 1372–1384. [CrossRef] [PubMed]
59. Sharick, J.T.; Walsh, C.M.; Sprackling, C.M.; Pasch, C.A.; Pham, D.L.; Esbona, K.; Choudhary, A.; Garcia-Valera, R.; Burkard, M.E.; McGregor, S.M.; et al. Metabolic Heterogeneity in Patient Tumor-Derived Organoids by Primary Site and Drug Treatment. *Front. Oncol.* **2020**, *10*, 1–17. [CrossRef]
60. Sachs, N.; de Ligt, J.; Kopper, O.; Gogola, E.; Bounova, G.; Weeber, F.; Balgobind, A.V.; Wind, K.; Gracanin, A.; Begthel, H.; et al. A Living Biobank of Breast Cancer Organoids Captures Disease Heterogeneity. *Cell* **2017**, *172*, 373–386.e10. [CrossRef] [PubMed]

61. Tiriac, H.; Belleau, P.; Engle, D.D.; Plenker, D.; Deschênes, A.; Somerville, T.D.D.; Froeling, F.E.M.; Burkhart, R.A.; Denroche, R.E.; Jang, G.H.; et al. Organoid Profiling Identifies Common Responders to Chemotherapy in Pancreatic Cancer. *Cancer Discov.* **2018**, *8*, 1112–1129. [CrossRef] [PubMed]
62. de Witte, C.J.; Espejo Valle-Inclan, J.; Hami, N.; Lõhmussaar, K.; Kopper, O.; Vreuls, C.P.H.; Jonges, G.N.; van Diest, P.; Nguyen, L.; Clevers, H.; et al. Patient-Derived Ovarian Cancer Organoids Mimic Clinical Response and Exhibit Heterogeneous Inter- and Intrapatient Drug Responses. *Cell Rep.* **2020**, *31*, 107762. [CrossRef] [PubMed]
63. Kopper, O.; de Witte, C.J.; Lõhmussaar, K.; Valle-Inclan, J.E.; Hami, N.; Kester, L.; Balgobind, A.V.; Korving, J.; Proost, N.; Begthel, H.; et al. An Organoid Platform for Ovarian Cancer Captures Intra- and Interpatient Heterogeneity. *Nat. Med.* **2019**, *25*, 838–849. [CrossRef] [PubMed]
64. Li, X.; Francies, H.E.; Secrier, M.; Perner, J.; Miremadi, A.; Galeano-Dalmau, N.; Barendt, W.J.; Letchford, L.; Leyden, G.M.; Goffin, E.K.; et al. Organoid Cultures Recapitulate Esophageal Adenocarcinoma Heterogeneity Providing a Model for Clonality Studies and Precision Therapeutics. *Nat. Commun.* **2018**, *9*, 1–13. [CrossRef]
65. Scognamiglio, G.; De Chiara, A.; Parafioriti, A.; Armiraglio, E.; Fazioli, F.; Gallo, M.; Aversa, L.; Camerlingo, R.; Cacciatore, F.; Colella, G.; et al. Patient-Derived Organoids as a Potential Model to Predict Response to PD-1/PD-L1 Checkpoint Inhibitors. *Br. J. Cancer* **2019**, *121*, 979–982. [CrossRef] [PubMed]
66. Saengwimol, D.; Rojanaporn, D.; Chaitankar, V.; Chittavanich, P.; Aroonroch, R.; Boontawon, T.; Thammachote, W.; Jinawath, N.; Hongeng, S.; Kaewkhaw, R. A Three-Dimensional Organoid Model Recapitulates Tumorigenic Aspects and Drug Responses of Advanced Human Retinoblastoma. *Sci. Rep.* **2018**, *8*, 1–13. [CrossRef]
67. Hubert, C.G.; Rivera, M.; Spangler, L.C.; Wu, Q.; Mack, S.C.; Prager, B.C.; Couce, M.; McLendon, R.E.; Sloan, A.E.; Rich, J.N. A Three-Dimensional Organoid Culture System Derived from Human Glioblastomas Recapitulates the Hypoxic Gradients and Cancer Stem Cell Heterogeneity of Tumors Found in Vivo. *Cancer Res.* **2016**, *76*, 2465–2477. [CrossRef]
68. Chadwick, M.; Yang, C.; Liu, L.; Gamboa, C.M.; Jara, K.; Lee, H.; Sabaawy, H.E. Rapid Processing and Drug Evaluation in Glioblastoma Patient-Derived Organoid Models with 4D Bioprinted Arrays. *iScience* **2020**, *23*, 101365. [CrossRef]
69. Loong, H.H.F.; Wong, A.M.; Chan, D.T.M.; Cheung, M.S.H.; Chow, C.; Ding, X.; Chan, A.K.Y.; Johnston, P.A.; Lau, J.Y.W.; Poon, W.S.; et al. Patient-Derived Tumor Organoid Predicts Drugs Response in Glioblastoma: A Step Forward in Personalized Cancer Therapy? *J. Clin. Neurosci.* **2020**, *78*, 400–402. [CrossRef] [PubMed]
70. Gatto, L.; Franceschi, E.; Di Nunno, V.; Tosoni, A.; Lodi, R.; Brandes, A.A. Liquid Biopsy in Glioblastoma Management: From Current Research to Future Perspectives. *Oncologist* **2021**, *26*, 865–878. [CrossRef]
71. Ivanova, E.; Ward, A.; Wiegmans, A.P.; Richard, D.J. Circulating Tumor Cells in Metastatic Breast Cancer: From Genome Instability to Metastasis. *Front. Mol. Biosci.* **2020**, *7*, 1–11. [CrossRef] [PubMed]
72. Lee, H.L.; Chiou, J.F.; Wang, P.Y.; Lu, L.S.; Shen, C.N.; Hsu, H.L.; Burnouf, T.; Ting, L.L.; Chou, P.C.; Chung, C.L.; et al. Ex Vivo Expansion and Drug Sensitivity Profiling of Circulating Tumor Cells from Patients with Small Cell Lung Cancer. *Cancers* **2020**, *12*, 3394. [CrossRef]
73. Ried, K.; Tamanna, T.; Matthews, S.; Eng, P.; Sali, A. New Screening Test Improves Detection of Prostate Cancer Using Circulating Tumor Cells and Prostate-Specific Markers. *Front. Oncol.* **2020**, *10*, 582. [CrossRef]
74. Jiang, M.; Jin, S.; Han, J.; Li, T.; Shi, J.; Zhong, Q.; Li, W.; Tang, W.; Huang, Q.; Zong, H. Detection and Clinical Significance of Circulating Tumor Cells in Colorectal Cancer. *Biomark. Res.* **2021**, *9*, 85. [CrossRef]
75. Lallo, A.; Schenk, M.W.; Frese, K.K.; Blackhall, F.; Dive, C. Circulating Tumor Cells and CDX Models as a Tool for Preclinical Drug Development. *Transl. Lung Cancer Res.* **2017**, *6*, 397–408. [CrossRef] [PubMed]
76. Smit, D.J.; Pantel, K.; Jücker, M. Circulating Tumor Cells as a Promising Target for Individualized Drug Susceptibility Tests in Cancer Therapy. *Biochem. Pharmacol.* **2021**, *188*, 114589. [CrossRef]
77. Lin, K.C.; Ting, L.L.; Chang, C.L.; Lu, L.S.; Lee, H.L.; Hsu, F.C.; Chiou, J.F.; Wang, P.Y.; Burnouf, T.; Ho, D.C.Y.; et al. Ex Vivo Expanded Circulating Tumor Cells for Clinical Anti-Cancer Drug Prediction in Patients with Head and Neck Cancer. *Cancers* **2021**, *13*, 6076. [CrossRef] [PubMed]
78. Wang, R.; Chu, G.C.Y.; Mrdenovic, S.; Annamalai, A.A.; Hendifar, A.E.; Nissen, N.N.; Tomlinson, J.S.; Lewis, M.; Palanisamy, N.; Tseng, H.R.; et al. Cultured Circulating Tumor Cells and Their Derived Xenografts for Personalized Oncology. *Asian J. Urol.* **2016**, *3*, 240–253. [CrossRef] [PubMed]
79. Teng, T.; Kamal, M.; Iriondo, O.; Amzaleg, Y.; Luo, C.; Thomas, A.; Lee, G.; Hsu, C.J.; Nguyen, J.D.; Kang, I.; et al. N-Acetyl-L-Cysteine Promotes Ex Vivo Growth and Expansion of Single Circulating Tumor Cells by Mitigating Cellular Stress Responses. *Mol. Cancer Res.* **2021**, *19*, 441–450. [CrossRef] [PubMed]
80. Müller Bark, J.; Kulasinghe, A.; Chua, B.; Day, B.W.; Punyadeera, C. Circulating Biomarkers in Patients with Glioblastoma. *Br. J. Cancer* **2020**, *122*, 295–305. [CrossRef] [PubMed]
81. Zhang, H.; Yuan, F.; Qi, Y.; Liu, B.; Chen, Q. Circulating Tumor Cells for Glioma. *Front. Oncol.* **2021**, *11*, 1–9. [CrossRef] [PubMed]
82. Lin, D.; Shen, L.; Luo, M.; Zhang, K.; Li, J.; Yang, Q.; Zhu, F.; Zhou, D.; Zheng, S.; Chen, Y.; et al. Circulating Tumor Cells: Biology and Clinical Significance. *Signal Transduct. Target. Ther.* **2021**, *6*, 404. [CrossRef] [PubMed]
83. Crook, T.; Gaya, A.; Page, R.; Limaye, S.; Ranade, A.; Bhatt, A.; Patil, S.; Kumar, P.; Patil, D.; Akolkar, D. Clinical Utility of Circulating Tumor-Associated Cells to Predict and Monitor Chemo-Response in Solid Tumors. *Cancer Chemother. Pharmacol.* **2021**, *87*, 197–205. [CrossRef]

84. Park, J.H.; Feroze, A.H.; Emerson, S.N.; Mihalas, A.B.; Keene, C.D.; Cimino, P.J.; de Lomana, A.L.G.; Kannan, K.; Wu, W.-J.; Turkarslan, S.; et al. A single-cell based precision medicine approach using glioblastoma patient-specific models. *bioRxiv* **2021**, 1–28. [CrossRef]
85. Bouzekri, A.; Esch, A.; Ornatsky, O. Multidimensional Profiling of Drug-Treated Cells by Imaging Mass Cytometry. *FEBS Open Bio* **2019**, *9*, 1652–1669. [CrossRef] [PubMed]
86. Sun, J.; Warden, A.R.; Ding, X. Recent Advances in Microfluidics for Drug Screening. *Biomicrofluidics* **2019**, *13*, 1–13. [CrossRef]
87. Straehla, J.P.; Hajal, C.; Safford, H.C.; Offeddu, G.S. A Predictive Micro Fl Uidic Model of Human Glioblastoma to Assess Traf Fi Cking of Blood—Brain Barrier-Penetrant Nanoparticles. *Proc. Natl. Acad. Sci. USA* **2022**, *119*, e2118697119. [CrossRef] [PubMed]
88. Stevens, M.M.; Maire, C.L.; Chou, N.; Murakami, M.A.; Knoff, D.S.; Kikuchi, Y.; Kimmerling, R.J.; Liu, H.; Haidar, S.; Calistri, N.L.; et al. Drug Sensitivity of Single Cancer Cells Is Predicted by Changes in Mass Accumulation Rate. *Nat. Biotechnol.* **2016**, *34*, 1161–1167. [CrossRef] [PubMed]
89. Stockslager, M.A.; Malinowski, S.; Touat, M.; Yoon, J.C.; Geduldig, J.; Mirza, M.; Kim, A.S.; Wen, P.Y.; Chow, K.H.; Ligon, K.L.; et al. Functional Drug Susceptibility Testing Using Single-Cell Mass Predicts Treatment Outcome in Patient-Derived Cancer Neurosphere Models. *Cell Rep.* **2021**, *37*, 109788. [CrossRef]
90. Kornauth, C.; Pemovska, T.; Vladimer, G.I.; Bayer, G.; Bergmann, M.; Eder, S.; Eichner, R.; Erl, M.; Esterbauer, H.; Exner, R.; et al. Functional Precision Medicine Provides Clinical Benefit in Advanced Aggressive Hematologic Cancers and Identifies Exceptional Responders. *Cancer Discov.* **2022**, *12*, 372–387. [CrossRef]
91. Skaga, E.; Kulesskiy, E.; Brynjulvsen, M.; Sandberg, C.J.; Potdar, S.; Langmoen, I.A.; Laakso, A.; Gaál-Paavola, E.; Perola, M.; Wennerberg, K.; et al. Feasibility study of using high-throughput drug sensitivity testing to target recurrent glioblastoma stem cells for individualized treatment. *Clin. Transl. Med.* **2019**, *8*, 33. [CrossRef] [PubMed]
92. Skaga, E.; Kulesskiy, E.; Fayzullin, A.; Sandberg, C.J.; Potdar, S.; Kyttälä, A.; Langmoen, I.A.; Laakso, A.; Gaál-Paavola, E.; Perola, M.; et al. Intertumoral heterogeneity in patient-specific drug sensitivities in treatment-naïve glioblastoma. *BMC Cancer* **2019**, *19*, 628. [CrossRef]
93. Grube, S.; Freitag, D.; Kalff, R.; Ewald, C.; Walter, J. Characterization of adherent primary cell lines from fresh human glioblastoma tissue, defining glial fibrillary acidic protein as a reliable marker in establishment of glioblastoma cell culture. *Cancer Rep.* **2021**, *4*, e1324. [CrossRef] [PubMed]
94. Oliva, M.A.; Staffieri, S.; Castaldo, S.; Giangaspero, F.; Esposito, V.; Arcella, A. Characterization of primary glioma cell lines derived from the patients according to 2016 CNS tumour WHO classification and comparison with their parental tumours. *J. Neurooncol.* **2021**, *151*, 123–133. [CrossRef] [PubMed]
95. Rapport, F.; Smith, J.; O'Brien, T.A.; Tyrrell, V.J.; Mould, E.V.; Long, J.C.; Gul, H.; Braithwaite, J. Development of an Implementation and Evaluation Strategy for the Australian "Zero Childhood Cancer" (Zero) Program: A Study Protocol. *BMJ Open* **2020**, *10*, e034522. [CrossRef] [PubMed]
96. Reed, M.R.; Lyle, A.G.; De Loose, A.; Maddukuri, L.; Learned, K.; Beale, H.C.; Kephart, E.T.; Cheney, A.; van den Bout, A.; Lee, M.P.; et al. A Functional Precision Medicine Pipeline Combines Comparative Transcriptomics and Tumor Organoid Modeling to Identify Bespoke Treatment Strategies for Glioblastoma. *Cells* **2021**, *10*, 3400. [CrossRef]
97. Shuford, S.; Lipinski, L.; Abad, A.; Smith, A.M.; Rayner, M.; O'Donnell, L.; Stuart, J.; Mechtler, L.L.; Fabiano, A.J.; Edenfield, J.; et al. Prospective Prediction of Clinical Drug Response in High-Grade Gliomas Using an Ex Vivo 3D Cell Culture Assay. *Neuro-Oncol. Adv.* **2021**, *3*, 1–13. [CrossRef]
98. Tanda, E.T.; Vanni, I.; Boutros, A.; Andreotti, V.; Bruno, W.; Ghiorzo, P.; Spagnolo, F. Current State of Target Treatment in BRAF Mutated Melanoma. *Front. Mol. Biosci.* **2020**, *7*, 154. [CrossRef] [PubMed]
99. Dardis, C.; Ashby, L.; Shapiro, W.; Sanai, N. Biopsy vs. extensive resection for first recurrence of glioblastoma: Is a prospective clinical trial warranted? *BMC Res. Notes* **2015**, *8*, 414. [CrossRef] [PubMed]
100. Alberter, B.; Klein, C.A.; Polzer, B. Single-Cell Analysis of CTCs with Diagnostic Precision: Opportunities and Challenges for Personalized Medicine. *Expert Rev. Mol. Diagn.* **2016**, *16*, 25–38. [CrossRef]

International Journal of *Molecular Sciences*

MDPI

Review

A Critical Overview of Targeted Therapies for Vestibular Schwannoma

Ryota Tamura * and Masahiro Toda

Department of Neurosurgery, Keio University School of Medicine, 35 Shinanomachi, Shinjuku-ku, Tokyo 160-8582, Japan; todam@keio.jp
* Correspondence: moltobello-r-610@keio.jp

Abstract: Vestibular schwannoma (VS) is a benign tumor that originates from Schwann cells in the vestibular component. Surgical treatment for VS has gradually declined over the past few decades, especially for small tumors. Gamma knife radiosurgery has become an accepted treatment for VS, with a high rate of tumor control. For neurofibromatosis type 2 (NF2)-associated VS resistant to radiotherapy, vascular endothelial growth factor (VEGF)-A/VEGF receptor (VEGFR)-targeted therapy (e.g., bevacizumab) may become the first-line therapy. Recently, a clinical trial using a VEGFR1/2 peptide vaccine was also conducted in patients with progressive NF2-associated schwannomas, which was the first immunotherapeutic approach for NF2 patients. Targeted therapies for the gene product of SH3PXD2A-HTRA1 fusion may be effective for sporadic VS. Several protein kinase inhibitors could be supportive to prevent tumor progression because merlin inhibits signaling by tyrosine receptor kinases and the activation of downstream pathways, including the Ras/Raf/MEK/ERK and PI3K/Akt/mTORC1 pathways. Tumor-microenvironment-targeted therapy may be supportive for the mainstays of management. The tumor-associated macrophage is the major component of immunosuppressive cells in schwannomas. Here, we present a critical overview of targeted therapies for VS. Multimodal therapy is required to manage patients with refractory VS.

Keywords: schwannoma; NF2; bevacizumab; VEGF; SH3PXD2A-HTRA1 fusion; molecular targeted therapy

Citation: Tamura, R.; Toda, M. A Critical Overview of Targeted Therapies for Vestibular Schwannoma. *Int. J. Mol. Sci.* **2022**, 23, 5462. https://doi.org/10.3390/ijms23105462

Academic Editors: Nives Pećina-Šlaus and Ivana Jovčevska

Received: 13 March 2022
Accepted: 12 May 2022
Published: 13 May 2022

1. Introduction

Schwann cells originate from neural crest cells, which migrate with growing neurites during nerve development. Schwann cells, which form the myelin sheath of an axon, support neuronal function and regeneration [1].

Schwannoma (Sch) is one of the common benign intracranial tumors with an incidence of 1 per 100,000 [2]. Sch often presents between the ages of 40 and 60 years [2]. Among these cases, 80–90% originate from the vestibular nerve. About 5–10% of vestibular Schs (VSs) are observed as bilateral in neurofibromatosis 2 (NF2) patients. A total of 95% of NF2 patients show bilateral VSs [3]. About 60% of unilateral VSs and 90% of bilateral VSs show NF2 gene mutation and the dysfunction of its transcription product, moesin–ezrin–radixin-like (merlin) protein [4].

Currently, the mainstays of management are observation, surgery, and radiosurgery. Surgery with facial and auditory monitoring remains the only curative treatment for growing VSs of all sizes. Stereotactic radiosurgery is considered as a widely accepted treatment option for small-sized VSs. For larger tumors, combined treatment strategies are mostly recommended. In particular, gamma knife radiosurgery (GKRS) has become an accepted treatment for VS [5]. However, additional treatment is needed for some refractory cases. Tumor volume $\geq$15 cm^3 is a significant factor predicting poor tumor control following GKRS [6]. There is no approved medical therapy for VS. For refractory VS with high risks of surgical treatment or GKRS, medical therapies that can slow tumor growth are urgently needed. Here, we review the molecular biology and its relevance to treatment for VS.

2. NF2 Gene

NF2 is an autosomal-dominant disease caused by a biallelic loss of the NF2 gene on chromosome 22. Although 50% of NF2 patients have an affected parent with the disease, the remaining 50% have de novo gene mutations [7].

Although 60% of patients with de novo NF2 show mosaic NF2, the actual diagnostic rate of this condition remains low at 20% because of the difficulties in detecting NF2 variants with a low variant allele frequency [8]. Teranishi et al. improved the diagnostic rate of mosaic NF2 using targeted deep sequencing of DNA. The mosaic NF2 phenotype was found to be different from that in the NF2 germline variant in terms of tumor growth and hearing outcome [8].

Differentiated Schwann cells become quiescent because merlin regulates this contact-dependent inhibition of proliferation. Merlin plays a significant role in regulating the actin cytoskeleton, adhesion junction formation, and cell proliferation [9]. Merlin can regulate multiple tumorigenic pathways, including retrovirus-associated DNA sequences (Ras)/rapidly accelerated fibrosarcoma (Raf)/mitogen extracellular signal-regulated kinase (MEK)/extracellular-signal-regulated kinases (ERK), and mammalian target of rapamycin complex 1 (mTORC1)/phosphoinositide 3-kinase (PI3K)/Akt [10,11].

3. SH3PXD2A-HTRA1 Fusion

In 2016, alternative tumorigenic mechanisms were proposed, including a recurrent in-frame fusion transcript of the HTRA1 and SH3PXD2A genes. The gene product of SH3PXD2A-HTRA1 fusion promotes proliferation and invasion. In a previous study, the frequency of this fusion gene was investigated [12]. The fusion gene SH3PXD2A-HTRA1, activating the MAPK pathway, has been associated with 10% of sporadic Schs. Agnihotri et al. suggested that SH3PXD2A-HTRA1 fusion promoted tumorigenesis and sensitivity to an MEK-ERK inhibitor [12]. Even though SH3PXD2A-HTRA1 fusion has been shown to be a driver of tumorigenesis, the fusion transcript was extremely rare in Norwegian sporadic VS patients [13]. Further investigation is warranted to elucidate the importance of this fusion gene.

4. Protein-Kinase-Related Pathway

4.1. VEGF-A/VEGFRs

The vascular endothelial growth factor receptor (VEGFR) family mainly includes VEGFR-1 (Flt-1), VEGFR-2 (Flk-1/KDR), and VEGFR-3 (Flt-4), which are important regulators of physiological and pathological angiogeneses [14]. Merlin deletion leads to the downregulation of the protein semaphorin 3F, which inhibits VEGF-mediated angiogenesis [15]. A previous study has shown that the concentrations of VEGF-A and VEGFR-1 are related to the growth rate of VS [16].

Tumor shrinkage and hearing improvement have been identified after the administration of bevacizumab (anti-VEGF-A antibody) in about 41% and 20% of progressive VSs in NF2 patients, respectively [17]. Bevacizumab may be considered as first-line medical therapy for rapidly growing VS. In a recent meta-analysis, the median treatment duration was 16 months [18,19]. Recently, the first phase III randomized clinical trial using bevacizumab was conducted in Japan [20,21]. Furthermore, progressive sporadic VS also exhibited significant tumor shrinkage after bevacizumab administration [22].

However, some aspects of bevacizumab treatment are problematic, such as the need for frequent parenteral administration, side effects, apparent drug resistance, and rebound tumor progression [23]. In the majority of published case series of bevacizumab usage for VS, their conclusions on efficacy were based on relatively short follow-ups. Long-term follow-up studies using a large number of patients are warranted. A clinical trial using a VEGFR-1/2 peptide vaccine was also conducted in patients with progressive NF2-derived

Schs, showing hearing improvement and tumor volume reduction. Memory cytotoxic T lymphocytes have the possibility to persist in the long-term [24]. This was the first immunotherapeutic approach for NF2 patients.

4.2. ErbB

The ErbB family's cell membrane receptors include the epidermal growth factor receptor (EGFR) (HER1/ErbB-1), HER2 (neu/ErbB-2), HER3 (ErbB-3), and HER4 (ErbB-4). MAPK/ERK and PI3K/Akt signaling pathways are considerably downstream of ErbB-2 activation [25]. ErbB receptors were activated in both sporadic and NF2-related VSs, and EGFR expression levels correlated with Sch size [26]. Furthermore, EGF was upregulated in NF2-related VS but not in sporadic VS, suggesting that an EGFR inhibitor might have efficacy in NF2 patients [27,28].

The predominant ErbB receptor dimerization patterns in VS are EGFR and ErbB2 heterodimers [29]. Trastuzumab, a humanized anti-ErbB2 monoclonal antibody, could significantly reduce tumor growth; however, this antibody did not induce significant cell death in VS xenografts [29].

Lapatinib is a potent and reversible tyrosine kinase inhibitor, showing a dual inhibitory effect on the EGF activation of EGFR/ErbB2 [30]. A phase II clinical trial showed that lapatinib has minor toxicity and the minor effects of reducing tumor volume and improving hearing in NF2-related progressive VS [30]. This treatment failure was due to the ErbB3 upregulation caused by the inhibition of ErbB2. Erlotinib is a reversible, small-molecule EGFR-specific tyrosine kinase inhibitor [30]. However, erlotinib was ineffective in NF2-related VSs for tumor shrinkage and improving hearing outcome. Bevacizumab has shown better benefits in the treatment of NF2 patients compared with lapatinib and erlotinib [31].

4.3. PDGFR

Platelet-derived growth factor (PDGF) regulated the migration of mesenchymal stem cells via PI3K signaling [32]. The PDGF receptor (PDGFR) family includes PDGFR-α, PDGFR-β, colony-stimulating factor1 receptor (CSF1-R), fetal liver kinase-2 (Flk-2), and c-kit [32]. Compared with normal nerves, the expressions of c-kit, PDGFR-α, and PDGFR-β are increased in sporadic and NF2-related VSs [33]. Imatinib mesylate (STI571) is an inhibitor of the BCR-ABL fusion kinase for chronic myelogenous leukemia [34]. Imatinib mesylate inhibits the activation of c-KIT, PDGFR-α, and PDGFR-β and their downstream signaling pathways, leading to increased apoptosis in the immortalized NF2-null VS cell line. Moreover, imatinib has an inhibitory effect for angiogenesis in both sporadic and NF2-related VSs [35].

Nilotinib (Bcr-Abl tyrosine kinase inhibitor) is 10–30-fold more potent than imatinib in inhibiting Bcr-Abl tyrosine kinase activity and proliferation [36]. Nilotinib also inhibited cell proliferation more effectively compared with imatinib in Sch cell lines. Anti-tumor effects were related to the inhibition of PDGFR-α and PDGFR-β, as well as their downstream signaling mediators, Akt and mTOR [36].

Ponatinib inhibits SRC, fibroblast growth factor receptor (FGFR), PDGFR, and VEGFR1–3, stimulating a robust G1 cell cycle arrest of merlin-deficient human Schwann cells [37]. However, in the clinical setting, targeting PDGF/PDGFR signaling did not show significant benefits in the treatment of NF2 patients.

4.4. HGFR

The hepatocyte growth factor receptor (HGFR), known as c-mesenchymal–epithelial transition (c-MET), is a glycosylated receptor tyrosine kinase and plays a role in driving tumorigenesis [38,39]. The activation of the HGF/c-MET pathway in sporadic VS can promote the inflammation network and cancer progression [40]. This pathway can also protect cells from apoptosis induced by chemotherapy or radiotherapy through PI3K/Akt signaling [41].

Therefore, crizotinib (a c-MET and anaplastic lymphoma kinase inhibitor) can enhance the radiation-induced DNA damage of NF2-related Sch cells, enhancing radio sensitivity. This effect leads to a reduction in radiation dose and protects hearing [42]. A phase II clinical trial using crizotinib for NF2 and progressive sporadic VSs in children and adults is ongoing (NCT04283669). The simultaneous use of the c-MET inhibitor "cabozantinib" and the Src inhibitor "saracatinib" can reduce the viability of human VS cells with he NF2 mutation, which is more effective than using either inhibitor alone [43].

There is a crosstalk between c-MET and VEGF-A in VSs. Sonam et al. found that c-MET and VEGF-A protein levels decreased using c-MET-targeted siRNA, while VEGF-A-targeted siRNA reduced c-MET expression. The combined inhibition of VEGF-A and c-MET may be an effective therapy [40].

4.5. PI3K/Akt/mTOR

PI3K/Akt/mTOR signaling contributes to a variety of processes that are critical in anabolic reactions and cell growth and survival. PI3K/Akt/mTOR signaling is elevated in VS [44]. Therefore, the PI3K/Akt pathway is also an attractive treatment target for VS [44].

OSU-03012 is an ATP-competitive inhibitor of PAK activity and suppresses the phosphorylation of Akt, which inhibits VS cell growth and promotes apoptosis [45]. Additionally, OSU-HDAC42 (AR-42), a novel phenylbutyrate-derived histone deacetylase inhibitors, can inhibit the downstream Akt expression of PI3K through protein phosphatase-1-mediated Akt dephosphorylation, showing the effect of G2 cell cycle arrest and cell apoptosis in a VS animal model [46].

mTOR is a downstream signal of the PI3K/Akt pathway [47]. A previous study has shown that an mTORC1 inhibitor (rapamycin) can inhibit the growth of merlin-deficient tumors in vivo. Rapamycin can lead to tumor shrinkage in NF2 patients with growing VSs [48].

Everolimus (RAD001), a derivative of rapamycin, can inhibit mTORC1 and reduce tumor angiogenesis. Although a phase II study has shown that everolimus is ineffective in progressive NF2-related VS patients [49], another study has shown that everolimus reduced the tumor volume in 55.6% of patients with NF2-related VS [50,51]. The effect of everolimus is still debatable.

5. Cytokines and Chemokines

C-X-C motif chemokine ligand 12 (CXCL12) binds to C-X-C chemokine receptor type 4 (CXCR4). The CXCL12/CXCR4 axis plays a pivotal role in tumor development, survival, angiogenesis, metastasis, and the tumor microenvironment. In addition, this chemokine axis promotes chemoresistance in cancer therapy. CXCR4 is also considered to be correlated with the tumorigenesis and functional disturbance of sporadic and NF2-related VSs [52]. CXCR4-directed positron emission tomography/computed tomography imaging with radiolabeled CXCR4-targeted ligand [68Ga]-Pentixafor was used to evaluate CXCR4 expression in VS patients [53]. These results provide a possibility for the use of Plerixafor (AMD3100) as a CXCR4-targeting drug [52,53].

Multiple cytokines and chemokines, including CXCL12, CXCL16, interleukin (IL)-1β, IL-6, IL-34, macrophage colony-stimulating factor (M-CSF), and tumor necrosis factor-α (TNF-α), are also associated with tumor progression and hearing disturbance [54].

In addition to the direct compression of auditory nerve fibers by tumors, in cases of NF2-associated deafness, detrimental paracrine substances, such as proinflammatory cytokines from tumors, have been proposed as a mechanism of cochlear hearing loss [55]. A novel therapeutic strategy targeting cytokines and chemokines may support other treatment strategies.

6. Tumor Microenvironment

Sch consists of different cell types, including tumorigenic Schwann cells, axons, macrophages, T cells, fibroblasts, blood vessels, and an extracellular matrix. The tumor microenvironment plays a relevant role in the development and progression of Sch. There are few studies regarding the tumor microenvironment in Sch [56,57].

Fast-growing VSs expressed high M-CSF and IL-34 levels that could regulate the chemotaxis of tumor-associated macrophages (TAMs). TAMs produce growth factors and anti-inflammatory cytokines to suppress the host immune response, resulting in tumor progression. VEGF in the hypoxic tumor microenvironment is a key factor for transitioning from the M1 to the M2 macrophage phenotype [58]. A greater TAM infiltration was found in growing sporadic VSs compared with non-growing sporadic VSs [54,59,60].

Programmed death-1 (PD-1) is expressed on CD8+T cells. Programmed death-ligand 1 (PD-L1) is expressed on tumor cells in numerous malignant tumors and binds to PD-1 to negatively regulate the immune response of CD8+T cells [58]. In 11 NF2-associated Schs, both high levels of programmed death-ligand 1 (PD-L1) expression and the presence of TAMs and T lymphocytes were identified in nearly all specimens [61]. In another study of 44 sporadic Schs, an increased presence of TAMs and an elevated PD-L1 expression were significantly associated with tumor progression [62].

Regulatory T cells (Tregs) (CD4 + CD25 + Foxp3+) play an active and significant role in the progression of tumors, and they play an important role in suppressing tumor-specific immunity [58]. In NF2 patients, the number of Foxp3-positive cells in Sch with a progressive course was significantly higher than in those without a progressive course, suggesting that growth may be associated with Foxp3-positive Tregs [24,63].

A previous study investigated the hypoxic tumor microenvironment of patients with NF2 Sch. Hypoxia was important for the shorter progression-free survival of NF2 Sch [59]. An immunotherapy that specifically targets the tumor microenvironment may emerge as a new class of Sch therapeutics.

7. Inflammation and Stress Reaction

7.1. COX2

The expression of cyclooxygenase 2 (COX-2) is associated with sporadic and NF2-related VS proliferation. Mutations in the NF2 gene can activate the Hippo pathway, in which YAP can promote the transcription of COX-2 for prostaglandin production. Prostaglandin E2 (PGE2) catalyzed by COX-2 has multiple roles in cell proliferation, apoptosis, angiogenesis, inflammation, and immune monitoring. COX-2 inhibitors may have the potential to inhibit the growth of VS [64,65].

A negative correlation between aspirin users and sporadic VS growth has been demonstrated [66,67]. In addition to inhibiting COX-2, aspirin can also suppress the activated NF-κB pathway in VS, which may be another potential mechanism. However, other studies demonstrated that there is no growth inhibitory effect for celecoxib on NF2-related VS or aspirin on sporadic VS [66,67]. Other studies have shown that NSAIDs, glucocorticoids, and other immunosuppressive drugs could not alter the expression of COX-2 in sporadic Sch [68].

7.2. Hsp90

Heat shock protein 90 (HSP90) is a ubiquitous molecule. The absence of Hsp90 results in proteasomal degradation [69]. The dysregulation of the Hippo pathway is necessary for schwannomagenesis, and MAPK signaling acts as a modifier for Sch formation. Furthermore, the pharmacological co-inhibition of YAP/TAZ transcriptional activity and MAPK signaling shows a synergistic size reduction in a mouse Sch model [70].

In a recent study, a novel small-molecule inhibitor compound of HSP90, NXD30001 (pochoxime A), was able to show reduced growth of NF2-deficient tumors in vivo. There are no current clinical trials using an HSP90 inhibitor [71].

The molecular patterns and mutations described for VS are summarized in Table 1.

Table 1. Molecular patterns and mutations currently described for VS.

	Targeted Pathway
	NF2 (merlin)-related pathway
1	Ras/Raf/MEK/ERK signaling
2	PI3K/Akt/mTORC1 signaling
	SH3PXD2A-HTRA1-fusion-related pathway
1	MAPK signaling
	Protein-kinase-related pathway
1	VEGF-A/VEGFR signaling
2	ErbB family signaling
3	PDGF/PDGFR signaling
4	HGF/HGFR (c-MET) signaling
	Cytokines and chemokines
1	CXCL12/CXCR4 signaling
2	IL-1β, IL-6, IL-34, M-CSF, TNF-α
	Tumor microenvironment
1	Tumor-associated macrophage
2	Regulatory T cell
3	PD-1/PD-L1
4	Hypoxia
	Inflammation and stress reaction
1	COX2
2	Hsp90

c-MET, c-mesenchymal–epithelial transition; COX2, cyclooxygenase 2; CXCL12, C-X-C motif chemokine ligand 12; CXCR4, C-X-C chemokine receptor type 4; ERK, extracellular-signal-regulated kinases; HGFR, hepatocyte growth factor receptor; Hsp90, heat shock protein 90; IL, interleukin; MAPK, mitogen-activated protein kinase; M-CSF, macrophage colony-stimulating factor; MEK, mitogen extracellular signal-regulated kinase; mTORC1, mammalian target of rapamycin complex 1; NF, neurofibromatosis; PDGFR, platelet-derived growth factor; PD-1, programmed death-1; PD-L1, programmed death-ligand 1; PI3K, phosphoinositide 3-kinase; Raf, rapidly accelerated fibrosarcoma; TNF-α, tumor necrosis factor-α; VEGF, vascular endothelial growth factor; VEGFR, vascular endothelial growth factor receptor.

8. Drug Repositioning

Mifepristone (RU486), a progesterone and glucocorticoid receptor antagonist that has already been approved for medical abortion, was chosen as the most promising candidate drug [72]. In a preclinical study, mifepristone reduced cellular proliferation in primary human VS cultures regardless of NF2 mutation. A phase II clinical trial on mifepristone in VS is currently being planned [72].

In VS, genes associated with NLRP3 were significantly upregulated in patients with poor hearing. NLRP3 mutation is associated with cochlear autoinflammation in conjunction with DFNA34-mediated hearing loss and age-rated hearing loss. The activation of NLRP3 triggers the production of IL-1β [73]. A recombinant human IL-1 receptor antagonist reversed the hearing loss observed in a family with sensorineural hearing loss and NLRP3 mutations [54].

9. Gene Therapy

Gene therapy offers the potential to treat a wide range of inherited and acquired human diseases. The direct modulation of affected genes in specific cell types represents the most powerful treatment strategy for NF2 patients. Delivery platforms typically include viral vectors, such as retroviruses, adenoviruses, and adeno-associated viruses (AAVs), as well as nonviral vectors, including nanoparticles and polymers [74].

A direct injection of an AAV serotype 1 vector encoding caspase-1 (ICE) under the Schwann-cell specific promoter led to the regression of Sch in a mouse model. Recently, a direct injection of AAV1 encoding the apoptosis-associated speck-like protein reduced tumor growth and resolved tumor-associated pain in a human xenograft Sch model [75].

Nonviral vectors, such as liposomal-, polymeric-, and peptide-based nanoparticles, offer an attractive alternative for gene delivery. Liposomes were used to deliver genome-editing agents to the cochlea of neonatal mice with dominant genetic deafness. By decorating the nanoparticle surface with a peptide targeting Schwann cells, peptide-based nanoparticles were used to deliver genetic materials, resulting in a decreased secretion of an ototoxic inflammatory cytokine from tumor cells [76].

10. Ongoing Clinical Trials

Table 2 shows ongoing clinical trials using multimodal treatment strategies for Sch. The superselective intraarterial infusion of bevacizumab is performed to control tumor progression (NCT01083966). Because of the promising results found with bevacizumab, it may be safely used by direct intracranial superselective intraarterial infusion up to a dose of 10mg/kg in order to enhance survival and hearing function. Another six trials are using medical treatment strategies. Crizotinib, AR-42 (OSU-HDAC42), everolimus, selumetinib (MEK 1/2 inhibitor), and tanezumab (a monoclonal antibody against nerve growth factor as a treatment for pain) are being evaluated in the trials. A previous meta-analysis suggests that there is insufficient evidence to recommend aspirin usage in patients with VS [77,78]. High-quality trials are warranted to determine the efficacy of aspirin in reducing VS growth (NCT03079999).

Table 2. Active and recruiting clinical trials using medical therapeutic approaches for schwannoma.

ClinicalTrials.Gov Identifier	ID	RP	EE	Age	TS
NCT01083966	8, 2011	Lenox Hill Brain Tumor Center	30	≥18	Superselective intraarterial intracranial infusion of bevacizumab
NCT04283669	2, 2020	University of Alabama at Birmingham	19	≥6	Crizotinib
NCT03079999	6, 2018	Massachusetts Eye and Ear Infirmary	300	≥12	Aspirin
NCT02282917	9, 2015	Massachusetts Eye and Ear	5	≥18	AR-42 (OSU-HDAC42)
NCT01345136	7, 2015	University of California	4	16–65	Everolimus
NCT03095248	5, 2017	Children's Hospital Medical Center	34	3–45	Selumetinib
NCT04163419	4, 2020	Massachusetts General Hospital	46	≥18	Tanezumab

ER, estimated enrollment; ID, initiation date; RP, responsible party; TS, treatment strategy.

11. Future Direction

Bevacizumab has recently been considered as the first-line medical therapy for rapidly growing VS. Furthermore, new therapeutic strategies targeting the SH3PXD2A-HTRA1 fusion gene, several protein kinases, and the tumor microenvironment may be supportive for the mainstays of management. An immunotherapeutic approach may also be needed to control multiple tumor progression in the long term. In addition to the standard treatment strategy, including surgery and radiotherapy, these targeted medical therapies are needed for multiple and large tumors of VS (Figure 1). Multimodal therapy is required to manage patients with refractory VS.

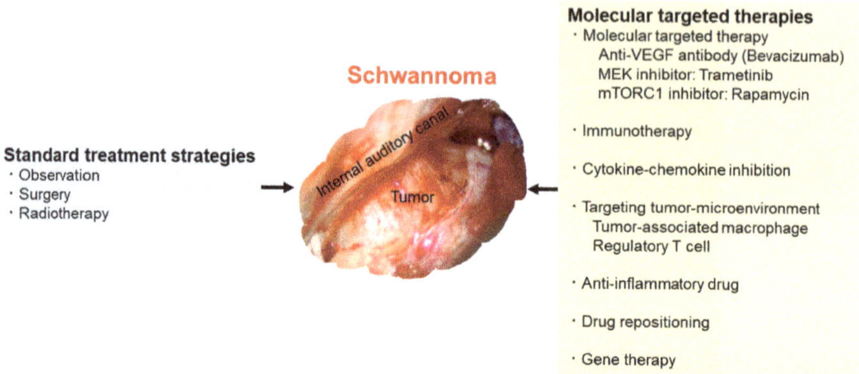

Figure 1. Multimodal treatment and management strategies.

The mainstays of management are observation, surgery, and radiation therapy. Bevacizumab has recently been considered as the first-line medical therapy for rapidly growing vestibular schwannomas. Furthermore, new therapeutic strategies targeting the SH3PXD2A-HTRA1 fusion gene, several protein kinases, and the tumor microenvironment may be supportive for the mainstays of management.

Author Contributions: Conception and Design: R.T.; Manuscript Writing: R.T.; Manuscript Reviewing: M.T.; Supervision: M.T. All authors have read and agreed to the published version of the manuscript.

Funding: This research received no external funding.

Institutional Review Board Statement: Not applicable.

Informed Consent Statement: Not applicable.

Data Availability Statement: Not applicable.

Conflicts of Interest: The authors declare no potential conflict of interest.

References

1. Jessen, K.R.; Mirsky, R.; Lloyd, A.C. Schwann Cells: Development and Role in Nerve Repair. *Cold Spring Harb. Perspect. Biol.* **2015**, *7*, a020487. [CrossRef] [PubMed]
2. Fisher, J.L.; Pettersson, D.; Palmisano, S.; Schwartzbaum, J.; Edwards, C.G.; Mathiesen, T.; Prochazka, M.; Bergenheim, T.; Florentzson, R.; Harder, H.; et al. Loud Noise Exposure and Acoustic Neuroma. *Am. J. Epidemiol.* **2014**, *180*, 58–67. [CrossRef] [PubMed]
3. Slattery, W.H. Neurofibromatosis type 2. *Otolaryngol. Clin. N. Am.* **2015**, *48*, 443–460. [CrossRef] [PubMed]
4. Evans, D.G.R.; Ramsden, R.T.; Shenton, A.; Gokhale, C.; Bowers, N.L.; Huson, S.M.; Pichert, G.; Wallace, A. Mosaicism in neurofibromatosis type 2: An update of risk based on uni/bilaterality of vestibular schwannoma at presentation and sensitive mutation analysis including multiple ligation-dependent probe amplification. *J. Med. Genet.* **2007**, *44*, 424–428. [CrossRef] [PubMed]
5. Yao, L.; Alahmari, M.; Temel, Y.; Hovinga, K. Therapy of Sporadic and NF2-Related Vestibular Schwannoma. *Cancers* **2020**, *12*, 835. [CrossRef] [PubMed]
6. Huang, C.W.; Tu, H.T.; Chuang, C.Y.; Chang, C.S.; Chou, H.H.; Lee, M.T.; Huang, C.F. Gamma Knife radiosurgery for large vestibular schwannomas greater than 3 cm in diameter. *J. Neurosurg.* **2018**, *128*, 1380–1387. [CrossRef]
7. Zhang, Z.-Y.; Wu, Y.-Y.; Cai, X.-Y.; Fang, W.-L.; Xiao, F.-L. Molecular Diagnosis of Neurofibromatosis by Multigene Panel Testing. *Front. Genet.* **2021**, *12*, 603195. [CrossRef]

8. Teranishi, Y.; Miyawaki, S.; Hongo, H.; Dofuku, S.; Okano, A.; Takayanagi, S.; Ota, T.; Yoshimura, J.; Qu, W.; Mitsui, J.; et al. Targeted deep sequencing of DNA from multiple tissue types improves the diagnostic rate and reveals a highly diverse phenotype of mosaic neurofibromatosis type 2. *J. Med. Genet.* **2021**, *58*, 701–711. [CrossRef]
9. Gladden, A.B.; Hebert, A.M.; Schneeberger, E.E.; McClatchey, A.I. The NF2 Tumor Suppressor, Merlin, Regulates Epidermal Development through the Establishment of a Junctional Polarity Complex. *Dev. Cell* **2010**, *19*, 727–739. [CrossRef]
10. Pećina-Šlaus, N. Merlin, the NF2 Gene Product. *Pathol. Oncol. Res.* **2013**, *19*, 365–373. [CrossRef]
11. Santarpia, L.; Lippman, S.M.; El-Naggar, A.K. Targeting the MAPK-RAS-RAF signaling pathway in cancer therapy. *Expert Opin. Ther. Targets* **2012**, *16*, 103–119. [CrossRef] [PubMed]
12. Agnihotri, S.; Jalali, S.; Wilson, M.R.; Danesh, A.; Li, M.; Klironomos, G.; Krieger, J.R.; Mansouri, A.; Khan, O.; Mamatjan, Y.; et al. The genomic landscape of schwannoma. *Nat. Genet.* **2016**, *48*, 1339–1348. [CrossRef] [PubMed]
13. Taule-Sivertsen, P.; Bruland, O.; Håvik, A.L.; Bratland, E.; Lund-Johansen, M.; Knappskog, P.M. The SH3PXD2A-HTRA1 fusion transcript is extremely rare in Norwegian sporadic vestibular schwannoma patients. *J. Neuro-Oncol.* **2021**, *154*, 35–40. [CrossRef] [PubMed]
14. Simons, M.; Gordon, E.; Claesson-Welsh, L. Mechanisms and regulation of endothelial VEGF receptor signalling. *Nat. Rev. Mol. Cell Biol.* **2016**, *17*, 611–625. [CrossRef] [PubMed]
15. Wong, H.K.; Shimizu, A.; Kirkpatrick, N.D.; Garkavtsev, I.; Chan, A.W.; di Tomaso, E.; Klagsbrun, M.; Jain, R.K. Merlin/NF2 regulates angiogenesis in schwannomas through a Rac1/semaphorin 3F-dependent mechanism. *Neoplasia* **2012**, *14*, 84–94. [CrossRef] [PubMed]
16. Uesaka, T.; Shono, T.; Suzuki, S.O.; Nakamizo, A.; Niiro, H.; Mizoguchi, M.; Iwaki, T.; Sasaki, T. Expression of VEGF and its receptor genes in intracranial schwannomas. *J. Neuro-Oncol.* **2007**, *83*, 259–266. [CrossRef]
17. Plotkin, S.R.; Duda, D.G.; Muzikansky, A.; Allen, J.; Blakeley, J.; Rosser, T.; Campian, J.L.; Clapp, D.W.; Fisher, M.J.; Tonsgard, J.; et al. Multicenter, Prospective, Phase II and Biomarker Study of High-Dose Bevacizumab as Induction Therapy in Patients with Neurofibromatosis Type 2 and Progressive Vestibular Schwannoma. *J. Clin. Oncol.* **2019**, *37*, 3446–3454. [CrossRef]
18. Shi, J.; Lu, D.; Gu, R.; Sun, H.; Yu, L.; Pan, R.; Zhang, Y. Reliability and toxicity of bevacizumab for neurofibromatosis type 2-related vestibular schwannomas: A systematic review and meta-analysis. *Am. J. Otolaryngol.* **2021**, *42*, 103148. [CrossRef]
19. Morris, K.A.; Golding, J.F.; Axon, P.R.; Afridi, S.; Blesing, C.; Ferner, R.E.; Halliday, D.; Jena, R.; Pretorius, P.M.; Evans, G.; et al. Bevacizumab in neurofibromatosis type 2 (NF2) related vestibular schwannomas: A nationally coordinated approach to delivery and prospective evaluation. *Neuro-Oncol. Pract.* **2016**, *3*, 281–289. [CrossRef]
20. Fujii, M.; Ichikawa, M.; Iwatate, K.; Bakhit, M.; Yamada, M.; Kuromi, Y.; Sato, T.; Sakuma, J.; Saito, K. Bevacizumab Therapy of Neurofibromatosis Type 2 Associated Vestibular Schwannoma in Japanese Patients. *Neurol. Med. Chir.* **2020**, *60*, 75–82. [CrossRef]
21. Fujii, M.; Kobayakawa, M.; Saito, K.; Inano, A.; Morita, A.; Hasegawa, M.; Mukasa, A.; Mitsuhara, T.; Goto, T.; Yamaguchi, S.; et al. Rationale and Design of BeatNF2 Trial: A Clinical Trial to Assess the Efficacy and Safety of Bevacizumab in Patients with Neurofibromatosis Type 2 Related Vestibular Schwannoma. *Curr. Oncol.* **2021**, *28*, 726–739. [CrossRef] [PubMed]
22. Karajannis, M.A.; Hagiwara, M.; Schreyer, M.; Haque, S. Sustained imaging response and hearing preservation with low-dose bevacizumab in sporadic vestibular schwannoma. *Neuro-Oncology* **2019**, *21*, 822–824. [CrossRef] [PubMed]
23. Tamura, R.; Tanaka, T.; Miyake, K.; Yoshida, K.; Sasaki, H. Bevacizumab for malignant gliomas: Current indications, mechanisms of action and resistance, and markers of response. *Brain Tumor Pathol.* **2017**, *34*, 62–77. [CrossRef]
24. Tamura, R.; Fujioka, M.; Morimoto, Y.; Ohara, K.; Kosugi, K.; Oishi, Y.; Sato, M.; Ueda, R.; Fujiwara, H.; Hikichi, T.; et al. A VEGF receptor vaccine demonstrates preliminary efficacy in neurofibromatosis type 2. *Nat. Commun.* **2019**, *10*, 5758. [CrossRef] [PubMed]
25. Wee, P.; Wang, Z. Epidermal Growth Factor Receptor Cell Proliferation Signaling Pathways. *Cancers* **2017**, *9*, 52. [CrossRef]
26. Ahmad, Z.K.; Brown, C.M.; Cueva, R.A.; Ryan, A.F.; Doherty, J.K. ErbB Expression, Activation, and Inhibition with Lapatinib and Tyrphostin (AG825) in Human Vestibular Schwannomas. *Otol. Neurotol.* **2011**, *32*, 841–847. [CrossRef]
27. Doherty, J.K.; Ongkeko, W.; Crawley, B.; Andalibi, A.; Ryan, A.F. ErbB and Nrg: Potential Molecular Targets for Vestibular Schwannoma Pharmacotherapy. *Otol. Neurotol.* **2008**, *29*, 50–57. [CrossRef]
28. Zhang, Y.; Long, J.; Ren, J.; Huang, X.; Zhong, P.; Wang, B. Potential Molecular Biomarkers of Vestibular Schwannoma Growth: Progress and Prospects. *Front. Oncol.* **2021**, *11*, 731441. [CrossRef]
29. Clark, J.J.; Provenzano, M.; Diggelmann, H.R.; Xu, N.; Hansen, S.S.; Hansen, M.R. The ErbB inhibitors trastuzumab and erlotinib inhibit growth of vestibular schwannoma xenografts in nude mice: A preliminary study. *Otol. Neurotol.* **2008**, *29*, 846–853. [CrossRef]
30. Ammoun, S.; Cunliffe, C.H.; Allen, J.; Chiriboga, L.; Giancotti, F.G.; Zagzag, D.; Hanemann, C.O.; Karajannis, M.A. ErbB/HER receptor activation and preclinical efficacy of lapatinib in vestibular schwannoma. *Neuro-Oncology* **2010**, *12*, 834–843. [CrossRef]
31. Plotkin, S.R.; Halpin, C.; McKenna, M.J.; Loeffler, J.S.; Batchelor, T.T.; Barker, F.G., 2nd. Erlotinib for progressive vestibular schwannoma in neurofibromatosis 2 patients. *Otol. Neurotol.* **2010**, *31*, 1135–1143. [CrossRef] [PubMed]
32. Salha, S.; Gehmert, S.; Brébant, V.; Anker, A.; Loibl, M.; Prantl, L.; Gehmert, S. PDGF regulated migration of mesenchymal stem cells towards malignancy acts via the PI3K signaling pathway. *Clin. Hemorheol. Microcirc.* **2018**, *70*, 543–551. [CrossRef] [PubMed]
33. Mukherjee, J.; Kamnasaran, D.; Balasubramaniam, A.; Radovanovic, I.; Zadeh, G.; Kiehl, T.-R.; Guha, A. Human Schwannomas Express Activated Platelet-Derived Growth Factor Receptors and c-kit and Are Growth Inhibited by Gleevec (Imatinib Mesylate). *Cancer Res.* **2009**, *69*, 5099–5107. [CrossRef] [PubMed]

34. O'Dwyer, M.E.; Druker, B.J. STI571: An inhibitor of the BCR-ABL tyrosine kinase for the treatment of chronic myelogenous leukaemia. *Lancet Oncol.* **2000**, *1*, 207–211. [CrossRef]
35. Yener, U.; Avsar, T.; Akgün, E.; Şeker, A.; Bayri, Y.; Kılıç, T. Assessment of antiangiogenic effect of imatinib mesylate on vestibular schwannoma tumors using in vivo corneal angiogenesis assay laboratory investigation. *J. Neurosurg.* **2012**, *117*, 697–704. [CrossRef] [PubMed]
36. Ammoun, S.; Schmid, M.C.; Triner, J.; Manley, P.; Hanemann, C.O. Nilotinib alone or in combination with selumetinib is a drug candidate for neurofibromatosis type 2. *Neuro-Oncology* **2011**, *13*, 759–766. [CrossRef]
37. Petrilli, A.M.; Garcia, J.; Bott, M.; Plati, S.K.; Dinh, C.T.; Bracho, O.R.; Yan, D.; Zou, B.; Mittal, R.; Telischi, F.F.; et al. Ponatinib promotes a G1 cell-cycle arrest of merlin/NF2-deficient human schwann cells. *Oncotarget* **2017**, *8*, 31666–31681. [CrossRef]
38. Fu, J.; Su, X.; Li, Z.; Deng, L.; Liu, X.; Feng, X.; Peng, J. HGF/c-MET pathway in cancer: From molecular characterization to clinical evidence. *Oncogene* **2021**, *40*, 4625–4651. [CrossRef]
39. Konstorum, A.; Lowengrub, J.S. Activation of the HGF/c-Met axis in the tumor microenvironment: A multispecies model. *J. Theor. Biol.* **2018**, *439*, 86–99. [CrossRef]
40. Dilwali, S.; Roberts, D.; Stankovic, K.M. Interplay between VEGF-A and cMET signaling in human vestibular schwannomas and schwann cells. *Cancer Biol. Ther.* **2015**, *16*, 170–175. [CrossRef]
41. Hu, S.-Y.; Duan, H.-F.; Li, Q.-F.; Yang, Y.-F.; Chen, J.-L.; Wang, L.-S.; Wang, H. Hepatocyte growth factor protects endothelial cells against gamma ray irradiation-induced damage. *Acta Pharmacol. Sin.* **2009**, *30*, 1415–1420. [CrossRef] [PubMed]
42. Zhao, Y.; Liu, P.; Zhang, N.; Chen, J.; Landegger, L.D.; Wu, L.; Zhao, F.; Zhao, Y.; Zhang, Y.; Zhang, J.; et al. Targeting the cMET pathway augments radiation response without adverse effect on hearing in NF2 schwannoma models. *Proc. Natl. Acad. Sci. USA* **2018**, *115*, E2077–E2084. [CrossRef] [PubMed]
43. Fuse, M.A.; Plati, S.K.; Burns, S.S.; Dinh, C.T.; Bracho, O.; Yan, D.; Mittal, R.; Shen, R.; Soulakova, J.N.; Copik, A.J.; et al. Combination Therapy with c-Met and Src Inhibitors Induces Caspase-Dependent Apoptosis of Merlin-Deficient Schwann Cells and Suppresses Growth of Schwannoma Cells. *Mol. Cancer Ther.* **2017**, *16*, 2387–2398. [CrossRef] [PubMed]
44. Welling, D.B.; Lasak, J.M.; Akhmametyeva, E.; Ghaheri, B.; Chang, L.-S. cDNA Microarray Analysis of Vestibular Schwannomas. *Otol. Neurotol.* **2002**, *23*, 736–748. [CrossRef] [PubMed]
45. Lee, T.X.; Packer, M.D.; Huang, J.; Akhmametyeva, E.M.; Kulp, S.K.; Chen, C.-S.; Giovannini, M.; Jacob, A.; Welling, D.B.; Chang, L.-S. Growth inhibitory and anti-tumour activities of OSU-03012, a novel PDK-1 inhibitor, on vestibular schwannoma and malignant schwannoma cells. *Eur. J. Cancer* **2009**, *45*, 1709–1720. [CrossRef]
46. Bush, M.L.; Oblinger, J.; Brendel, V.; Santarelli, G.; Huang, J.; Akhmametyeva, E.M.; Burns, S.S.; Wheeler, J.; Davis, J.; Yates, C.W.; et al. AR42, a novel histone deacetylase inhibitor, as a potential therapy for vestibular schwannomas and meningiomas. *Neuro-Oncology* **2011**, *13*, 983–999. [CrossRef]
47. Porta, C.; Paglino, C.; Mosca, A. Targeting PI3K/Akt/mTOR Signaling in Cancer. *Front. Oncol.* **2014**, *4*, 64. [CrossRef]
48. James, M.F.; Stivison, E.; Beauchamp, R.L.; Han, S.; Li, H.; Wallace, M.R.; Gusella, J.F.; Stemmer-Rachamimov, A.O.; Ramesh, V. Regulation of mTOR Complex 2 Signaling in Neurofibromatosis 2–Deficient Target Cell Types. *Mol. Cancer Res.* **2012**, *10*, 649–659. [CrossRef]
49. Lane, H.A.; Wood, J.M.; McSheehy, P.M.; Allegrini, P.R.; Boulay, A.; Brueggen, J.; Littlewood-Evans, A.; Maira, S.-M.; Martiny-Baron, G.; Schnell, C.R.; et al. mTOR Inhibitor RAD001 (Everolimus) Has Antiangiogenic/Vascular Properties Distinct from a VEGFR Tyrosine Kinase Inhibitor. *Clin. Cancer Res.* **2009**, *15*, 1612–1622. [CrossRef]
50. Goutagny, S.; Raymond, E.; Esposito-Farese, M.; Trunet, S.; Mawrin, C.; Bernardeschi, D.; Larroque, B.; Sterkers, O.; Giovannini, M.; Kalamarides, M. Phase II study of mTORC1 inhibition by everolimus in neurofibromatosis type 2 patients with growing vestibular schwannomas. *J. Neuro-Oncol.* **2015**, *122*, 313–320. [CrossRef]
51. Karajannis, M.A.; Legault, G.; Hagiwara, M.; Giancotti, F.G.; Filatov, A.; Derman, A.; Hochman, T.; Goldberg, J.D.; Vega, E.; Wisoff, J.H.; et al. Phase II study of everolimus in children and adults with neurofibromatosis type 2 and progressive vestibular schwannomas. *Neuro-Oncology* **2014**, *16*, 292–297. [CrossRef] [PubMed]
52. Breun, M.; Schwerdtfeger, A.; Martellotta, D.D.; Kessler, A.F.; Perez, J.M.; Monoranu, C.M.; Ernestus, R.-I.; Matthies, C.; Löhr, M.; Hagemann, C. CXCR4: A new player in vestibular schwannoma pathogenesis. *Oncotarget* **2018**, *9*, 9940–9950. [CrossRef] [PubMed]
53. Breun, M.; Monoranu, C.M.; Kessler, A.F.; Matthies, C.; Löhr, M.; Hagemann, C.; Schirbel, A.; Rowe, S.P.; Pomper, M.G.; Buck, A.K.; et al. [68Ga]-Pentixafor PET/CT for CXCR4-Mediated Imaging of Vestibular Schwannomas. *Front. Oncol.* **2019**, *9*, 503. [CrossRef] [PubMed]
54. Hannan, C.J.; Lewis, D.; O'Leary, C.; Donofrio, C.A.; Evans, D.G.; Roncaroli, F.; Brough, D.; King, A.T.; Coope, D.; Pathmanaban, O.N. The inflammatory microenvironment in vestibular schwannoma. *Neuro-Oncol. Adv.* **2020**, *2*, vdaa023. [CrossRef] [PubMed]
55. Fujioka, M.; Okano, H.; Ogawa, K. Inflammatory and immune responses in the cochlea: Potential therapeutic targets for sensorineural hearing loss. *Front. Pharmacol.* **2014**, *5*, 287. [CrossRef] [PubMed]
56. Helbing, D.-L.; Schulz, A.; Morrison, H. Pathomechanisms in schwannoma development and progression. *Oncogene* **2020**, *39*, 5421–5429. [CrossRef] [PubMed]
57. Tamura, R.; Tanaka, T.; Yamamoto, Y.; Akasaki, Y.; Sasaki, H. Dual role of macrophage in tumor immunity. *Immunotherapy* **2018**, *10*, 899–909. [CrossRef]

58. Tamura, R.; Tanaka, T.; Akasaki, Y.; Murayama, Y.; Yoshida, K.; Sasaki, H. The role of vascular endothelial growth factor in the hypoxic and immunosuppressive tumor microenvironment: Perspectives for therapeutic implications. *Med. Oncol.* **2019**, *37*, 2. [CrossRef]
59. Tamura, R.; Morimoto, Y.; Sato, M.; Kuranari, Y.; Oishi, Y.; Kosugi, K.; Yoshida, K.; Toda, M. Difference in the hypoxic immunosuppressive microenvironment of patients with neurofibromatosis type 2 schwannomas and sporadic schwannomas. *J. Neuro-Oncol.* **2020**, *146*, 265–273. [CrossRef]
60. Nisenbaum, E.; Misztalm, C.; Szczupak, M.; Thielhelm, T.; Peña, S.; Mei, C.; Goncalves, S.; Bracho, O.; Ma, R.; Ivan, M.E.; et al. Tumor-Associated Macrophages in Vestibular Schwannoma and Relationship to Hearing. *OTO Open.* **2021**, *5*, 2473974X211059111. [CrossRef]
61. Li, Z.; Liu, X.; Guo, R.; Wang, P. TIM-3 plays a more important role than PD-1 in the functional impairments of cytotoxic T cells of malignant Schwannomas. *Tumor Biol.* **2017**, *39*, 1010428317698352. [CrossRef] [PubMed]
62. Wang, S.; Liechty, B.; Patel, S.; Weber, J.S.; Hollmann, T.J.; Snuderl, M.; Karajannis, M.A. Programmed death ligand 1 expression and tumor infiltrating lymphocytes in neurofibromatosis type 1 and 2 associated tumors. *J. Neuro-Oncol.* **2018**, *138*, 183–190. [CrossRef] [PubMed]
63. Jacobs, J.F.; Idema, A.J.; Bol, K.F.; Nierkens, S.; Grauer, O.M.; Wesseling, P.; Grotenhuis, J.A.; Hoogerbrugge, P.M.; de Vries, I.J.; Adema, G.J. Regulatory T cells and the PD-L1/PD-1 pathway mediate immune suppression in malignant human brain tumors. *Neuro Oncol.* **2009**, *11*, 394–402. [CrossRef] [PubMed]
64. Nakanishi, M.; Rosenberg, D.W. Multifaceted roles of PGE2 in inflammation and cancer. *Semin. Immunopathol.* **2013**, *35*, 123–137. [CrossRef] [PubMed]
65. Finetti, F.; Travelli, C.; Ercoli, J.; Colombo, G.; Buoso, E.; Trabalzini, L. Prostaglandin E2 and Cancer: Insight into Tumor Progression and Immunity. *Biology* **2020**, *9*, 434. [CrossRef] [PubMed]
66. Hong, B.; Krusche, C.A.; Schwabe, K.; Friedrich, S.; Klein, R.; Krauss, J.K.; Nakamura, M. Cyclooxygenase-2 Supports Tumor Proliferation in Vestibular Schwannomas. *Neurosurgery* **2011**, *68*, 1112–1117. [CrossRef]
67. Kandathil, C.K.; Dilwali, S.; Wu, C.-C.; Ibrahimov, M.; McKenna, M.J.; Lee, H.; Stankovic, K.M. Aspirin Intake Correlates with Halted Growth of Sporadic Vestibular Schwannoma In Vivo. *Otol. Neurotol.* **2014**, *35*, 353–357. [CrossRef]
68. Mackeith, S.; Wasson, J.; Baker, C.; Guilfoyle, M.; John, D.; Donnelly, N.; Mannion, R.; Jefferies, S.; Axon, P.; Tysome, J.R. Aspirin does not prevent growth of vestibular schwannomas: A case-control study. *Laryngoscope* **2018**, *128*, 2139–2144. [CrossRef]
69. Whitesell, L.; Lindquist, S.L. HSP90 and the chaperoning of cancer. *Nat. Rev. Cancer* **2005**, *5*, 761–772. [CrossRef]
70. Chen, Z.; Li, S.; Mo, J.; Hawley, E.; Wang, Y.; He, Y.; Brosseau, J.-P.; Shipman, T.; Clapp, D.W.; Carroll, T.J.; et al. Schwannoma development is mediated by Hippo pathway dysregulation and modified by RAS/MAPK signaling. *JCI Insight* **2020**, *5*, e141514. [CrossRef] [PubMed]
71. Tanaka, K.; Eskin, A.; Chareyre, F.; Jessen, W.; Manent, J.; Kawakita, M.; Chen, R.; White, C.; Vitte, J.; Jaffer, Z.M.; et al. Therapeutic Potential of HSP90 Inhibition for Neurofibromatosis Type 2. *Clin. Cancer Res.* **2013**, *19*, 3856–3870. [CrossRef] [PubMed]
72. Sagers, J.E.; Brown, A.S.; Vasilijic, S.; Lewis, R.M.; Sahin, M.I.; Landegger, L.D.; Perlis, R.H.; Kohane, I.S.; Welling, D.B.; Patel, C.J.; et al. Computational repositioning and preclinical validation of mifepristone for human vestibular schwannoma. *Sci. Rep.* **2018**, *8*, 5437. [CrossRef] [PubMed]
73. Nakanishi, H.; Kawashima, Y.; Kurima, K.; Chae, J.J.; Ross, A.M.; Pinto-Patarroyo, G.; Patel, S.K.; Muskett, J.A.; Ratay, J.S.; Chattaraj, P.; et al. *NLRP3* mutation and cochlear autoinflammation cause syndromic and nonsyndromic hearing loss DFNA34 responsive to anakinra therapy. *Proc. Natl. Acad. Sci. USA* **2017**, *114*, E7766–E7775. [CrossRef] [PubMed]
74. Ren, Y.; Landegger, L.D.; Stankovic, K.M. Gene Therapy for Human Sensorineural Hearing Loss. *Front. Cell. Neurosci.* **2019**, *13*, 323. [CrossRef]
75. Prabhakar, S.; Taherian, M.; Gianni, D.; Conlon, T.J.; Fulci, G.; Brockmann, J.; Stemmer-Rachamimov, A.; Sena-Esteves, M.; Breakefield, X.O.; Brenner, G.J. Regression of Schwannomas Induced by Adeno-Associated Virus-Mediated Delivery of Caspase-1. *Hum. Gene Ther.* **2013**, *24*, 152–162. [CrossRef]
76. Ahmed, S.G.; Abdelnabi, A.; Maguire, C.A.; Doha, M.; Sagers, J.E.; Lewis, R.M.; Muzikansky, A.; Giovannini, M.; Stemmer-Rachamimov, A.; Stankovic, K.M.; et al. Gene therapy with apoptosis-associated speck-like protein, a newly described schwannoma tumor suppressor, inhibits schwannoma growth in vivo. *Neuro-Oncology* **2019**, *21*, 854–866. [CrossRef]
77. Ignacio, K.H.D.; Espiritu, A.I.; Diestro, J.D.B.; Chan, K.I.; Dmytriw, A.A.; Omar, A.T., 2nd. Efficacy of aspirin for sporadic vestibular schwannoma: A meta-analysis. *Neurol. Sci.* **2021**, *42*, 5101–5106. [CrossRef]
78. Tamura, R. Current Understanding of Neurofibromatosis Type 1, 2, and Schwannomatosis. *Tamura R. Int. J. Mol. Sci.* **2021**, *22*, 5850. [CrossRef]

MDPI
St. Alban-Anlage 66
4052 Basel
Switzerland
Tel. +41 61 683 77 34
Fax +41 61 302 89 18
www.mdpi.com

International Journal of Molecular Sciences Editorial Office
E-mail: ijms@mdpi.com
www.mdpi.com/journal/ijms

www.ingramcontent.com/pod-product-compliance
Lightning Source LLC
Chambersburg PA
CBHW040820250926
51850CB00007BA/1224
* 9 7 8 3 0 3 6 5 6 6 4 4 3 *